CONTAMINATED RIVERS

Contaminated Rivers

A Geomorphological-Geochemical Approach to Site Assessment and Remediation

By

Jerry R. Miller
Department of Geosciences & Natural Resources Management,
Western Carolina University, NC, USA

and

Suzanne M. Orbock Miller
Haywood County Schools, NC, USA

 Springer

A C.I.P. Catalogue record for this book is available from the Library of Congress.

ISBN-10 94-017-7631-8
ISBN-13 978-94-017-7631-8
DOI 10.1007/978-1-4020-5602-4
ISBN-10 1-4020-5602-8 (eBook)
ISBN-13 978-1-4020-5602-4 (eBook)

Published by Springer,
P.O. Box 17, 3300 AA Dordrecht, The Netherlands.

www.springer.com

Printed on acid-free paper

To Mary and Rebecca

TABLE OF CONTENTS

PREFACE

By the end of the 1960s, it became acutely apparent that major problems existed with the quality of both surface and subsurface waters on a world-wide scale. In response to these discoveries numerous legislative initiatives were enacted in most developed countries to limit the introduction of contaminants to the environment. It quickly became apparent, however, that not only was there a need to reduce the quantity of contaminants introduced to surface and subsurface waters, but previously contaminated resources had to be remediated to reduce the potential risks on human and ecosystem health. Effective remediation proved to be a difficult task that required an improved understanding of the transport and fate of contaminants in aquatic environments. This fact resulted in a wide range of analyses regarding contaminant transport and cycling in riverine environments during the past several decades. Nonetheless, in comparison to the enormous efforts which have been made to characterize, assess, and remediate contaminated soils and groundwater, contaminated rivers have received relatively little attention. This is in spite of the fact that polluted reaches may cover tens of kilometers of stream channel and the adjacent valley floor. Progress, however, in the soils and groundwater arena has recently produced a shift in emphasis from the subsurface to the surface environment, particularly with regards to cleaning up contaminated rivers.

Rivers and their associated drainage basins tend to be geological, hydrological, and geochemically more variable than either soil or groundwater systems. Perhaps more importantly, geomorphic processes play a much larger role in controlling the dispersal, distribution, and geochemical cycling of contaminants. Thus, the direct application of site assessment and remediation protocols developed over the past several decades for soils and groundwater to rivers is generally inappropriate. In fact, post-remedial reviews demonstrate that the application of approaches created for soil and groundwater contaminated sites to fluvial (river) systems often results in less than successful cleanup programs that have required unnecessarily large financial expenditures. What is needed is a more thorough incorporation of the catchment's geomorphology and surficial processes into the utilized site character-ization, assessment, and remediation strategies.

The primary purpose of this book is to provide students and professionals with an introductory understanding of fluvial geomorphic principles, and how these principles can be integrated with geochemical data to cost-effectively assess and remediate contaminated rivers. We stress the importance of needing to understand both geomorphic and geochemical processes. A process-oriented approach is required because it goes beyond the simple description of the river channel and its associated drainage basin to enhance the predictive capabilities of models used in the investigation of riverine environments. Thus, the overall presentation is first an analysis of physical and chemical processes and, second, a discussion of how an understanding of these processes can be applied to specific aspects of site assessment and remediation. We also emphasize the need to take a catchment-scale approach when conducting site investigations, and the potential for changes in process rates through time as a result of both natural and anthropogenic disturbances. Such analyses provide the basis for a realistic prediction of the kinds of environmental responses that might be expected, for example, during future changes in climate or land-use.

Although much of the discussion is derived from work in the United States with which we are most familiar, case studies from many other parts of the world have also been included, particularly studies from the U.K. We purposely include a large number of literature citations in the text to make it easier for those wishing to pursue a topic in more depth to begin their literature review. The hope is that the discussion provided herein, combined with the literature at large, will lead to more effective cleanup programs, and to innovate ways to assess and remediate contaminated rivers.

A number of individuals reviewed parts of the manuscript, and we extend our thanks to all of those who graciously devoted their time and effort to do so. They include Michael Amacher (Chapter 2), Douglas Boyle (Chapter 3), Dru Germanoski (Chapter 5), Karen Hudson-Edwards (Chapter 2), Mark Lord (Chapter 1), Peter Richards (Chapter 4), and several anonymous reviewers (Chapter 8, 9, 10). Special thanks are given to Dale (Dusty) Ritter for critically reviewing nearly all of the manuscript, and for providing much needed focus to parts of the discussion. Thanks are also given to numerous colleagues and friends with whom we have spent countless hours discussing the topics in the book. All shortcomings and errors in the book are, of course, ours.

Jerry R. Miller
Suzanne M. Orbock Miller

CHAPTER 1

CONTAMINATED RIVERS: AN OVERVIEW

1.1.   INTRODUCTION

Approximately 75% of the Earth's surface is covered in water. The vast majority ($\sim 97\%$) of it is saline, and cannot be used for domestic, agricultural, or industrial needs. Of the remaining 3% that is fresh, nearly all is inaccessible as it is either frozen in glaciers or locked in deep geological strata as groundwater (Smol 2002). Only about 0.003% of all freshwater is found in rivers. Nevertheless, the importance of river water to sustaining civilizations can hardly be overstated. Rivers provide water for irrigation, industries, and domestic purposes, serve as fisheries and the source of other food stuffs, act as routes of exploration and travel, and provide a wealth of recreational opportunities.

For millennia humans have also used rivers as a form of natural sewer system whereby waste products were disposed of within, and carried away by, the river's flow. In spite of the introduction of these substances, aquatic degradation was initially minimal and localized as their adverse affects on water quality were negated by dilution. The cliché "dilution is the solution to pollution" held true for most, but certainly not all, locations. That began to change with the onset of industrialization at the beginning of the 19th century (Dunnette 1992), and an accompanying increase in global population from approximately 1.5 billion in 1800 to over 6 billion in 1999 (UNPD 2001).

The widespread effects of chemical contaminants on human and ecological health were not immediately recognized. In fact, an awakening to the potential effects of chemical substances did not begin to materialize on a global scale until the 1960s, by which time it was often difficult to ignore. For example, petroleum-based products floating on the surface of Cuyahoga River near Cleveland, Ohio were ignited in 1969 sending flames shooting nearly 20 m high (Garnett 2002). The absurdity of a river on fire had a significant impact on the public's perception of the Nation's (if not the planet's) water quality. The Cuyahoga, however, was only one of many sites around the world that began to be recognized as having a significant problem.

The increasing concern for high-quality aquatic resources led to the development of state and federal regulatory agencies and the implementation of stringent environmental regulations in most developed countries. Much of the legislation was aimed

1

at reducing the introduction of chemical pollutants to surface and subsurface waters, particularly from points of contaminant discharge. Such regulations have greatly improved water quality along many river systems in the developed world (USEPA 1997; Lomborg 2001). In fact, today the Cuyahoga River is lined by condominiums and restaurants, and its waters, once devoid of fish, now possess abundant quantities of walleye, pike and smallmouth bass (Garnett 2002). Nonetheless, as we will see below, national surveys on river health indicated that there is still much to be done. Significant riverine problems exist in every region of the planet, primarily as a result of modifications in flow regimes by dams and diversions, physical alterations in river and riparian habitats (e.g., associated with channelization), and perhaps most importantly, the historic or ongoing introduction of contaminants.

In light of the above, it is clear that improvements in river health will require a continued effort to decrease contaminant influx to rivers and other aquatic environments. It will also depend in no small way on our ability to cost effectively assess and remediate currently contaminated rivers.

The term *assessment* has been defined in a multitude of ways, depending on the objective of the investigator or agency performing the analysis. Here, it is defined rather broadly (following the terminology put forth in the Australian National Environmental Protection Act of 1994) as the means for determining the nature, extent and levels of existing contamination at a site and the actual or potential risk(s) that the contaminant poses to human or ecosystem health. A *contaminant* represents a chemical substance that can potentially threaten human health or the environment. It differs from a *pollutant* in that a pollutant is thought to have already had an adverse affect on humans or other biota. The basic elements of site assessment commonly include the quantification of the chemical forms, concentrations, and distribution of contaminants within water, sediment, and biota of a river and its associated sedimentary deposits as well as the analysis of the rates and mode of transport, the constraints on bioaccumulation, the toxicity of the contaminants in question, and the potential exposure pathways and health risks to humans and the environment (see, for example, USEPA 1988). It is important to recognize that assessment is the *process* through which data from a diversity of fields are brought together, manipulated, and interpreted to determine what, if any, action should be taken to remediate or restore the site.

The precise use of the term remediation has become complicated in recent years as it has often been used interchangeably in both legislation and academic reports with restoration, reclamation, rehabilitation, and mitigation (NRC 1992). Rather than expend unnecessary effort attempting to sort out the nomenclature, *remediation* will be defined for our purposes as the application of any method that prevents, minimizes, or mitigates the damage to human health or the environment by a contaminant.

The assessment and remediation of contaminated sites is an extremely expensive process. For example, if the U.S. maintains its current trends, the costs of site characterization, assessment, and remediation are projected to amount to $750 billion from 1990 to 2020 (as reported in 1990 dollars) (Russell et al. 1991). Data regarding the nationwide costs of cleaning up contaminated rivers are currently unavailable. Nevertheless, contaminated rivers, particularly those impacted by historic mining operations, represent some of the most costly remediation projects on record. As an example, the Record of Decision issued by the USEPA to tackle the impacts of mining on the Coeur d'Alene River basin in Idaho contains a "final remedy" to address metal-related human health risks and an "interim remedy" to initiate the process of addressing ecological risks. Completion of the program is estimated to cost \$360 million over 30 years, and a review by the National Research Council suggests that even this effort is insufficient to limit ecological risks to an acceptable level (NRC 2005).

In the following pages, we will examine a cost-effective, geomorphological-geochemical approach to site assessment and remediation. However, before we can effectively discuss the approach we must examine the types and sources of contamination in riverine environments, the general modes through which contaminants are dispersed, and the current quality of river waters.

1.2. TYPES OF CONTAMINANTS

A wide range of contaminants can be found in river water, sediment, and biota. Subdividing them into specific types is a useful process in understanding their behavior in aquatic environments, their effects on humans and other biota, or their primary source(s) from which they are derived. Their classification, however, is not as easy you might think as any given contaminant may be categorized in many different ways, depending on the goal of the classification system. Take, for instance, pesticides. We are all aware that these represent a group of substances that were created to control, repel, or kill pests. They may be further classified according to the pests that they were designed to specifically address. From this perspective, specific types of pesticides may include fungicides, herbicides, insecticides, or rodenticides. An alternative means of classifying pesticides is on the basis of their chemical composition, in which case they may be categorized as organochlorines, organophosphates, carbamates, etc. The point is that a given contaminant may fall within several different categories depending on the criteria used in the classification.

A common practice that has been extensively used in the analysis of sediment and water quality is to mix categories of contaminants, defined using different criteria, on charts and graphs. This practice is followed to focus attention on groups of substances which are of most importance to a particular resource. We will take a similar approach and describe the organic and inorganic contaminants that are of most importance to riverine environments. Our discussion, then, does not include all of the potential contaminants that may be found in surface waters, only the most common.

1.2.1. Organic Contaminants

At a broad scale, organic contaminants can be subdivided into pathogens and other toxic organic substances. *Pathogens* of most importance include bacteria, viruses, protozoa, and parasitic worms which are responsible for a variety of water-borne diseases including gastroenteritis, malaria, river blindness (onchocerciasis), schistosomiasis, cholera, and typhoid fever (Kubasek and Silverman 1997). As we will see later in the chapter, pathogens affect more river miles than any other type of contaminant in the U.S. Nonetheless, fatalities resulting from water-borne pathogens in most developed countries are minimal in comparison to the estimated 25 million deaths per year attributed to water-borne diseases in developing countries (GTC 1990).

The enormous diversity of pathogens found in natural waters makes it impossible to routinely monitor for specific organisms. Thus, waters (particularly drinking water) are usually analyzed for groups of bacteria and other microorganisms. Perhaps the most important of these is a group of bacteria referred to as *fecal coliform*, the most common of which is *Escherichia coli, or E. coli*. The presence of fecal coliform in river waters is an important finding because it indicates that the water has been contaminated by human or animal wastes (i.e., fecal materials). The extensive use of *E. coli* as an indirect indicator of other contaminants is related to a lack of routine, cost-effective methods to analyze for a wide range of other pathogens (e.g., viruses).

In addition to pathogens and other microorganisms, most river water contains minor amounts of dissolved or particulate bound organic substances. The description of these organic substances is made difficult by the complex binding capabilities of carbon; it can form single, double, or triple bonds with other carbon atoms, and up to four covalent bonds with other elements. As a result, there are literally millions of possible organic compounds that can be found in water, sediment, and soils, each with its own distinct set of physical and chemical properties. Some of these compounds, such as the naturally occurring humic and fulvic acids, pose little threat to human or ecosystem health. This is not necessarily the case for the more than two million different kinds of man-made organic substances known to exist (Giger and Roberts 1977).

Domenico and Schwartz (1998) provided an effective means of classifying organic contaminants commonly found in aquatic environments. Their classification system includes 16 major categories of organic substances, and is based on the categorization of selected functional groups composed C, H, O, S, N, or P (Fig. 1.1). While organic substances from all of these groups are found in surface- and ground-water, several kinds of hydrocarbons (of which the basic building blocks are C and H) are particularly important. One such group is the *polycyclic aromatic hydro-carbons* (PAHs). PAHs actually include more than 100 different kinds of closely related organic substances which are known to pose a potential threat to human and ecosystem health. They were, in fact, one of the first groups of substances identified as a potential carcinogen.

PAHs are formed from the natural and anthropogenic combustion of organic matter, particularly during the burning of fossil fuels at power plants, in automobiles, or in residential, commercial, or industrial heating systems. They are also found in a number of commonly used commercial products such as creosote and roofing tar. Some forms of PAHs are volatile and can enter the air from contaminated soil, sediment, or water, upon which they are broken down by sunlight and other chemicals within a period of days to weeks. PAHs in soils and water are degraded by microbes but at a slower rate than in the atmosphere, perhaps within a few weeks or months.

1. Miscellaneous Nonvolatile Compounds

2. Halogenated Hydrocarbons

Aliphatic	Aromatic
Trichloroethylene	Chlorobenzene

3. Amino Acids

Basic Structure   Aspartic acid

4. Phosphorous Compounds

Basic Structure   Malathion

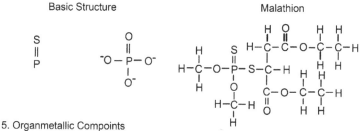

5. Organmetallic Compoints

Tetraethyllead

6.Carboxylic Acid

Basic Structure

Acetic acid

Figure 1.1. Classification of organic compounds as presented by Domenico and Schwartz (1998) (From Domenico and Schwartz 1998)

7. Phenols

Basic Structure   Cresol

8. Amines

Basic Structure

Aromatic   Aliphatic

$$CH_3 - N - H$$
with CH_3 above N

Dimethylamine

9. Ketones

Basic Structure   Acetone

$$R - \overset{\overset{\displaystyle O}{\|}}{C} - R'$$

$$CH_3 - \overset{\overset{\displaystyle O}{\|}}{C} - CH_3$$

10. Aldehydes

Basic Structure   Formaldehyde

$$R - \overset{\overset{\displaystyle O}{\|}}{C} - H$$

$$H - \overset{\overset{\displaystyle O}{\|}}{C} - H$$

11. Alcohols

Basic Structure   Methanol

$$R - OH$$

$$CH_3 - OH$$

12. Esters

Basic Structure   Vinyl Acetate

$$R - \overset{\overset{\displaystyle O}{\|}}{C} - OR'$$

$$H_2C = CH - O - \overset{\underset{\displaystyle O}{\|}}{C} - CH_3$$

13. Ethers

Basic Structure   1,4 Dioxane

$$C - O - C$$

14. Polynuclear Aromatic Hydrocarbons

Phenanthrene

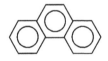

Figure 1.1. (Continued)

15. Aromatic Hydrocarbons

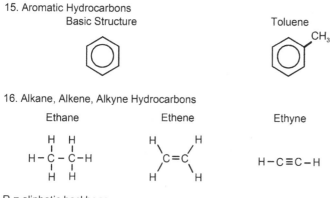

Basic Structure

Toluene

16. Alkane, Alkene, Alkyne Hydrocarbons

Ethane

Ethene

Ethyne

$H-C\equiv C-H$

R = aliphatic backbone

Figure 1.1. (Continued)

One of the largest and most important groups of organic contaminants, particular for groundwater, is the halogenated hydrocarbons. *Halogenated hydrocarbons* are hydrocarbons consisting of one or more atoms of Cl, Br, or F. Many of the halogenated hydrocarbons are associated with solvents, cleansers and degreasers used in industrial processing. Examples include tetrachloroethylene (PCE), trichloroethylene (TCE), dichloroethylene (1,2 DCE), and chloroform. Other important contaminants that fall within this general group include a number of pesticides (e.g., DDD, DDE, DDT, 2,4-D and 2,4,5-T) as well as polychlorinated biphenyls (PCBs). The latter group of substances, PCBs, have been found to be of particularly importance in alluvial (river) deposits.

PCBs are mixtures of chlorinated hydrocarbons that are made by substituting 1 to 10 Cl atoms onto a biphenyl aryl structure (Fig. 1.2). Thus, the mixtures may contain 209 different compounds. PCBs were extensively used for decades in electrical equipment, hydraulic fluids, and a variety of other commercial products because of their chemical, thermal, and biological stability, their high boiling point and their excellent insulating properties. It is these same characteristics, however, that have allowed them to accumulate to exceedingly high levels in many river systems around the world, in spite of the fact that PCB production was terminated in the 1970s in most developed countries. The modern occurrence of PCBs is related to

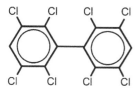

Figure 1.2. Example of a polychorinated biphenyl compound (PCB). The number of attached Cl atoms may vary from one to ten

two factors. First, enormous quantities were produced ($\sim$ 680 million kg in the U.S. alone) prior to the end of production. Second, equipment containing PCBs is still being utilized and thus, they continue to enter the environment as industrial or electrical equipment fails. In addition, once PCBs are in the environment, they can persist for long-periods of time without degrading to non-toxic or less-toxic forms. They are therefore classified, along with a number of other organic contaminants, as *Persistent Organic Pollutants* (POPs).

In addition to degrading slowly in the environment, POPs are particularly hazardous because they possess (1) a strong propensity to accumulate in biota, (2) a tendency for concentrations to increase up the food chain (i.e., biomagnify), (3) an ability to be transported over long distances, and (4) a high potential to be toxic to both humans and other animals at very low concentrations. POPs have been linked to disruptions of the immune system, cancer, reproductive problems, and nervous system dysfunction.

Because of their potentially hazardous nature, POPs were recently addressed by the Stockholm Convention, a United Nations Treaty that seeks to reduce or completely eliminate the production, use, and/or release of the 12 most important POPs. In addition to PCBs, the targeted POPs include pesticides such as aldrin, chlordane, DDT, dieldrin, endrin, mirex, toxaphane, and heptachlor as well as dioxins, hexachlorobenzene, and furans. Dioxins are a family of highly toxic compounds that are produced as a byproduct of various industrial processes involving chlorine such as waste incineration, chemical and pesticide manufacturing, and pulp and paper bleaching. Furans are often associated with dioxins as they are also byproducts of combustion, and like dioxins, are toxic. Because of their widespread production, dioxins and furans are common in riverine environments.

1.2.2.   Inorganic Contaminants

There are a large number of inorganic contaminants that affect the environment, but three groups are of significant importance to river systems: nutrients, metals and metalloids, and radionuclides. The effects of nutrients are primarily manifested through a process referred to as eutrophication. *Eutrophication* refers to a condition in which lakes, reservoirs, and, to a lesser degree, rivers exhibit excess algal or plant growth which leads to severe degradation of the water body (Manahan 2000). Some investigators view eutrophication as a negative feedback loop in which the first step in the process is the input of nutrients, which causes excessive rates of growth. Upon death, the biomass accumulates in the bottom sediment where it decays. Decay not only recycles the nutrients for additional plant growth, but leads to anoxic conditions that eliminates favorable biota, encourages undesirable species, and results in objectionable odors and tastes.

In actuality, eutrophication is much more complicated than described above and is in need of additional research. Most ongoing research focuses on *cultural eutrophication*, rather than natural eutrophication. The former is driven by the input of nutrients, primarily phosphorous, nitrogen, and potassium from human activities.

Recent studies have shown that in most aquatic systems the majority of the essential elements for plant growth are available in sufficient quantities. Thus, they do not limit the rate of biomass production. The exceptions include P, N, and occasionally K. In freshwater P is most often the rate limiting nutrient, and is the primary culprit in cultural eutrophication. An important source of P in river waters has been household detergents and, thus, much effort has been devoted to reducing their phosphate content and in removing phosphate from water in sewage treatment plants. Nitrogen can be the rate limiting nutrient in certain situations, particularly in marine environments or where the influx of animal wastes is significant.

In addition to its potential effects through eutrophication, N can directly impact human health. High levels of nitrate, which is the most common form of dissolved N in surface- and groundwater, can cause methemoglobinemia (or blue baby disease) in infants. Once entering the body, nitrate interferes with the ability of the blood to carrier oxygen (turning the blood a shade of blue), which may then inhibit the normal development of the central nervous system.

Trace metals represent another important group of inorganic contaminants. The term trace metal as used here refers to any metal found in very low concentrations, generally less than a few mg/kg, in the river system. The most important of these from a contaminant point of view include Ag, Cd, Cr, Cu, Hg, Fe, Mn, Pb, and Zn. Some of these elements are essential for plant and animal life at low concentrations, but become toxic at higher concentrations (Table 1.1). Others serve no biological purpose.

Much of literature has historically focused on the so called *heavy metals* (e.g., Pb, Hg, Zn), most of which are transition elements. These elements are often considered as the most harmful to aquatic ecosystems, and are of particular concern to site assessments because they can be extremely toxic to humans at high concentrations (Manahan 2000). The specific effects of these metals on human and biotic health vary with the element and the form of the element as it exists in the environment. Many of them, however, interfere with important enzyme functions or bind to cell membranes, thereby hindering transport of materials through the cell wall.

The term heavy metal has recently come under attack because it has been defined in different ways by different people (Duffus 2002; Hodson 2004). For example, heavy metals are often defined according the density of the metal; densities that have been cited in the literature range from $> 3.5\,g/cm^3$ to $> 7\,g/cm^3$. In addition, it has been pointed out that a number of the utilized definitions are contradictory. Thus, Hodson (2004) has argued that investigators should avoid use of the term. While we will follow Hodson's suggestion, the term is so ingrained in the literature that it is unlikely go away any time soon.

In addition to trace metals, a number of metalloids have also been found to contaminate river systems. The most important of these are As, Se, and Sb. Arsenic (As) is particularly important as the ingestion of no more than 100 mg of the element can result in acute As poisoning (Manahan 2000). Ingestion of smaller amounts over long periods of time can result in chronic poisoning. Because of its extreme

Table 1.1. Trace elements found as contaminants in natural waters

Element	Sources	Effect and significance
Arsenic	Mining byproduct; chemical waste	Toxic; possibly carcinogen
Beryllium	Coal; industrial waste	Toxic
Boron	Coal; detergents; wastes	Toxic
Chromium	Metal plating	Essential as Cr(III); toxic as Cr (VI)
Copper	Metal plating; mining; industrial waste	Essential trace element; toxic to plants and algae at higher levels
Fluorine (F^-)	Natural geological sources; wastes; water additive	Prevents tooth decay at around 1 mg/L; toxic at higher levels
Iodine (I^-)	Industrial wastes; corrosion; acid mine water microbial action	Prevents goiter
Iron	Industrial waste; mining; fuels	Essential nutrient; damages fixtures by staining
Lead	Industrial wastes; mining; fuels	Toxic, harmful to wildlife
Manganese	Industrial wastes; acid mine water; microbial action	Toxic to plants; damages fixtures by staining
Mercury	Industrial waste; mining, coal	Toxic; mobilized as methyl compounds by anaerobic bacteria
Molybdenum	Industrial wastes; natural source	Essential to plants; toxic to animals
Selenium	Natural sources; coal	Essential at lower levels; toxic at higher levels
Zinc	Industrial waste; metal plating; mining; plumbing	Essential element; toxic to plants at higher levels

From Manahan 2000

toxicity, it has historically been used in a number of insecticides and herbicides as well as in various agents for chemical warfare. Arsenic has also been used in the production of certain metal alloys and various pigments. Its use for most of these purposes has generally been replaced by synthetic organic compounds. Nevertheless, arsenic continues to be a problem in many areas because of past uses, its continued release from ore mining and processes facilities, and its natural occurrence in the environment (Freeze and Cherry 1979).

Radionuclides (or radioactive isotopes) are produced by the fission of relatively heavy elements such as uranium, thorium, and plutonium. In contrast to stable nuclei, they emit ionizing radiation which consists of alpha particles, beta particles, and gamma rays. Ionizing radiation, in turn, produces ions in the materials which it contacts, thereby disrupting the material's existing chemical structure. Alpha particles are composed of two neutrons and two protons, or a helium nuclei. In contrast, beta particles consist of positively or negatively charged particles referred to as positrons or electrons, respectively. Gamma rays are a true form of electro-magnetic radiation that is found at the highest energy end of the electromagnetic spectrum. In general, the depth to which each of the types of radiation can penetrate increases from alpha particles, to beta particles, and finally to gamma rays, but the amount of ionization cause by each follows the reversed order. Thus, while alpha particles cannot penetrate most matter to any significant depth, they can cause a large amount of ionization along their pathway. It follows, then, that radionuclides emitting alpha particles are not a major concern when they are found outside the body, but can be an extremely hazardous if ingested (Manahan 2000). Both beta and gamma radiation can penetrate the skin, and are therefore of concern even while outside the body.

Radionuclides are found naturally in aquatic environments, but artificially produced radionuclides are usually of more importance in terms of aquatic contamination. Sources of radioactive materials include medical and industrial wastes, weapons production (particularly during World War II), uranium ore processing and enrichment, and to a lesser degree, power plant releases (such as Chernobyl). In contrast to most other contaminants, the amount of radionuclides found in water, sediment, etc., is measured in a quantity known as picocuries, where one picocurie (pCi) represents 3.7×10^{-2} disintegrations per second. The primary health effect of ionizing radiation is that it initiates a variety of adverse chemical reactions in biotic tissues. In the most severe case where radiation poisoning has occurred, the production of red blood cells is reduced as bone marrow is destroyed. The potential for ionizing radiation to result in genetic damage over longer periods of time is also a concern.

1.3.   NATIONAL ASSESSMENTS OF RIVER HEALTH

In the previous section, we examined the type of contaminants frequently found in rivers. We can now examine the significance of these contaminants in river water and sediment, and the primary sources from which they are derived. One way of effectively doing this is to analyze the results obtained from national assessments of river health.

Most developed countries have one or more program in place to conduct large-scale (national) assessments of the physical and chemical conditions of river systems. These assessments are utilized for a variety of purposes, particularly (1) the identification of important contaminant types and sources, and (2) the delineation of rivers or river reaches where contaminants may pose a threat to human health or the environment, and are therefore in need of further investigation. Most are also designed to collect data over a period of years so that decadal-scale trends in riverine health can be developed and used to determine, if the existing legislation and management strategies are having the desired affect.

In general, national assessments of riverine health involve the collection of physical and geochemical data from multiple sites and sources, its integration into a single database, and ultimately its interpretation using procedures designed to answer a specific set of questions. Although the integration of records from multiple sources may seem to be a rather simple task, in reality the uncertainties (errors) associated with geochemical data collected at a specific location are compounded as they are combined with information from other sites (Dunnette 1992). In addition, the collection of chemical and physical data is an extremely expensive undertaking. Therefore, to reduce costs, data are often obtained from sites that were established for other purposes (e.g., to characterize the contaminant load downstream of a contaminanted site). Thus, the data contained in most national assessments are biased by the geographical distribution of the sampling locations.

Although large-scale assessments of water and sediment quality are not perfect, the derived information provides important insights into the types and sources of contaminants found in freshwater systems and, to a lesser degree, the extent and severity of freshwater contamination. Comparison of the results from different countries reveals that water and sediment quality varies from country to country and from region to region within any given country. The types and sources of riverine contamination, however, are surprisingly similar. In the following sections, we will briefly examine three assessments of riverine health conducted in the U.S. in order to more closely examine the nature of national assessment programs, and to develop a better understanding of the importance of the various contaminants described above to riverine environments. In doing so, we will also gain insights in to the current conditions of river systems in the U.S.

1.3.1.   The National Water Quality Inventory

Section 305b of the Clean Water Act requires states, U.S. territories, and other jurisdictions (e.g., interstate river commissions and Indian tribes) to assess the quality of their surface- and groundwater and report their findings on a biennial basis to the U.S. Environmental Protection Agency (USEPA). The assessments involve the collection of data from selected water bodies and its subsequent comparison to water quality standards that have been defined by the jurisdiction performing the analysis. Water quality standards, which must be approved by the USEPA, consist of three elements: a designation of the primary uses of the water (e.g., to support aquatic life or recreation), a set of criteria that are intended to protect aquatic life and humans from the adverse effects of contaminants or other forms of stress, and some form of antidegradation policy aimed at preventing degraded waters from getting worse, and high quality waters from deteriorating from their current condition. Most jurisdictions utilize a wide variety of parameters in the analysis, including biological, chemical, and physical measurements of the water, chemical analysis of fish and sediment, land-use records, and predictive models. The results of the analysis are then reported according to the degree to which the river meets the water quality standards, so that the assessed river miles can be classified as:

    Good – fully supporting all of their uses, or fully supporting of all uses but are
        threatened for one or more;

    Impaired – partially or not supporting of one or more use;

    Not attainable – not able to support one or more use.

The results of the most recent findings of the National Water Quality Inventory have been summarized by the USEPA (2000). With regards to rivers, 19% of the 3,692,946 miles of stream channels in the reporting jurisdictions were assessed. Some form of pollution, or degradation, impairs 39% of the assessed river miles to the extent that one or more of its designated uses were not fully supported (Fig. 1.3). Note, however, that ten states did not include the effects of a statewide advisory on the consumption of fish as a result of Hg contamination, and New York did not include a statewide advisory on fish consumption associate with PCB, chlordane, mirex, and DDT pollution. If the above, statewide advisories had been included in the analysis, as much as 67% of the assessed river miles may have been listed as partially or fully impaired. As stated earlier, the data clearly show that there is still much to be done to improve the quality of riverine environments.

For those waters that are classified as impaired, an attempt is made to determine the primary causes of degradation, and the sources of any identified pollutant. Figure 1.4 presents the leading causes of degradation identified in the USEPA (2000) report. An important conclusion that can be drawn from the figure is that several of the causes of impairment are not related exclusively to chemical contaminants, but also to physical forms of impairment including alterations in habitat and flow conditions, thermal modifications, and siltation. Thus, river health is dependent on both physical and chemical stressors. It is also important to recognize that more

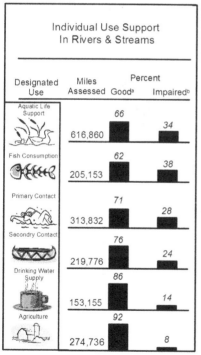

Guidelines Provided for Designated Uses to be Fully Supportive

Aquatic Life: *Water quality is good enough to support a healthy, balanced community of aquatic organisms including fish, plants, insects and algae*

Fish Consumption: *People can safely eat fish caught in the river or stream*

Primary Contact Recreation *(swimming): People can make full body contact with water without risk to their health*

Secondary Contact Recreation: *No risk to public health from recreational activities on the water, such as boating, that expose the public to minimal contact with the water*

Drinking Water Supply: *The river provides a safe supply of water with standard treatment*

Agricultural Uses: *The water can be used for irrigating fields and watering livestock*

Figure 1.3. Summary of impaired river miles according to designated uses from the 2000 National Water Quality Inventory (Modified from USEPA 2000)

than one cause of degradation may have been identified and reported for any given reach. As a result, the percentages shown on Fig. 1.4 do not add up to 100%.

1.3.2.   The National Sediment Quality Survey

The National Sediment Quality Survey represents the most comprehensive, nationwide assessment of the extent and severity of contaminated river bed sediment in the U.S. (USEPA 1997). Produced by the USEPA as a requirement of the Water Resources Development Act of 1992, the National Sediment Inventory, upon which the analysis is based, consists of approximately 2 million records of both sediment chemistry and related biological data, collected at 21,000 monitoring stations across the country. The assessment focuses on the probable risk that contaminated channel bed sediment poses to benthic organisms as well as to humans who consume aquatic organisms that have been directly or indirectly exposed to contaminated materials. The assessment differs significantly from the National Water Quality Inventory in

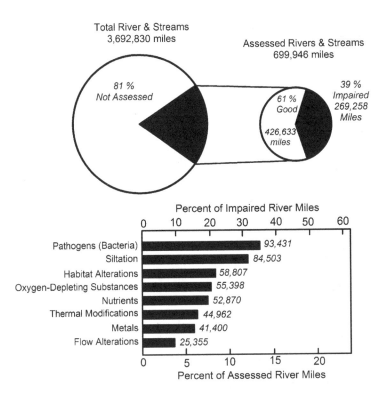

Figure 1.4. Leading causes of degradation of impaired rivers. Data based on the 2000 National Water Quality Inventory. Note that leading causes include both physical and chemical factors. Percentages do not sum to 100% because more than one pollutant or source of stress may impair any given river reach (Modified from USEPA 2000)

that it focuses on the potential effects of contaminated bed sediment, and it does not extensively include the analysis of physical stressors on the aquatic environment.

The probability of adverse effects is determined by a rather complex set of analyses that involve multiple parameters (see USEPA 1997 for details). These analyses ultimately allow each of the monitoring stations to be placed into one of three categories:

Tier 1 – associated adverse effects on aquatic life or human health are probable;

Tier 2 – associated adverse effects on aquatic life or human health are possible, but infrequently expected;

Tier 3 – no indication of associated adverse effects on aquatic life or human health.

Of the 21,000 sampling stations evaluated, 26% were classified as Tier 1, and 49% were classified as Tier 2. It follows, then, that adverse effects on aquatic life or human health were considered to be either possible or probable at 75% of the sites. These statistics, however, are somewhat misleading because the stations are not uniformly distributed across the country, but are more frequently located along rivers that are suspected to have been contaminated. Thus, the statistics presented above do not provide an accurate assessment of the overall chemical condition of channel bed sediment in the U.S. Another commonly cited estimate, based on the volume of material contained within the upper 5 cm of the channel bed, is that approximately 10% (or $9 \times 10^8 \, m^3$) of the sediment underlying surface water bodies in the U.S. is sufficiently polluted to pose a threat to fish or humans who eat contaminated fish (USEPA 1998). The justification for focusing on the upper 5 cm of sediment in the channel bed is that these materials control the interchange of contaminants with the water column, and represent the materials with which benthic organisms are most likely to come in contact.

It may seem odd that the assessment cannot be used to assess the general quality of river sediment on a national scale. However, this was never its intent. Rather, its primary objective is to identify regions of pollution that may warrant further, more detailed investigation of the potential risks of contaminants to ecosystem and human health. To accomplish this objective, the assessment has delineated *areas of probable concern*, defined as watersheds in which ten or more of the sampling stations were categorized as Tier 1, and in which at least 75% of all sampling sites were categorizes as either Tier 1 or Tier 2. A total of 488 watersheds nationwide contain ten or more sampling stations and were therefore eligible to be classified as an area of probable concern. Of these, 96 or roughly 20% were delineated as areas of probable concern.

A significant outcome of the National Sediment Quality Survey is that it delineates types of chemicals that pose a risk to aquatic ecosystems or human health within the areas where problems are likely to exist. In doing so, Hg was examined separately from the other metals because of its widespread impacts on sediment quality and its unique geochemical behavior in the environment, including the occurrence of both inorganic and organic chemical forms (USEPA 1997). The data revealed that PCBs were the most commonly observed contaminant at Tier 1 sampling stations, followed by pesticides, Hg, and PAHs (Fig. 1.5). When considering both Tier 1 and Tier 2 stations, metals become the most frequently encountered contaminant (Fig. 1.5).

The increase in the importance of metals as a contaminant from Tier 1 to Tier 2 stations is a function of the criteria used to define the probability of adverse effects to human or aquatic health. In order for a sampling station to be defined as a Tier 1 site on the basis of metal contamination, acid-volatile sulfide concentration (AVS) data must be available. The logic behind the criteria is that sulfide, when available in the sediment, will react with divalent metal cations (e.g., Cd, Cu, Pb, Ni, and Zn) to form highly insoluble metal species that are not accumulated in biota. Only when the sulfide is depleted are metals likely to be taken up by biota. AVS is a measure of the availability of sulfide in the sediment; however, it is a recently developed analysis and AVS data are generally lacking for most sampling stations. Thus, most sites cannot be categorized as Tier 1 on the basis of metal contamination and could not be considered in the analysis (Fig. 1.5). The point is that metals are likely to represent a more important Tier 1 contaminant than is portrayed by the current data set.

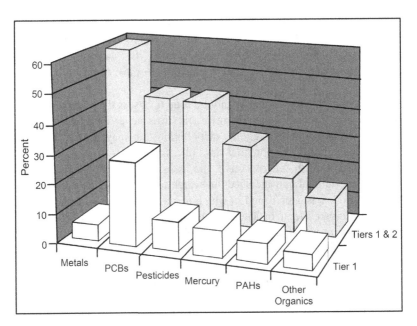

Figure 1.5. Average percent of sampling sites contaminated by each of the major chemical classes designated in the 1997 National Sediment Inventory for watersheds containing areas of probable concern (From USEPA 1997)

1.3.3. The National Water-Quality Assessment Program

A significant shortcoming of many national evaluations of water quality is the consistency with which data are collected, analyzed and interpreted across large areas. The National Water-Quality Assessment Program (NAWQA), conducted by the U.S. Geological Survey, is intended to address this issue by obtaining consistent and comparable data on water quality on a national scale which is capable of supporting watershed management and policy decisions. To accomplish this objective, data pertaining to streams, groundwater and aquatic ecosystems are collected from rivers and aquifers on a watershed basis. Thus far, more than 50 watersheds and aquifer systems have been instrumented and monitored since 1991. The program differs from other assessments in the U.S. in that the study design, including protocols for sampling and analysis, are consistent so that water quality conditions can be compared between watersheds and through time. In addition, it focuses much more heavily on assessing the predominant controls on water quality by addressing specific questions such as (1) what is the current condition of water resources in the U.S., (2) how is water quality changing over time, and (3) what are the natural and human controls on stream and groundwater quality (USGS 2001)?

Between 1991 and 2001 most of the effort was aimed at documenting water-quality conditions within the watersheds selected for study. Since 2001, the focus has changed to examine the sources of contaminants that degrade water quality, the contaminant transport mechanisms operating in streams and groundwater, and the effects of contaminants and other disturbances on humans and aquatic ecosystems. At the time of this writing, summary reports in the form of USGS Circulars had been published for 36 of the studied watersheds. The results from these documents are too extensive to present in the space that is available here. However, data concerning pesticides and nutrients have been summarized on a national scale (USGS 1999), and some of the major conclusions from this work are worth noting.

In agricultural areas, pesticides were detected in more than 95% of the stream samples, but the concentrations were generally below the levels of concern for human health set by the USEPA (USGS 1999). Approximately 66% of the samples contained five or more pesticides, demonstrating that they commonly occur as chemical mixtures. In spite of the fact that certain pesticides, such as DDT, were banned in the1970s, they were still present in both waters and sediment of agricultural streams. Similar results were found for urban watersheds. Nearly 80% of the stream samples contained five or more pesticides, and some pesticides, such as malathion and chlorpyrifos, were found more frequently and at higher concentrations than in agricultural areas. The concentrations, however, rarely exceeded USEPA drinking water standards. Historic pesticides, such as DDT, were also found in the channel bed sediment of urban streams. With regards to nutrients, average annual concentrations of phosphorous in more than 70% of the samples from urban and agricultural areas exceeded the USEPA goal for preventing nuisance plant growth in rivers.

Some regional statistics are also available for the assessment of trace metals. In urban areas, concentrations of selected trace elements including Cd, Pb, Hg, and Zn are elevated above the naturally occurring concentrations found in the area. Moreover, in an analysis of data from 20 of the studied basins, median concentrations measured for seven trace metals (Cd, Cr, Cu, Pb, Hg, Ni, and Zn) in materials from urban environments exceeded those found in agricultural and forested terrains (Rice 1999). In fact, concentrations of Cu, Hg, Pb, and Zn increased within increasing population density.

1.3.4.   Sources of Contamination

The toxic substances described above are released to water bodies from a wide variety of sources. These sources can be differentiated into two broad categories on the basis of the spatial extent over which the materials are released. *Point sources* refer to the discharge of contaminants from a specific location, such as the end of a pipe or canal. In contrast, *non-point sources* refer to contaminants derived from a diffuse area, such as an agricultural field or an urban center. It is not always easy to classify a particular source as either a point or nonpoint source of pollution. For example, mining districts are commonly characterized by multiple waste piles from which metal contaminated particles can be eroded and transported to an adjacent river. Individually, the waste piles represent a point source of pollution, but collectively, they represent a form of nonpoint source pollution in that the contaminants are derived from multiple sites spread over a broad region.

Identifying the types of pollutants causing degradation to aquatic ecosystems is a relatively straight forward process, based on field observations, the collection of sediment, water or biotic samples, and their subsequent analyses. However, delineating and ranking the primary sources of the degrading chemicals is not a simple task. Source ascription is frequently complicated by the occurrence of multiple point and nonpoint sources which may deliver similar contaminants to the river in unknown quantities and at varying rates through time.

Table 1.2 lists the potential sources of selected contaminants found in river systems. The sources, in this case, have been subdivided into six categories. Four of these categories represent nonpoint sources, including (1) the deposition of contaminants from the atmosphere, (2) runoff from agricultural lands, (3) runoff

Table 1.2. Sources associated with selected classes of contaminants

Source/chemical class	Mercury	Metals	PCBs	PAHs	Pesticides	Other organics
Agricultural croplands[1]					*	
Mine Sites[1]	*	*				
Atmospheric Deposition[1]	*	*	*	*	*	*
Urban Sources[1]	*	*		*	*	*
Industrial Discharges	*	*	*	*	*	*
Municipal Discharges	*	*	*	*	*	*

Modified from USEPA 1997

[1] Classified as non-point source pollution

from urban areas, and (4) drainage from mine sites. The two remaining categories, industrial and municipal discharges, represent point sources of contamination. Data on other source types, such as landfills and recreational activities, are generally lacking, and their relative importance as a contaminant source to rivers is unknown. Thus, they are not listed in Table 1.2. In addition, Table 1.2 does not include natural sources of pollution, such as might be important for trace metals or radionuclides.

Atmospheric deposition is most commonly associated with PCBs, PAHs, and trace metals, particularly Pb and Hg (USEPA 1997). It is also a significant source of dioxins and furans, listed as "other organics" in Table 1.2. Anthropogenic sources of metals within the atmosphere include coal combustion, the burning of leaded gasoline and other petroleum products, primary and secondary metal production, waste incineration and the application of paints, among a host of others. In general, atmospheric sources are not considered as important as other sources of anthropogenic contaminants in terms of their direct input to rivers. This follows because most of the materials are deposited over the terrestrial landscape and remobilized in runoff to surface water bodies. As a result, a number of pollutants found, for example, in urban runoff may have originally been derived from atmospheric sources, but were considered in water quality assessments as part of the urban load.

A number of recent studies have found that Hg is one of the most important atmospheric pollutants (Fitzgerald et al. 1991; Sorensen et al. 1990). In fact, some investigators have suggested that the atmospheric deposition of Hg can account for nearly all of the Hg in fish, water and sediment in secluded, temperate lakes of the mid-continental U.S. (Fitzgerald et al. 1991). Similar arguments have been put forth for relatively secluded areas of Canada and Sweden.

Agricultural lands encompass more than 50% of the continental U.S. and even larger portions of some other countries. The application of chemicals to enhance crop production, combined with the release of materials from livestock operations, has commonly transformed these areas into significant sources of contaminants to riverine ecosystems. In fact, the National Water Quality Inventory considers agricultural lands the leading source of pollutants to river systems, affecting nearly 129,000 miles of channel in the 2000 survey (Fig. 1.6). The primary contaminants of concern include nutrients (P and N), pesticides, and herbicides.

In comparison to agricultural lands, urban areas cover a relatively small percentage of the landscape (< 5% of the continental U.S.). Historically, they were not considered a significant source of chemical pollutants (USGS 2001). This view is changing, however, as the distribution of water quality data increases. In the U.S., the 2000 National Water Quality Inventory estimated that approximately 35,000 miles of impaired river channel is affected by urban runoff and storm sewers, and an additional 28,000 miles were impacted by municipal point sources. The pollutants include pesticides and nutrients, primarily used to enhance lawns and gardens, as well as a trace metals, oil, PAHs, and other petroleum hydrocarbons (USEPA 1992). Data from municipal sewage treatment plants and major industrial facilities indicate that they discharge all six of the chemical classes shown in Table 1.2.

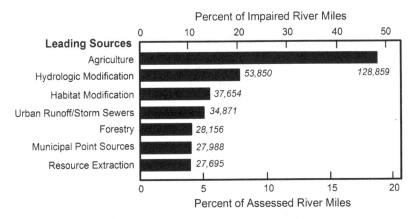

Figure 1.6. Leading sources of river impairment based on the 2000 National Water Quality Inventory. Percentages do not sum to 100% because more than one pollutant source may impair any given river reach (Modified from USEPA 2000)

The largest quantities of contaminants from these facilities were metals, and they were released at the most number of sites (USEPA 1997).

Much of our data concerning the physical and chemical transport of trace metals in rivers comes from sites contaminated by mining activities. In developed countries, the effects of ore extraction and processing are primarily related to historic mining operations. In the U.S., for example, metal flux from historic mining has resulted in the listing of several river basins on the USEPA Superfund Program's National Priorities List. These include the Carson River, Nevada, Whitewood Creek and the Belle Fourche River, South Dakota, the Clark Fork River basin, Montana, the Coeur d'Alene River basin, Idaho, Tar Creek, Oklahoma, Iron Mountain, California, and the Arkansas River and its tributaries near Leadville, Colorado (USEPA 1997). Not only is the severity of contamination significant, but the total area impacted is large, currently exceeding 27,500 miles of impaired channel (USEPA 2000) (Fig. 1.6). In many developing countries, metal releases to the environment from mining continue at an ever increasing rate.

One of the most frequently cited metal contaminants from mining is Hg which was (and continues to be) used to recover gold and silver from placer and ore bodies. In both cases, precious metal extraction involves Hg amalgamation, a process during which Hg is mixed with ore containing fine gold or silver particles. The Hg binds with the Au and Ag creating an amalgam. The dense amalgam grains are subsequently separated from the rest of the material and heated, driving Hg off as a vapor and leaving a pure form of the precious metal. While Hg amalgamation is no longer used in the U.S. and most developed countries, it persists in the environment as a legacy of past mining operations at a surprisingly large number of localities. Basham et al. (1996), for example, found that Hg may have been used for the extraction of gold and silver in the U.S. at more than 1600 sites in 21 different states during the late 1800s and early 1900s. Approximately

6.20×10^7 kg of Hg may have been released to the environment at these sites between 1820 and 1900 (Nriagu 1994). In addition, Hg amalgamation is still widely used in many developing parts of the world. Malm (1998), for example, estimated that 240 tons of Hg are released annually into the environment in the Brazilian Amazon, Venezuela, Colombia, and the Pando Department of Bolivia. Hg amalgamation mining is also occurring in such places as Ecuador, French Guiana, Guyana, Suriname, Tanzania, Indonesia, the Philippines, and Vietnam (Jernelov and Ramel 1994).

1.4.   THE DISSOLVED VERSUS PARTICULATE LOAD

Once chemical contaminants enter a river channel, they can found in two predo-minant forms: (1) as a dissolved constituent, or (2) attached to sediment which is suspended within the water column. The dissolved constituents are generally thought to be more bioavailable than those attached by various mechanisms to particulate matter. By saying it is *bioavailable*, we mean that it can be taken up by and potentially accumulated in biota. Thus, the dissolved constituents pose the largest risk to human and ecosystem health (Salomons and Förstner 1984). It is for this reason that regulations are heavily biased toward the concentration of contaminants dissolved within river waters. The question arises, then, as to why any site assessment program should be concerned with the geochemistry of the particulate load. The answer rests partly on the fact that sediment is a primary sink for reactive, hydrophobic contaminants (e.g., trace metals), and the nature (mineralogy, crystallinity, size, abundance, etc.) of the alluvial materials directly influences the concentrations of aqueous species within surface waters. Moreover, in many, if not most, rivers, the sediment which is suspended within the water column and that forms the channel bed and banks, exhibit concentrations that are orders of magnitude higher than those associated with the aqueous (dissolved) load. Gibbs (1977), for example, demonstrated that while the Yukon and Amazon Rivers drain very different terrains, Cr, Mn, Fe, Co, Ni, and Cu concentrations in the suspended sediment of both systems ranged from 6,000 to more than 10,000 times greater than their dissolved concentrations. Such large differences in concentration in such diverse environments have led to the argument that trace element transport in most rivers is dominated by the movement of particulate matter (Fig. 1.7). In fact, Meybeck and Helmer (1989) argue that for a great majority of trace metals more than 90% of the load is associated with the physical movement of sediment (Table 1.3).

Horowitz (1991) cautions that care must be taken when examining data sets such as those presented above because contaminant transport in rivers is not only dependent on the differences between their concentrations in the dissolved and suspended loads, but the quantity of suspended particles within the water column. Take, for instance, a gravel bed stream in which the suspended sediment concen-tration (i.e., the weight of suspended sediment per unit volume of water) is low. Even though contaminant concentrations associated with the suspended particles may

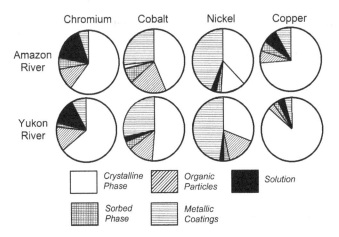

Figure 1.7. Percentages of selected metals transported in the dissolved and particulate phases within the Amazon and Yukon Rivers. The percentage transported as a dissolved species is generally less than 10% (Data from Gibbs 1977)

exceed that of the dissolved load by several orders of magnitude, there may be vastly more water than sediment moving through the channel and the transport of contaminants in association with particulate matter may be rather modest. Nonetheless, there is little argument that the physical transport of particulate matter is a significant process through which trace metals and other hydrophobic contaminants are dispersed through riverine environments.

In light of the above, let us return to our original question: why must a site assessment focus on the particulate load? The answer is because the particulate load, and sediment in generally, significantly influences contaminant concentrations in the water column, contaminant bioavailability, contaminant dispersal processes, and the distribution of contaminant storage, or hotspots, within the river valley. If we were to ignore the particulate phase, we could not possibly assess the risks associated with the contaminants of concern, or provide an accurate analysis of

Table 1.3. Ratio between dissolved and total elemental transport in rivers. Higher percentages indicates a greater proportion within the dissolved phase

Percentage	Elements
90 − 50%	Br, I[1], S[1], Cl[1], Ca, Na, Sr
50 − 10%	Li, N[1], Sb, As, Mg, B, Mo, F[1], Cu, Zn, Ba, K
10 − 1%	P, Ni, Si, Rb, U, Co, Mn, Cr, Th, Pb, V, Cs
1 − 0.1%	Ga, Tm, Lu, Gd, Ti, Er, Nd, Ho, La, Sm, Tb, Yb, Fe, Eu, Ce, Pr, Al

Adapted from Martin and Meybeck 1979
[1] Estimates based on elemental contents in shales

how to effectively remediate the riverine environment. Thus, our characterization, assessment, and remediation strategy must include an analysis of both the dissolved and particulate load.

1.5. SITE CHARACTERIZATION, ASSESSMENT, AND REMEDIATION

The primary purpose for investigating a contaminated river is to determine the nature and extent of contamination and to evaluate whether the contaminant(s) pose an unacceptable risk to human health or the environment (USEPA 2005). If the level of risk is considered excessive, it may be necessary to remediate the site to reduce the identified risks to within acceptable values. The investigation, then, can be envisioned as consisting of two distinct, but closely related components: site assessment and remedial action (Soesilo and Wilson 1997). The primary intent of site assessment is to characterize the river system in terms of the types of contaminants and their associated sources, the history of contaminant release, the areal extent and magnitude of contamination, and the rates of contaminant migration, among a host of other things. These data are subsequently used in various types of assessments to evaluate whether the contaminant(s) pose a threat to human health and/or the environment if no actions are taken to address the problem. Site assessment, then, goes beyond characterization in that it manipulates and interprets data obtained during site characterization in order to answer questions related to site remediation.

Numerous definitions have been put forth for site characterization, but in general it is the process of collecting data that quantitatively describes the river in terms of its physical, chemical, and biological conditions both now and in the past, including the flux of materials between each of the system's components. Much of the literature on characterization consists of extensive lists of parameters which are important for one reason or another, along with short descriptions of the process that a specific parameter, such as sediment size, may influence (Table 1.4). Such lists tend to be more extensive for rivers than for any other environment because characterization must not only focus on the site itself, but on the entire watershed. Upstream changes in land-use, for example, have the potential to change the hydrologic regime of the river, which may subsequently alter the river's ability to erode and transport contaminated particles.

It is easy when examining such extensive lists of parameters to get lost in the process. Characterization can seem like an overwhelming task requiring the collection of huge amounts of data. It is therefore essential to develop a sound understanding of why each part of the characterization program is being conducted, and how the collection of a particular type of data will lead to the program's overall success. One way that this can be done is to formulate a set of objectives that must be accomplished with the collected information. These objectives will

Table 1.4. Abbreviated list of site characterization parameters for metal contaminated rivers

Physical parameters	Chemical parameters
Channel Characteristics & Stability	*General*
• Width, depth, gradient, planimetric configuration, and pattern of channel	• Types of contaminants in channel bed or other alluvial deposits, in water, and biota
• Magnitude and rate of changes in channel morphology through time	• Metal speciation and bioavailability
• Bank erosion rates	• Physical partitioning as a function of grain size and mineralogy
• Depths of scour & fill	• Sediment toxicity in relation to Sediment Quality Guidelines
• Current rates of aggradation or degradation	• Background concentrations in specific media
Channel Bed Sediments	*Channel bed sediments*
• Sediment grain-size distribution & mineralogy	• Concentrations as a function of macroscale bedforms
• Type and nature of macroscale bed forms	• Concentrations as a function of depth
	• Total organic content; Fe & Mn oxyhydroxide content
Holocene Depositional Units within floodplain and terraces	*Holocene Depositional Units within floodplain and terraces*
• Horizontal and vertical extent	• Variations in concentration both horizontally and with depth; data stratified by delineated depositional units
• Age	
• Sediment size distribution and mineralogy	• Oxidation-reduction variations in sediment cores
• Bearing strength and bulk density	
• Current depositional or erosional rates	• pH profile in cores
Flow & Load Characteristics	*Flow Chemistry*
• Hydrologic regime as measured by flood frequency and magnitude analysis	• Changes in dissolved and sediment-bound concentrations with discharge
• Characterization of stream power and tractive force	• Characterization of hysterises affects
• Suspended sediment concentrations, loads, and variations	• Typical water quality parameters (pH, DOC, temperature, Eh, electrical conductivity)
• Bedload transport rates	• Non-contaminant substance concentrations that may affect contaminant mobility
• Water temperature and turbidity	
Groundwater	*Groundwater Chemistry*
• Location and extent of influent and effluent flow	• Groundwater flow paths and rates
• Rates of water losses and gains to channel	• Contaminant concentrations and loadings from groundwater
• Aquifer characteristics (e.g., hydraulic conductivity, storativity, porosity, confined, unconfined)	• Non-contaminant substance concentrations that may affect contaminant mobility
Basin Characteristics	*Basin Characteristics*
• Area, relief, drainage morphometry	• Weathering rates and products
• Underlying bedrock composition	• Soil chemistry and sorption potentials
• Upland erosion rates and flow patterns	

vary depending on the location and complexity of the river being investigated, the contaminant of concern, and the level of potential risk at the site, among a host of other factors. There are, however, a number of general, task oriented objectives that are applicable to most sites and are therefore worth noting. They are (as modified from the USEPA 2005) to:

(1) Identify the type of contaminants in the various media (water, sediment, biota, etc.) that are of concern;
(2) Determine the three-dimensional (vertical and horizontal) distribution of contaminants within the channel bed, floodplain, and terrace deposits, and determine how this distribution changes through time;
(3) Identify and characterize all sources that are currently or may have historically contributed contaminants to the site, including those located beyond the site boundaries;
(4) Quantify the key geomorphological processes that can remobilize and disperse contaminated particles within the river system;
(5) Examine and quantify the primary chemical and biological processes that affect the transport, fate, and bioavailability of the contaminants;
(6) Determine the pathways through which both human and ecological receptors may become exposed to the contaminants; and
(7) Obtain information to evaluate the potential effectiveness of natural processes in cleaning up the site.

The above objectives clearly demonstrate that the magnitude of investigation required for site assessment can be enormous. Because of its complexity, it is important to begin the assessment process by developing a conceptual model of the physical, chemical, and/or biological system which is being evaluated. Conceptual models describe the primary processes which operate within the system and which therefore require characterization in quantitative terms (Fig. 1.8). In addition, they help to develop hypotheses that require testing, identify processes that require quantification, and provide insights into areas of uncertainty that must be addressed (USEPA 2005). Put differently, conceptual models provide a means of determining not only what processes may be important to site assessment, but what types of data need to be collected and why.

It is important to recognize that separate conceptual models may need to be developed for specific contaminants or groups of contaminants. In contrast to Pb, for example, the bioavailability of Hg is strongly dependent on the degree to which inorganic Hg is converted into organic forms of Hg by sulfate-reducing bacteria. Trying to apply a conceptual model developed to describe the biogeo-chemical cycling of Pb to Hg would likely lead to a seriously flawed investigation. It is also possible that different conceptual models will need to be developed for distinct parts or reaches of the river, even when only one contaminant is of concern.

In contrast to site characterization and assessment, remedial action involves the identification, evaluation, and selection of technologies that are technically feasible, cost-effective, and socially acceptable to reduce the risk posed by a contaminant to

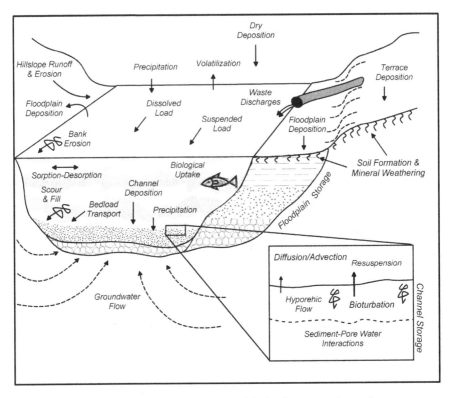

Figure 1.8. Hypothetical conceptual model of sediment-water interactions

an acceptable level. Remedial action is based on the results of site assessment and should serve as a logical continuation of the assessment process. In practice, both are often conducted concurrently in an iterative manner; data from site assessment is used to develop remediation alternatives, and the identified alternatives dictate the types of additional data that need to be collected as part of the assessment process.

In most countries, highly formalized programs have been developed to cleanup sites contaminated by hazardous wastes. One of the largest, and most complex, of these is the remedial investigation and feasibility study (RI/FS) process associated with the U.S. Superfund program. The basic components of the RI/FS process are shown in Fig. 1.9 and discussed in detail by the USEPA (1988). Our intent in the following chapters is not to provide an analysis of how to conduct the RI/FS, or any other program, on river systems. Rather it is to provide an understanding of the interaction between geomorphic and geochemical processes, as inherent in the geomorphological-geochemical approach, that are essential to effectively conduct site assessments and remediation.

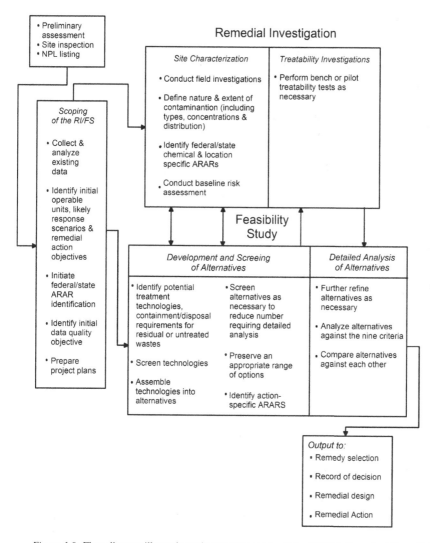

Figure 1.9. Flow diagram illustrating primary components of the USEPA Superfund Remedial Investigation and Feasibility Study (From USEPA 1988)

1.6. THE GEOMORPHOLOGICAL-GEOCHEMICAL APPROACH

Before the 1970s, relatively little was known about contaminated environments. Our knowledge base has improved dramatically during the past three decades and data are now accumulating faster than their meaning can be assimilated. Much of the collected data pertain to the assessment and remediation of contaminated groundwater or areas of contaminated soil. In fact, it seems fair to say that the methodologies used by regulatory agencies in the U.S. and elsewhere were primarily derived by learning from the successes and failures of cleaning up contaminated soils and subsurface waters. Riverine environments represent significantly different settings in that they are much more dynamic and tend to be geologically, hydrologically, and geochemically more variable than either soil or groundwater systems. As a result, what may represent an excellent approach to the assessment and remediation of soils or groundwater may be completely inadequate for rivers.

In the following chapters, we will examine a geomorphological-geochemical approach to the assessment and remediation of metal contaminated rivers. The term was first coined by Macklin et al. (1999) while determining the effects of metal mining on river basins. The approach is based, however, on a multitude of investigations conducted during the past several decades, including those by Macklin and his colleagues. The *geomorphological-geochemical approach* represents the combined and integrated use of geomorphic and geochemical principles to gain an in-depth understanding of the transport and fate of contaminants in riverine environments.

An important characteristic of the geomorphological-geochemical approach is its focus on processes. A *process* can be defined as the action involved when a force induces a physical or chemical change in the materials at or near the Earth's surface (Ritter 1978). In other words, it is the means through which one thing is produced from something else. A process-oriented approach is needed because it goes beyond the simple description of the river channel and its associated drainage basin. Rather it provides for an understanding of the cause and effect relationships between the river's components. These relationships are required to enhance the predictive capabilities of the conceptual or mathematical models which are used to characterize contaminated riverine environments.

A focus on processes necessarily requires that we view rivers as systems. A *system* is nothing more than a collection of related components (Ritter 1978). The systems approach in geomorphology has become extremely sophisticated in recent years (Chorley and Kennedy 1971), and a variety of system types have been identified and used in the geomorphic and hydrological analysis of rivers and river basins (Schumm 1977). For our purposes, the river (or fluvial) system will generally be defined as the drainage basin and all of its component parts (e.g., hillslopes, floodplains, channel networks, etc.). It would make no sense, for example, to remediate a river channel if the adjacent floodplain or hillslopes remain contaminated. Subsequent floods would simply remobilize the hillslope or floodplain materials and re-contaminate the remediated reach. Thus, we typically need to focus on the entire watershed and all of its components, although some parts of our analysis will undoubtedly require investigations at much smaller scales.

The geomorphological-geochemical approach is most applicable to rivers contaminated by reactive, hydrophobic contaminants (e.g., trace metals) which tend to be bound to sediment. The reason for this is that because most of the contaminants are associated with particulate matter, geomorphic processes governing the erosion, transport and deposition of sediment play a key role in determining how contaminants are moved through and distributed within river valleys. In addition, geomorphic alterations in channel form can significantly alter geochemical processes, thereby affecting such things as contaminant bioavailability and uptake.

The key principles inherent in the geomorphic-geochemical approach include the following:

Contaminant concentrations, transport, and fate within a river channel are controlled by basin-wide processes: Rivers are a product of their drainage basin characteristics and processes. The amount and composition of the water and sediment within the channel, for example, reflect hillslope processes occurring beyond the boundaries of the river channel. Thus, site characterization and assessment must consider processes operating over a range of spatial scales, up to and including the entire catchment.

Hydrophobic contaminants are transported largely with the particulate load: The strong affinity of reactive contaminants for particulate matter means that the transport of trace metals and other hydrophobic contaminants are predominantly driven by the movement of contaminated particles. Thus, the rate at which sediment-borne trace metals are eroded from the channel bed and banks and/or moved downstream is controlled by basin and river processes. These process rates can be determined in part using sediment transport functions developed during the past 50 years for non-contaminated particles. In addition, physical analyses can be used to determine what fraction of the total volume or weight of material transported by the river is moved in solution, in suspension or along the channel bed.

Sediment-borne contaminants are partitioned into specific depositional units: Contaminants are not uniformly distributed within alluvial (river) sediment, but are associated with specific types of particles possessing certain physical and geochemical properties. As a result, sediment-borne contaminants are partitioned into discrete sedimentary packages ranging in size from layers a few millimeters in thickness, to macroform depositional features on the channel bed, to large-scale floodplain and terrace deposits. Documenting the means through which this partitioning occurs allows contaminant distributions to be determined by means of mapping alluvial features or sedimentary packages. In addition, it allows for a prediction of chemical hotspots using an understanding of the means through which contaminants are disseminated through the river system.

Channel response to disturbance alters the physiochemical environment: Rivers are not static features, but adjust their form in response to tectonic activity, changes in climate or land-use and a variety of other factors. When these adjustments are significant, they can dramatically alter the physiochemical setting and, thus, such factors as the chemical concentrations within the water column and/or the rates of contaminant transport and storage along the river valley. Geomorphic analyses of the site's history is therefore required to determine how past instabilities have affected contaminant distributions, and the potential for future geomorphic adjustments to alter contaminant transport rates and availability, among other things.

Alluvial (river) deposits contain a record of contaminant influx to the river and the rates of river recovery. A time series of depositional events are preserved within alluvial deposits. When combined with geochemical data, deposit dating allows chronologies of contaminant influx to the river to be constructed over time scales which often exceed that of direct water quality monitoring. In addition, an analysis of the temporal changes in sediment-borne contaminant concentrations following the removal of a contaminant source(s) provides valuable insights into the rate of river recovery.

The successful use of many remediation technologies on contaminated rivers is influenced by geomorphic processes: Fluvial (river) processes strongly influence the likelihood of contaminant remobilization by physical and chemical processes as well as the potential for sites to become re-contaminated during subsequent flood events. For example, where in situ remediation technologies are used to contain the contaminants in place, containment will depend on the erosional and depositional processes that occur during large flood events. Thus, the selection of remedial technologies must be based on a geomorphic as well as geochemical understanding of the remediated site.

Channel reconstruction and restoration are both influenced by geomorphic processes, and are affected by them: Many remediation techniques, such as excavation or dredging, can significantly disrupt riverine ecosystems. Thus, some form of river rehabilitation or restoration is often required following the application of the site remediation strategy to improve riparian or aquatic habitat. The successful development of these restoration methods requires a detailed analysis of the geomorphic processes operating along the remediated reach, and where the contaminated sediments are left in place, their potential to remobilize contaminated particles.

Each of the primary principles of the geomorphological-geochemical approach listed above will be discussed in detail in the following chapters. Given the diversity of organic and inorganic contaminants that exist, and the extreme variability in their biogeochemistry, it is simply impossible to examine all the various contaminant types herein. Therefore, to focus our discussion, we will limit our analysis to rivers contaminated by trace metals and metalloids. Nonetheless, it should be recognized that the basic tenets of our analysis are applicable to rivers contaminated by other reactive, hydrophobic substances.

1.7.   SUMMARY

There is a wide variety of natural and anthropogenic contaminants found in river water and sediment. These contaminants can be broadly separated into organic and inorganic forms. One of the most important groups of organic contaminants are the POPs (Persistent Organic Pollutants) which include a number of pesticides, PCBs, dioxion, and furans. Other organic contaminants of concern are PAHs and various pathogens (e.g., bacteria, viruses, protozoa, and parasitic worms). The inorganic pollutants of common concern include nutrients (e.g., N and P), trace metals and metalloids (e.g., As, Cd, Hg, Pb, and Zn), and radionuclides. In the U.S., data from the National Sediment Quality Survey suggests that trace metals were the most commonly observed contaminant to pose a threat to human or aquatic life (Tier 1 plus Tier 2 stations), followed by PCBs, pesticides, and PAHs. These pollutants are derived from multiple sources, two of the most important of which are agricultural lands and point and nonpoint sources in urban areas. Other significant sources include the atmosphere, industrial discharges, mining operations, and landfills. Similar types and sources of contaminants are found in most other developed countries which conduct some form of national water quality assessment.

The primarily purpose for investigating a contaminated river is to determine the nature and extent of contamination and to evaluate whether the contaminant poses an unacceptable risk to human health or the environment (USEPA 2005). Where unacceptable risks are encountered, it may be necessary to remediate the site to reduce the identified risks to within acceptable values. Thus, site investigation can be viewed as consisting of two closely related components: site assessment and remedial action (Soesilo and Wilson 1997). The primary intent of site assessment is to characterize the river system in terms of the types of contaminants and their associated sources, the history of contaminant release, the areal extent and magnitude of contamination, and the rates of contaminant migration, among a host of other things. These data are subsequently used in determining various types of risk. Neither of these components is easy to accomplish as past investigations have demonstrated the enormous complexities inherent in the cycling of pollutants in riverine environments. In this book, we will apply a geomorphological-geochemical approach to site investigation which (1) uses geomorphic principles to gain a more in-depth understanding of the transport and fate of contaminants in rivers than can be obtained by studying biogeochemical processes alone, and (2) applies this knowledge to the development of site assessment, remediation, and restoration strategies.

1.8.   SUGGESTED READINGS

Dunnette, DA (1992) Assessing global river water quality: overview and data collection. In: Dunnette DA, O'Brien RJ (eds) The science of global change, the impact of human activities on the environment. America Chemical Society Symposium Series 483:240–259.

USEPA (1997) The incidence and severity of sediment contamination in surface waters of the United States, vol 1, The national sediment quality survey. EPA-823-R-97-006.

USEPA (2000) National water quality inventory: 2000 report. EPA-841-R-02-001.

CHAPTER 2

SEDIMENT-TRACE METAL INTERACTIONS

2.1. INTRODUCTION

The affinity of trace metals and many other contaminants for particulate matter ultimately results in the storage of potentially toxic substances within channel, floodplain and terrace deposits. It is, perhaps, tempting to interpret the potential environmental impacts of these constituents solely on the basis of their total concentration within the alluvial materials. From this simplistic perspective, deposits with the highest concentration pose the greatest risk to ecosystem or human health. Unfortunately, the risks posed by sediment having similar trace metal concentrations are not necessarily the same in every river system. Trace elements in one deposit may be bound to different materials (substrates) and by different mechanisms than another, and while the elements may be loosely bound and bioavailable in the former, they may be tightly bound and non-available in the latter (Horowitz 1991). Moreover, the likelihood that a trace element will be released back into the water column as a result of future changes in the physiochemical environment is directly dependent on how it occurs in the sediment and the mechanism through which it is bound to individual particles. Thus, prediction of the impacts of contaminants on human or ecosystem health requires an analysis of the chemical partitioning of contaminants, in our case trace metals, within any given alluvial deposit.

Much of the available information regarding sediment-trace metal interactions is derived from the investigation of processes that are not directly associated with rivers. These include the formation of ore deposits, the development of soils and soil profiles, and the retention and release of contaminants from groundwater moving through porous or fractured media. In the following pages, we will utilize data from all of these topical areas to examine the relationships between sediment and trace metals in aqueous environments, particularly the physical and chemical partitioning of trace metals within sedimentary deposits. This format is well suited to our geomorphic approach in that the partitioning can be directly related to the physical processes responsible for contaminant transport in rivers as well as the potential for trace metals to be remobilized by chemical processes operating within any given depositional site.

Solution-particle interactions are extremely complex; thus, we will only be able to examine the most important aspects of the topic in the following pages. Those interested in greater detail are referred to a number of excellent discussions, from which much of the following text is derived, including Horowitz (1991), Krauskopf and Bird (1995), Langmuir (1997), and Deutsch (1997).

2.2.   PHYSICAL PARTITIONING OF TRACE METALS IN SEDIMENT

The chemical and physical composition of alluvial sediment generally reflects the underlying geology of the watershed and the mechanical and chemical processes involved in rock weathering. Most rivers exhibit a wide range of particle types, each of which can be described by a suite of physical parameters such as the dimensions of their long-, intermediate, and short-axes (i.e., grain size), their surface area, their specific gravity, and their shape as defined by sphericity and rounding. If one could collect and analyze these particles individually for trace metals, it would be apparent that they do not exhibit the same elemental concentrations, nor do they have an equal propensity to retain metal contaminants. Part of the difference in their chemical nature rests on the fact that reactions responsible for the collection, concentration, and retention of trace metals occur on the particle's surface. Thus, the nature and surface area of the particles, which can be highly variable, is an important control on trace metal accumulation. In general, trace metal accumulation is enhanced by an increase in surface area. Thus, a single, large particle could potentially accumulate more metals than a single, smaller grain. However, given an equal mass of material, there are many times more smaller-diameter particles than larger particles and, thus, surface area increases with decreasing grain size of the sediment (Fig. 2.1). Take, for example, sediments which are represented by perfect spheres. The surface area for very coarse sand, calculated using the equation $4\pi r^2$, is $11.3\,\text{cm}^2/\text{g}$. In contrast, the surface area for very fine clay is $45,280\,\text{cm}^2/\text{g}$. It follows, then, that finer-grained alluvial materials with a higher surface area per unit mass have a larger area over which surface reactions can proceed. In addition, the added surface area provides more space for metals to accumulate, even when chemical reactions between the substrate and the metal constituent do not occur (Jenne 1976).

Actual measurements on natural sediments demonstrate that surface area is not only dependent on grain-size, but a host of other factors as well. One of the most important of these is grain composition. Förstner and Wittmann (1981), for example, show that for sediment $< 2\,\mu\text{m}$ in diameter, surface areas can vary from 10 to $1,900\,\text{m}^2/\text{g}$, depending on the composition of the material and its influence on particle shape (Table 2.1). Mineral composition also has an important influence on the type and extent of particle irregularities such as cracks, ridges, micropores, and depressions, all of which can add greatly to the surface area of an individual grain.

In many rivers, surface area of the sediment is not only controlled by the nature of the particle, but by coatings of reactive substances that partially or entirely cover

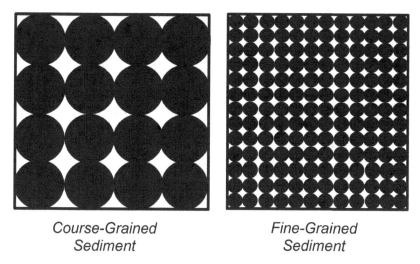

Course-Grained Sediment   Fine-Grained Sediment

Figure 2.1. Surface area of a large, individual particle is clearly larger than that of a smaller grain. However, per unit mass, smaller particles exhibit more surface area than larger grains because of the large number of smaller grains that exist (From Horowitz 1991)

it. One of the most detailed studies of the influence of grain coatings on surface area was conducted by Horowitz and Elrick (1988). Their data from rivers in highly diverse geologic and hydrologic settings demonstrated that the influence of grain coatings on surface area and, therefore, elemental concentrations is grain size dependent. For sand-sized and larger sediment, coatings composed of carbonates, Fe and Mn oxides and hydroxides, and organic matter are rough in comparison to the surface of the underlying particles (dominated by quartz and primary silicate minerals). Thus, while the coatings constitute a relatively minor percentage of the sample by weight (<5%), their irregularities contribute significantly to the total

Table 2.1. Variations in surface area as a function of particle composition

Grain composition	Surface area (m^2/g)
Clay Minerals	
Kaolinite	10–50
Chlorite	30–80
Montmorillonite	50–150
Fe Hydroxide (freshly precipitated)	300
Calcite ($<2\,\mu m$ in diameter)	12.5
Humic Acids from Soils	1900

Data compiled by Förstner and Wittmann (1979); table after Horowitz (1991)

surface area of the sediment. In contrast, coatings on silt-sized and finer particles fill in surface irregularities such as cracks and micropores, reducing the overall surface area of the material. Coatings have also been found to cement fine particles together, creating larger agglomerated grains, thereby reducing the sediment's surface area further.

In practice, relatively few investigations have attempted to relate metal concentrations to the surface area of alluvial (river) sediments. Where studies have been conducted, the expected relationship of increasing metal concentration with increasing surface area has generally been observed (Oliver 1973; Horowitz and Elrick 1988), although the relationship appears to be logarithmic rather than linear (Fig. 2.2). Most studies attempting to quantify the nature of the physical partitioning of trace metals in aquatic environments have examined the relationship between grain size and metal content. These investigations have shown that there is a tendency for metal concentrations to increase with decreasing grain size, a trend that undoubtedly reflects the close association between grain size and surface area. Horowitz and Elrick (1988) argue that the lack of surface area data is surprising given the theoretical basis for its use. This deficiency may be related to the perception that surface area and grain size data are interchangeable, or to the greater number of laboratories that are capable of analyzing sediment for grain size distributions rather than surface area. Nonetheless, surface area analyses have several advantages over typically utilized methods for particle size determinations. The most important advantages are their non-destructive and non-contaminating nature that allows the materials to be used in subsequent geochemical analyses. When working with larger volumes of sediment, this consideration may not be important, but it can be significant when sample sizes are small, such as those involving suspended sediment.

Quantification of the relationship between grain size and metal concentrations within alluvial deposits is typically carried out using one of two methods. First, individual grain size fractions can be collected, analyzed and compared to determine elemental distributions within the sample. These types of data generally show a

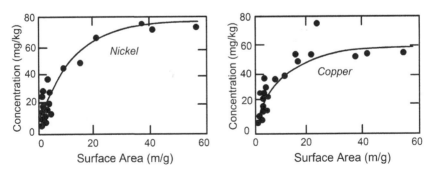

Figure 2.2. Influence of surface area per unit mass on nickel and copper concentrations within the Ottawa and Rideau Rivers, Canada (Data from Oliver 1973; figure from Förstner and Wittmann 1979)

semi-systematic increase in metal concentrations with decreasing grain-size below about 63 μm (Fig. 2.3). Elevated metal concentrations within the finer sediments are thought to be due to (1) the influences of surface area on metal content (as noted above), (2) larger quantities of reactive substances such as organic matter or Fe and Mn oxides and hydroxides in the silt- and clay-sized sediment fraction, and (3) differences in grain mineralogy. In contrast to fine sediment, sand-sized particles are generally dominated by non-reactive minerals such as quartz, feldspar and other primary silicates.

There are some data to suggest that concentrations of metals in extremely fine materials are slightly lower than those found in the rest of the fine-grained sediment fraction (Fig. 2.3). These lower concentrations probably result from influences of mineral composition and grain coatings on surface area as described above. Many investigations have also observed slightly elevated metal concentrations within various sand-sized fractions, particularly within contaminated rivers. For fine-sand, the elevated metal contents may be related to an abundance of metal-enriched detrital minerals (Förstner and Wittmann 1979); within the medium to coarse sand fraction elevated metal concentrations are likely to be related to the introduction of coarse, metal-enriched sediment to the river from pollutant sources (Salomons and Förstner 1984).

The second and more commonly used method of quantifying the relationship between grain size and metal concentrations within alluvial deposits is to correlate the percent of the deposit composed of a selected grain size fraction to the concentration of a given metal measured in the sample (Fig. 2.4). A high correlation coefficient is assumed to indicate a strong degree of physical partitioning within the materials. More importantly, the grain size fraction with the highest correlation coefficient is generally regarded as the chemically active phase, and will represent the size fraction in which most of the metals are found in the riverine environment.

Not all alluvial sediments are characterized by increasing metal concentrations with decreasing particle size. Moore et al. (1989) found that within the Clark

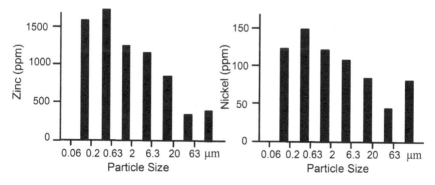

Figure 2.3. Nickel and zinc concentrations for selected grain size fraction within sediments from the Lower Rhine River (Modified from Förstner and Wittman 1979)

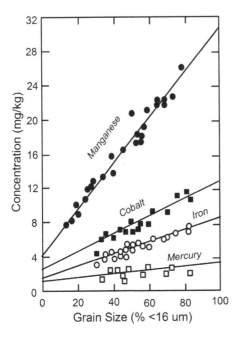

Figure 2.4. Relationship between selected elemental concentrations and the percent of sediment <16 μm in size for samples from the River Ems, Germany (From de Groot et al. 1982)

Fork River, Montana, all of the size fractions significantly contributed to the high metal concentrations observed in the alluvial deposits. The distribution of metals within the deposits results from the introduction of crushed mica and quartz, primary sulfides (ore), and smelter slag glasses to the river from mining operations, and to the fact that the waste materials not only contained large amounts of metals, but had a wide, and unnatural, range of size distributions. Other studies have come to similar conclusions for rivers impacted by mining (Miller et al. 1998, 2003).

It is important to recognize that the above discussion focuses on the *concentration* of metals within a given grain-size fraction. It is possible that while the highest concentrations are associated with fine sediment, there is so little of it in river deposits that most of the metals by weight are actually associated with coarse materials. For example, in the Carson River of west-central Nevada, Miller and Lechler (1998) found that Hg concentrations within channel bed sediment were an order of magnitude higher in the <63 μm fraction than the sand fraction. However, the channel bed deposits generally contained <5% silt and clay. Thus, most of the Hg per total mass of material was associated with fine sand-sized Ag and Au amalgam grains. These observations demonstrate that it is not only important to determine the physical partitioning that occurs in alluvial deposits of the river of interest, but the sedimentology of the deposits in general.

2.3. CHEMICAL PARTITIONING OF TRACE ELEMENTS IN SEDIMENT

In the previous section, we examined the partitioning of trace elements in alluvial deposits as a function of the physical characteristics of the sediment. Chemical partitioning also occurs within riverine environments, and it plays a significant role in the accumulation and retention of trace elements in alluvial deposits. The analysis of chemical partitioning has historically followed one of two approaches (Horowitz 1991): (1) an assessment of the mechanisms through which trace elements are bound to particulate matter, or (2) a determination of the phase (or composition) of the material with which the constituents are associated. In reality, binding mechanisms are compositionally dependent and, thus, most studies of chemical partitioning utilize a combination of the two techniques (Horowitz 1991).

2.3.1. Mechanistic Associations

Along any given reach of a river, water and sediment interactions are occurring in a host of different environments. Within the channel, solute-particle reactions are occurring between sedimentary grains and the water in which they are suspended. River water is also in contact with the materials that form the channel bed and banks, and may temporally move into pore spaces within the deposits with which they are comprised. The pores of floodplain and terrace deposits may also be filled with waters that foster chemical reactions. These pore waters may migrate toward the channel, or may carry contaminants away from the channel where they may affect local groundwater resources. Thus, they are an important part of the overall picture in terms of the assessment of contaminant transport and cycling.

Constituents dissolved within waters of any of these environments may accumulate on particle surfaces through a combination of mechanisms collectively referred to as *sorption*. Sorption strongly influences metal mobility, and includes three primary processes (Krauskopf and Bird 1995): (1) the *precipitation* of a surface coating that is compositionally different from that of the underlying host grain, (2) incorporation of ions into the crystal structure of the mineral by a set of mechanisms collectively referred to as *absorption*, which may involve diffusion as well as dissolution and reprecipitation reactions, and (3) *adsorption*. The latter process refers to the removal of dissolved species from solution and their accumulation on a particle surface without the formation of a distinct, three-dimensional molecular structure. In natural systems it is often difficult to ascertain the degree to which these three processes are involved in the removal of species from solution. In fact, their importance will change spatially along the river and through time with changes in physiochemical conditions. Nonetheless, it is generally agreed that the most important controls on metal concentrations and, thus their mobility, include some combination of the processes of precipitation/dissolution and adsorption/desorption.

2.3.1.1.   Precipitation and dissolution

Rivers can be thought of as containing a combination of solid phases (consisting of minerals, noncrystalline (amorphous) solids, and organic material) and an aqueous phase (including water and all of its dissolved constituents). River waters may also contain one or more gaseous phases, although for our purposes, gases will be ignored. *Precipitation* and *dissolution* reactions involve the transfer of mass from the solid to the aqueous phase and vice versa (Deutsch 1997). The amount of solid material that dissolves in a solution is described by the mineral's solubility. All minerals are soluble to some degree in pure water and much effort has been expended in the laboratory to quantify their solubility under equilibrium conditions at 25 °C and 1 atm of pressure. At equilibrium, both dissolution and precipitation may continue to occur, but the two processes will balance one another so the effective concentrations of the dissolved constituents remain constant through time.

As an example, let's look at the dissolution of barite, which may be found in rivers draining basins that contain ore deposits of Pb and Zn. We will start by placing a small crystal of barite ($BaSO_4$) in a beaker of pure water at 25 °C. The dissolution reaction for barite is:

(1) $BaSO_4 \leftrightarrow Ba^{2+} + SO_4^{2-}$

The above reaction will continue until the water is saturated with the dissociated components of the solid phase, at which point equilibrium concentrations of the constituents in the water have been reached. The ratio of the activities (effective concentrations) of the components involved in the chemical reaction can be expressed by an equilibrium constant (K). The equilibrium constant is determined by multiplying together the chemical activities of the constituents produced by the reaction, and dividing that product by the activity of the reactants. For the case of barite, at 25 °C,

(2) $K = \dfrac{(a_{Ba^{2+}})(a_{SO_4^{2-}})}{(a_{BaSO_4})} = 10^{-9.98}$

The activity of the solid (barite) is set at one as the amount that is present does not influence the outcome of the reaction (K) as long as some of the solid remains. Thus, the equation simplifies to:

(3) $K = (a_{Ba^{2+}})(a_{SO_4^{2-}}) = 10^{-9.98}$

In this rather simple laboratory experiment, the only source of Ba^{2+} and SO_4^{2-} to the beaker of water is the barite, and the effective concentrations for both constituents is $10^{-4.99}$ moles per liter when the system is at equilibrium. This follows because the dissolution of one barite molecule adds equal proportions of the two constituents to the water and their product must equal $10^{-9.98}$. In natural systems, however, multiple sources of the two reaction products are likely to be present and, thus, the amount of barite that is dissolved (i.e., its solubility) is reduced because some is already present within the water. The lowering of the solubility of a mineral as a result of the occurrence of one of its reaction products from another source in the solution is referred to as the *common-ion effect*.

Values of K are commonly compared to the ion activity product (IAP) to determine whether surface- or groundwaters are saturated with respect to a given mineral. The expression for determining the IAP for a mineral is equivalent to that used to calculate the equilibrium constant, but the chemical activities used in the equation are based on solution concentration data obtained from water samples. Thus, the expression used to calculate the IAP for barite is:

(4) $IAP_{barite} = (a_{Ba}^{2+})(a_{SO_4}^{2-})$

A convenient method of comparing the equilibrium constant (K) to the IAP is through the use of the mineral saturation index (SI), defined as (Deutsch 1997):

(5) $SI = \log_{10} \dfrac{IAP}{K_{mineral}}$

When,

$SI = 0$, the solution is saturated and in equilibrium with the mineral of interest;
$SI < 0$, the solution is undersaturated with respect to the mineral;
$SI > 0$, the solution is oversaturated with respect to the mineral, and the mineral is non-reactive, perhaps because reaction rates are too slow to limit dissolved ion concentrations within the water.

As seen in Eqs. 2 through 4, the determination of the equilibrium constant and the IAP is based on the activities of the *uncomplexed* constituents dissolved within the water rather than their total dissolved concentrations. This distinction may seem subtle, but it is an extremely important difference because the activity (effective concentration) of the free ions in solution can be quite different from their total concentrations. The differences in total and effective concentrations primarily result from two processes: ion shielding and the formation of aqueous complexes.

The theory behind ion shielding is that cations and anions in solution will attract ions of opposite charge dissolved within the solution. These oppositely charged particles will form a cloud or shield around the cations and anions of interest, thereby inhibiting their participation in chemical reactions (Deutsch 1997). Thus, the activity of a dissolved species is typically less than their total concentration within the solution. In practice, the two quantities are related to one another according to the following expression:

(6) $a_i = \gamma_i c_i$

where a_i and c_i are the activity (effective concentration) and total concentration, respectively, of the dissolved ion of interest, and γ_i is the activity coefficient.

The degree to which ion shielding occurs is directly related to the ionic strength of the solution (I), defined as:

(7) $I = 0.5 \, \Sigma \, c_i \, (z_i)^2$

where z_i is the charge (valence) of the dissolved species i, and c_i is the total dissolved concentration of that species in the solution. When the ionic strength of the solution is known, there are several methods that can be applied to determine the activity coefficient (γ_i) used in Eq. 6, allowing for an estimation of the ion's activity (Deutsch 1997).

The second factor that affects the activity of an ion is aqueous complexation. As mentioned above, the total concentration of the constituent used in Eq. 6 to determine its activity is based only on the concentration of the *uncomplexed* species. Many ions, however, will form multiple organic and inorganic complexes in water, and the formation of these complexes will lower the activity of the species beyond that caused by ion shielding. For example, the dissolution of $CaCO_3$ is expressed as:

(8) $CaCO_3 \leftrightarrow Ca^{2+} + CO_3^{2-}$

However, carbonate not only occurs in its uncomplexed form (CO_3^{2-}), but as a complexed ion with hydrogen to form HCO_3^- and H_2CO_3. In addition, if Mg is present, it may form additional complexes such as $MgHCO_3^-$. Because the activity of the ion of interest is based on its uncomplexed form, in this case, CO_3^{2-}, the dissolved concentrations of the other complexes must be determined and subtracted from its total (carbonate) concentration in the solution to determine the concentration of the uncomplexed ion. The uncomplexed concentration (CO_3^{2-}) can then be utilized in Eq. 6 to determine its activity. In nature, the large number of dissolved ions and ion complexes that exist in natural systems make it extremely difficult to perform these types of calculations "by hand". Thus, numerous specially designed computer models have been developed during the past decade to aid in the calculations of chemical activities (Deutsch 1997).

It is important to recognize that elemental solubility is not only dependent on the concentration of the complexes that are present, but also the *type* of complexes that exist in the solution. In both surface waters and the pore fluids of alluvial sediments, chloride, sulfate, bicarbonate, and, under reducing conditions, sulfide anions are common. In general, the chlorides and sulfates of most trace metals are highly soluble in comparison to those associated with carbonates, hydroxides, and sulfides, which dissolve only slightly in most natural waters (Förstner and Wittmann 1979).

The solubility of the solid phase is also dependent on a number of other factors that reflect the physical and chemical conditions of the water, including its temperature, pH, and Eh (Deutsch 1997). With regards to temperature, most minerals exhibit higher solubilities as the temperature of the water increases. It is for this reason that equilibrium constants (K) are usually reported for solutions at 25 °C.

A notable exception occurs for calcium and magnesium carbonate minerals, which become less soluble as water temperatures rise (Bathurst 1975).

The pH of the water affects the formation of aqueous complexes and for those constituents that form strong complexes, including most trace metals, it can have a large affect on ion solubility (Deutsch 1997). In addition, pH can influence the ease with which oxidation and reduction reactions occur. Oxidation and reduction reactions involve the transfer of electrons from one element to another, and result in a change in the charge (valence) of the primary constituents. Of significance to our discussion, some elements, including many trace metals, are relatively insoluble in one valence state and more easily dissolved in the other. For example, Fe^{2+} can be oxidized to Fe^{3+} if it is contacted by oxygenated waters or oxidizing agents. Fe^{2+} is significantly more mobile, and exhibits a higher solubility, than its oxidized form, Fe^{3+}. Thus, the oxidative/reductive state of the system, described by its redox potential (Eh) or alternatively electron activity (pe), will not only dictate the relative proportions of the two forms that exist in the system, but the overall mobility of Fe within the aquatic environment. When Eh is highly positive, more oxidizing valence states of the element will occur, whereas lower Eh values will favor more reduced valence states. It is important to recognize that Eh values are relative measures of oxidizing and reducing conditions because what may be oxidizing for one element may be reducing for another (Deutsch 1997).

Because of the influence of pH and Eh (pe) on mineral solubility and aqueous complexation, Eh-pH (or pe-pH) diagrams are commonly used to assess the types of precipitates and aqueous species that are likely to occur given the physiochemical state of the system. Discussion of Eh-pH diagrams can be found in almost any geochemical textbook, but in general, they define the stability fields for both the solid and aqueous phases of an element under varying Eh (pe) and pH conditions. Generalized pe-pH diagrams for several elements are presented in Fig. 2.5. In pH neutral or mildly alkaline waters that are oxygenated or slightly reducing (positive or slightly negative Eh values), the solubility of many heavy metal cations is low because they precipitate as oxides, hydroxides or carbonate minerals. For example, minimum solubility values for a variety of trace metal hydroxides falls between 9 and 12, whereas a decrease in pH below four may lead to the complete dissolution of the solid phase (Förstner and Wittmann 1979). Trace metal solubility is also low for a wide range of pH values in waters characterized by moderately to strongly reducing conditions, but only when enough sulfur is present to allow for the formation of sulfide minerals (Fig. 2.5).

2.3.1.2. *Adsorption-desorption processes*

Earlier it was stated that nearly all of the reactions pertinent to our discussion occur at the interface between a mineral and the surrounding solution. With respect to adsorption and desorption processes, the accumulation of trace metals is strongly dependent on the sign, strength, and cause of the electrical charge that occurs at the

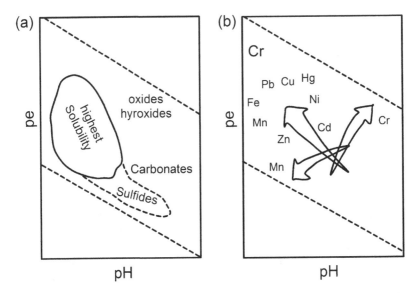

Figure 2.5. Generalized pe-pH diagrams showing (a) stability fields for substrates that influence trace metal solubility, and (b) trends in increasing solubility for selected trace metals. Dashed line in (a) indicates formation of insoluble sulfide minerals when sufficient sulfur is available (From Bourg and Loch 1995)

surface of a particle. The total electrical charge on the mineral surface ($\sigma_{min.\ surface}$) can be expressed as (Krauskopf and Bird 1995):

$$(9)\qquad \sigma_{min.\ surface} = \sigma_{psc} + \sigma_{reaction}$$

where σ_{psc} is the permanent structural charge and $\sigma_{reaction}$ is the charge due to chemical reactions that occur between the exposed ions in the mineral and the solution.

The permanent structural charge is independent of the composition of the surrounding solution, but, as the name implies, depends on crystallographic structure (and defects in the structure) of the mineral. One way in which it can be formed is through the isomorphic substitution of ions with different charges for ions within the mineral's crystalline structure. For example, Al^{3+} can replace Si^{4+} in the tetrahedral layer, and Mg^{2+} can substitute for Al^{3+} in the octahedral layer of clay minerals, producing a net negative charge. A negative charge can also be produced when octahedral Al^{3+} or interlayer K^+ ions are missing (see the section on clay minerals later in the chapter for a discussion on clay mineralogy). It should be no surprise, then, that clay minerals such as smectites, illites, and vermiculites exhibit a signif-icant permanent structural charge. In contrast, the permanent structural charge is close to zero for hydrous oxide minerals and some simple layer clay minerals where isomorphic substitutions are limited (e.g., kaolinite) (Krauskopf and Bird 1995).

The electrical charge produced by chemical reactions varies significantly with the composition of the solution and is particularly important for silicate minerals, metal oxide and hydroxides, and organic matter. The driving factor in this case is the occurrence of charged adsorption sites associated with exposed metal (M) and oxygen (O) ions that are not fully coordinated at the surface as they are within the interior of the crystal (Deutsch 1997). One important kind of reaction involves the addition or removal of a hydrogen (H^+) atom from a hydroxyl (OH^-) ion that is attached to a metal cation exposed at the mineral surface (Fig. 2.6). For example, hydroxyl groups bound to metal cations may gain a hydrogen atom according to the following reaction:

(10) $M\text{-}OH + H^+ \leftrightarrow MOH_2^+$

The reaction produces a net positive charge on the mineral surface, and, because of the need for H^+ ions, is favored by acidic solutions. Hydroxyls may also be lost from the surface in low-pH solutions leaving a positively charged cation exposed, as expressed by the following equation:

(11) $M\text{-}OH \leftrightarrow M^+ + OH^-$

In more basic solutions, H^+ ions tend to be lost from the hydroxyl groups to form water, as in:

(12) $M\text{-}OH + OH^- \leftrightarrow M\text{-}O^- + H_2O$

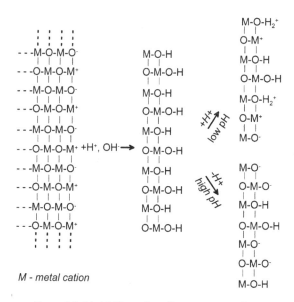

M - metal cation

Figure 2.6. Model illustrating affects of pH on H^+ and OH^- driven adsorption (Adapted from Deutsch 1997)

The important point to be gained from this discussion is that the charge on the mineral surface is variable and can change sign depending on the pH of the surrounding solution. At low pH values, mineral surfaces are positively charged and, therefore, tend to adsorb anions, whereas at high pH, the surface is negatively charged and tends to adsorb cations (Fig. 2.7). The pH at which the concentration of positively charged surface sites balances the concentration of negative sites is referred to as the *pristine point of zero charge* (PPZC). It is important to note, however, that both positively and negatively charged sites will be present at any give time, but one will predominant over the other (Fig. 2.7).

The accumulation (adsorption) of anions or cations on the mineral surface tends to reduce the net charge exhibited by the mineral. However, it is not uncommon for the adsorbed ions themselves to carry a charge that has not been completely neutralized by the reactions at the mineral surface. The net effect is the formation of an electric double layer that can be viewed as an inner layer of ions that are attached to the mineral and a diffuse outer layer of ions that are free to move around in the surrounding solution (Fig. 2.8). In the diffuse outer layer, the electrical charge decreases away from the mineral surface until it reaches zero. Various models also have been applied to describe the nature of this decreasing trend in charge, perhaps the most common one denoting an exponential decline (Fig. 2.8).

One effect of the interrelations between pH and surface charge is the amount of a cation (or anion) that can be adsorbed changes dramatically with alterations in the pH of the solution. The changes that occur for specific trace metals have been extensively examined in the laboratory, generating curves such as those presented in Fig. 2.9. In general, pH and cation adsorption are directly related because the negatively charged sites at the mineral-solution interface increases as pH rises. In most cases, major changes in cation adsorption occur over a very narrow range of pH values. The pH range over which this change occurs is known as the *pH edge*. A decrease in the number of adsorption sites, either by an increase in

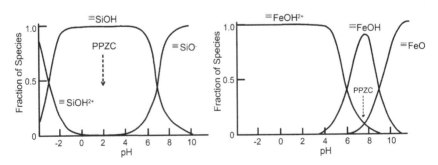

Figure 2.7. Changes in the distribution of chemical species on silica gel (a) and ferrihydrite (b) as a function of pH. The pristine point of zero charge (PPZC) is the pH at which the concentration of negatively charged surface sites are balanced by the positively charged sites (From Langmuir 1997; based on Healy 1974)

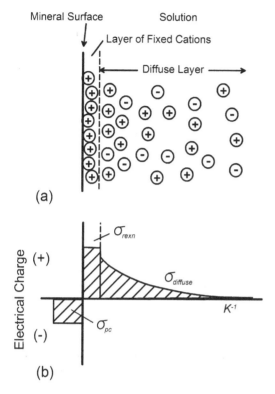

Figure 2.8. (a) Illustration of the electric double layer, showing ions fixed to the mineral surface (the Stern layer), and diffuse ions within the Gouy layer; (b) Net charge observed at the mineral-solution interface. K^{-1} represents distance from the mineral surface (From Krauskopf and Bird 1995)

solution concentration of the adsorbed cations (sorbate), or by a reduction in the amount of absorbing materials (sorbent) that is present, causes the pH edge to shift toward higher pH values (Deutsch 1997). Anion adsorption exhibits a trend similar to that of cations, although the trend is reversed; anion adsorption increases as pH values decline. Anion adsorption-pH plots, however, commonly exhibit a "plateau" at high sorbate/sorbent ratios due to a saturation of anion adsorption sites.

2.3.1.2.1.   Distribution coefficients, adsorption isotherms, and complexation models: An important question for many site assessment programs, particularly those associated with the migration of trace metals through the subsurface, is how far a contaminant can migrate from the point of release before sorption by the surrounding geological materials lowers its concentration below developed action levels. The answer to this question rests partly on the affinity of a given trace metal for the sediment at a site. One way to address this question is to examine the

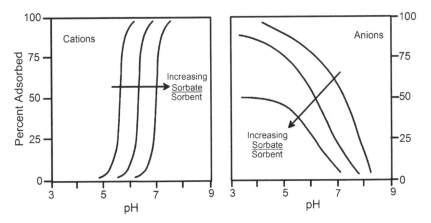

Figure 2.9. Alterations in cation and anion adsorption with variations in pH (From Deutsch 1997)

partitioning that occurs between the water (solution) and the particles for a given metal, where the sorption reaction can be written as:

$$(13) \qquad C_{aq} + X \leftrightarrow C_{ab}$$

where X is the weight of adsorbent (sediment), C_{aq} is the concentration of a trace metal in solution, and C_{ab} represents the concentration of the metal on the sediment, X. The relationship between the constituents of the reaction can be expressed by a distribution coefficient (Kd) that reflects the partitioning of the substance between the solution and the particulate phases. The distribution coefficient for the above reaction is:

$$(14) \qquad Kd = \frac{(C_{ab})}{(X)(C_{aq})}$$

If it is assumed that X is very large in comparison to the amount of the metal that can be absorbed and, thus, the adsorption sites cannot be filled, X is set to one and the distribution coefficient simplifies to:

$$(15) \qquad Kd = \frac{(C_{ab})}{(C_{aq})}$$

It is important to note that Kd is determined from empirical data that have been generated through laboratory experiments. One type of experiment used to determine Kd involves the use of column tests where sediment from the site of interest is placed in a vertically oriented tube and a solution with a known concentration of a contaminant is passed through the material. The Kd value can then be determined using the expression:

$$(16) \qquad Kd = \frac{[(C_i - C_f)/C_f]}{(V/M)}$$

where C_i and C_f represent the initial and final concentrations of the metal in solution, respectively, V is the volume of solution passed through the column, and M is the mass of the sorbent within the column. A second approach is to use batch experiments where a known amount of sorbent (sediment) from the site is mixed in a vessel with a selected amount of water (also collected from the site) until the amount of contaminant in solution and absorbed on the sediment reaches equilibrium. The amount of material adsorbed can subsequently be determined by subtracting the initial concentration of the constituent in the solution from the final concentration. Langmuir (1997) notes that the data from column and batch experiments often disagree. The causes of the inconsistencies are complex, but are related in part to differences in the experimental conditions between the two methods. In batch experiments, the material is stirred or shaken and, thus, diffusion gradients are reduced or eliminated. Moreover, desorbed ions remain in the solution surrounding the sediment rather than being carried away. The net result is that steady state conditions can be more readily attained since the desorbed products accumulate and diffusion is reduced or eliminated as a rate-controlling factor. In contrast, diffusion gradients are an important consideration in column experiments because the experimental flow conditions more closely match what occurs when stream waters move through alluvial sediment. In addition, the desorbed ions are removed from the immediate vicinity of the sediment surface by the moving water. It follows, then, that column tests, when properly designed, tend to provide more realistic in situ results of sorption to soils and sediments (Langmuir 1997).

These types of laboratory studies have demonstrated that as the initial concentration of the metal in solution increases, the amount of it that will be sorbed also increases. In the simplest case, the final concentration of the dissolved species versus the concentration of the species held on the surface of the solid plots as a straight line (Fig. 2.10). The line is referred to as an *isotherm*, because the experiments are carried out at a constant temperature. The Kd value is equal to the slope of the line, or C_{ads} divided by C_{aq}. High Kd values reflect metal species with a strong affinity for particulate matter, and which therefore have a preference to be adsorbed to the surface of the solid rather than to be dissolved within the surrounding solution. It is important to recognize that Kd is a parameter that includes the removal of dissolved species from solution by multiple sorption processes including precipitation, coprecipitation, and adsorption. It, therefore, tells us little about the mechanisms through which the dissolved species have been removed.

Linear isotherms indicate that the adsorption processes are not altered by changes in solute concentrations. This suggests that the capacity of the sediment to adsorb trace metals is essentially unlimited (Deutsch 1997). However, in natural systems, this is rarely the case and the adsorption capacity of the material will decrease as the adsorption sites become filled. Thus, two additional types of isotherms including the Freundlich isotherm and Langmuir isotherm (Fig. 2.10) have been developed. Both exhibit linear trends at low concentrations of the species on the adsorbing materials. At higher concentrations, the Freundlich isotherm becomes

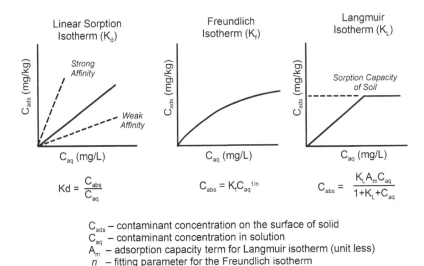

$$Kd = \frac{C_{abs}}{C_{aq}} \qquad C_{abs} = K_f C_{aq}^{1/n} \qquad C_{abs} = \frac{K_L A_m C_{aq}}{1 + K_L + C_{aq}}$$

C_{ads} – contaminant concentration on the surface of solid
C_{aq} – contaminant concentration in solution
A_m – adsorption capacity term for Langmuir isotherm (unit less)
n – fitting parameter for the Freundlich isotherm

Figure 2.10. Adsorption isotherm models (Adapted from Deutsch 1997)

non-linear, reflecting a decrease in sorption as the sorption sites become filled. At some point, the sorption sites essentially become saturated and the dissolved species can no longer be sorbed to the sediment. The point at which this occurs is referred to as the "break through" concentration of the sorbate. Determining the break through concentration is important in that it provides an estimate of the capacity of the material to reduce the mobility of contaminants in solution. Once break through is reached, the sediment is incapable of hindering further migration of contaminants from the point of release. The Langmuir isotherm differs from the Freundlich isotherm in that it possesses a term for maximum sorption capacity in its definition, thereby assuming that a finite number of sorption sites exist (Deutsch 1997). Once that capacity is reached, no additional adsorption can occur, no matter what the concentration is of the dissolved species in the surrounding solution.

In practice, adsorption isotherms must be determined for very specific physio-chemical conditions, including pH, Eh, and competing species concentrations. If any of these parameters change, the nature of the isotherms will likely be altered. Moreover, because the mechanisms through which sorption is occurring is unknown, it is not possible to precisely predict how changes in physiochemical conditions will affect the isotherms that have been previously developed. To overcome this problem, some investigators have turned to surface complexation models (James and Healy 1972; Davis and Leckie 1978; Davis and Kent 1990; Stumm 1992). These models simultaneously consider multiple parameters including pH, ionic strength, the formation of aqueous complexes, and the complexing properties of the solids (Langmuir 1997), thereby allowing for an estimation of the adsorption of a dissolved constituent over a large range of physiochemical conditions. A disadvantage is that the approach requires a set of intrinsic adsorption constants for the sorbents which are currently lacking for many geological materials (Deutsch 1997). Nonetheless, the application of this approach will probably increase as these data become available.

2.3.1.3. *Ion exchange*

Ion exchange is a poorly understood phenomenon in which ions in solution are substituted for those held by a mineral grain. While anionic substitution can occur, it is not as prevalent as cation exchange in aquatic systems with waters exhibiting the typically observed values of Eh and pH. Therefore, our discussion will focus primarily on cation exchange mechanisms.

In its simplest form, the cation exchange reaction can be expressed as:

$$(17) \qquad \text{Ma-substrate} + \text{Mb}^+ = \text{Mb-substrate} + \text{Ma}^+$$

where Ma and Mb represent different metal cations.

The exchange process may involve ions adsorbed to the mineral surface, and thus, can be viewed as a type of adsorption/desorption phenomenon. However, substitution is not limited to the ions adsorbed to the mineral surface, but can involve the replacement of an ion in solution for one held within the mineral's crystalline structure. As a result, a distinction is generally made between adsorption processes and ion exchange processes.

While all minerals exhibit some capacity for ion exchange, the most important substances in alluvial sediment include clay minerals, organic matter, and Fe and Mn oxides and hydroxides. The propensity for a given substance to exchange cations is determined by measuring the number of exchange sites that are present, and is referred to as the material's *cation exchange capacity* (CEC). The CEC is typically expressed as the number of milliequivalents per 100 grams of sediment (meq/100 g), and is most often defined by the uptake and release of cations from a solution. For example, a commonly utilized method measures the uptake of ammonium (NH_4^+) ions from a 1M solution of ammonium acetate at a pH of seven. The CEC is a function of both the mineral's surface charge and its surface area (Table 2.2). Thus, organic substrates, such as humic acids, and Mn oxides can exhibit an exceptionally high CEC.

The exchange capacity of a give substrate is not a constant, but varies as a function of the composition and pH of the solution as well as the type of ions occupying the exchange sites. Take, for example, a clay mineral. At a pH less than about six, hydrogen ions tend to be bound to oxygen atoms at the edges of the crystal. Cations cannot be exchanged with these positively charged sites. However, as pH increases, some of the hydrogen ions are released to the solution, thereby creating new negatively charged exchange sites that increase the clay's CEC. Thus, as was the case for other sorption processes, changes in pH will alter the magnitude to which cation exchange will occur.

Table 2.2. Cation exchange capacity of selected substrates

Grain composition	Cation exchangecapacity (meq/100 g)
Clay Minerals	
Kaolinite	3–15
Illite	10–40
Chlorite	20–40
Montmorillonite	80–120
Smectites	80–150
Vermiculites	120–200
Mn Oxides	200–300
Fe Hydroxides	10–25
Humic Acid (soil)	170–590

Data Compiled by Horowitz (1991)

When more than one type of cation is present in the solution, they will all compete for the available exchange sites. The outcome of this competition is governed by numerous parameters, one of which is the relative concentrations of the cations dissolved within the solution; as the concentration of the cation increases, the amount of it exchanged (or sorbed) to the mineral also increases. However, laboratory studies have shown that when the concentrations of the different cations are the same, some cations have a stronger propensity for exchange than others. These differences are dictated in part by the valence state (or charge) of the cation because the exchange processes is essentially a phenomenon driven by electrostatic forces. Cations with a greater charge tend to have a higher affinity for a given substrate, or (Förstner and Wittmann 1979):

$$M^{3+} > M^{2+} > M^{+}$$

where M represents a metal cation. The radius of the hydrated cation is also important, because the closer it can approach the mineral surface, the greater the attraction between the ion and the substrates surface (once again assuming equal concentrations of the cations in solution) (Deutsch 1997). In addition, ions whose bonds have a strong covalent character tend to be more readily exchanged than those whose bonds are predominantly ionic.

Various attempts have been made to develop selectivity series that define the relative affinity of metal cations for a mineral surface. These *selectivity series* are important because they provide a means of assessing the extent of metal accumulation by a particular substrate (particle) and the potential influence of one ion on the sorption of another in natural geological systems (Krauskopf and Bird 1995). Unfortunately, laboratory tests have shown that the generalized rules for controlling the affinity of a cation for a mineral or non-crystalline substance (e.g., Fe and Mn oxides), as described above, are commonly violated (Krauskopf and Bird 1995). Nonetheless, some useful selectivity series have been put forth.

One such series for divalent cations was presented by Deutsch (1997), using data from Appelo and Postma (1994):

$$Pb^{2+} > Ba^{2+} = Sr^{2+} > Cd^{2+} = Zn^{2+} = Ca^{2+} > Mg^{2+} = Ni^{2+}$$
$$= Cu^{2+} > Mn^{2+} > Fe^{2+} = Co^{2+}$$

2.3.2. Chemically Reactive Substrates

The composition of alluvial sediment varies greatly both between watersheds and within them, depending, in part, on the underlying geological units, the weathering regime, and the transport processes that move materials off hillslopes and redeposit them along the channel. Nonetheless, it seems fair to say that sand-sized and coarser particles tend to be dominated by quartz, feldspars, and carbonates, all of which tend to be chemically inert. Other types of substrates are, of course, mixed with these primary minerals, and it is these less abundant materials that provide alluvial sediments with their capacity to accumulate trace metals. Horowitz (1991) argues that clay minerals, organic matter, and Fe and Mn oxides and hydroxides are the most prevalent of these reactive substrates in aquatic environments. The reactive nature of these particles results from a high surface charge distributed over a relatively large surface area, and from possessing high cation exchange capacities. They also tend to exhibit an amorphous or cryptocrystalline structure that is unstable in the near surface environment. It is important to note that while the factors that produce the reactive nature of these substrates are similar, the release of trace metals from each of them occurs under different physiochemical conditions (as will be described in more detail below).

2.3.2.1. *Clay minerals*

Clay minerals are an indicator of the degree and character of chemical weathering, are a common component of clastic sedimentary rocks, and are an important hydrologic and geochemical barrier to contaminant migration in the subsurface. As a result, clay mineralogy has developed into a separate scientific discipline, the details of which are beyond the scope of this book. Nevertheless, no discussion of reactive substrates would be complete without briefly examining the nature of clay minerals and the various types that exist in riverine environments.

Clay minerals are composed of tetrahedral or octahedral sheets which are bound together to form a layered atomic structure. The tetrahedral layers are composed of individual tetrahedron which consist of either a Si^{4+} or Al^{3+} ion that is surrounded by four negatively charged O atoms in such as way that if forms a four-faced structure (Fig. 2.11). The octahedral layers are composed of an interconnected series of octahedrons, each of which consists of six negatively charged O or OH ions that surround a central Al, Mg, or Fe ion in an octahedral (eight-sided) arrangement (Fig. 2.11). Most clay minerals are either 1:1 layered silicates, or

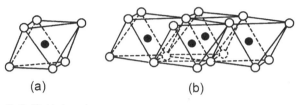

○ & ◌ Hydroxyls
● Aluminum, Manesium, etc.

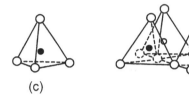

Figure 2.11. Illustration of an individual octahedron (a) and tetrahedron (c) and their combining to form octahedral (b) and tetrahedral (d) layers in the clay mineral structure (From Grim 1968)

2:1 layered silicates, depending on the number of octahedral or tetrahedral layers that are combined into repetitive sheets that are themselves fixed to one another through a variety of mechanisms. In 1:1 clays, one tetrahedral layer is fixed to one octahedral layer. In most riverine environments, kaolinite, and to a lesser degree, illite represent the most important 1:1 clay minerals. As shown in Fig. 2.12, the octahedral and tetrahedral layers form tightly bound sheets by the sharing of Al and Si atoms in the separate layers. The 2:1 clay mineral structure is characterized by an ocatahedral layer that is positioned (sandwiched) between two tetrahedral layers. There is a wide variety of minerals exhibiting the 2:1 structure, perhaps the most important of which for our discussion are the smectites and micas (Fig. 2.12).

The abundance of 1:1 versus 2:1 clay minerals is dependent in large part on the local climatic regime and its influence on weathering processes. In warm, wet climates characterized by intense weathering, kaolinite tends to be the predominant mineral, whereas smectites are often dominant in drier regions.

The reactive nature of the various clay mineral groups is highly variable. For example, adsorption processes are a function of the negative charge on the broken edges of the layered sheets that result from the isomorphic substitution of Al^{3+} for Si^{4+} in the tetrahedral layer, and Mg^{2+} or Fe^{2+} for Al^{3+} in the octahedral layers. In the case of kaolinite, very little substitution occurs. Thus, its net negative surface charge, and its potential to adsorb cations to external surfaces, is limited. In addition, the sheets (consisting of an octahedral and tetrahedral layer) are tightly bound together by hydrogen bonds, thereby preventing a change in the spacing between the sheets during wetting. As a result, cations and water cannot penetrate into kaolinite's crystalline structure. This factor explains its relatively low CEC because the exchange process is limited to external surfaces. In contrast, isomorphic

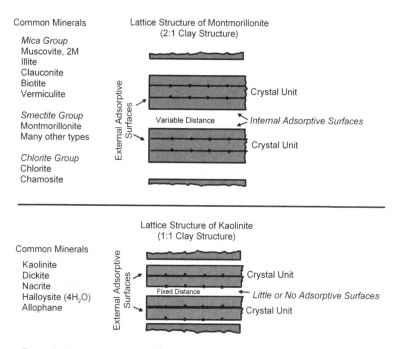

Figure 2.12. Basic structure of common clay minerals (Modified from Buckman and Brady 1960)

substitution can be extensive in the 2:1 smectite clay mineral group, particularly within the octahedral layers. They therefore exhibit a relative high net surface charge that promotes cation adsorption. Moreover, smectites tend to exhibit a relatively high CEC because the individual sheets are weakly bound together, and cations and water can penetrate between the sheets and participate in exchange processes (Fig. 2.12).

There is some debate about the significance of clay minerals in aquatic environments as a direct chemical collector of trace metals (Jenne 1976; Horowitz and Elrick 1988). Jenne (1976), for example, has proposed that clay minerals primarily serve as high surface area substrates for the precipitation and accumulation of organic matter and Fe and Mn hydroxides, and that these surface coatings actually accumulate the trace metals commonly associated with clays. Whether Jenne (1976) is entirely correct has yet to be demonstrated, but there seems to be a general consensus that clay minerals are not as important as either organic matter or oxides as trace metal collectors in most aquatic ecosystems.

2.3.2.2.   Al, Fe, and Mn oxides and hydroxides

Amorphous oxides and hydroxides of Al, Fe, and Mn are significant scavengers of trace metals in aquatic environments and, thus, play an important role in controlling the partitioning of trace metals between the dissolved and particulate phase in river systems. Fe and Mn hydroxides are particularly important because they possess a greater abundance of binding sites than is typically found on Al hydroxides (Jenne 1976). Moreover, unlike the oxides of Al, they are redox sensitive; thus, while they constitute trace metal sinks under oxidizing conditions, they can become a major source of dissolved metals under reducing conditions. This sensitivity to changes in Eh allows them to play an important role with regards to trace metal cycling in aquatic environments.

The accumulation of trace metals on Fe and Mn oxides and hydroxides can occur by both coprecipitation reactions and adsorption. Laboratory studies have suggested that metal adsorption on oxide surfaces is characterized by a two step process (Axe and Anderson 1998). The first step involves rapid adsorption of trace metals from the surrounding solution onto external surfaces of the mineral grains. This step is governed by the parameters described in the above section on adsorption/desorption processes. The second step entails slow diffusion of sorbed trace metals into the oxides through gaps in the structure, a step that may allow the trace metals to become isolated from the bulk solution (Axe and Anderson 1998). The rate of metal uptake by the oxides is ultimately limited by the latter process.

In natural systems, Fe and Mn oxides and hydroxides are produced by the precipitation of porous, X-ray amorphous substrates, either as coatings on grains or as finely dispersed particles. The very poorly crystalline nature of the Fe and Mn oxides is attributed in large part to their rapid precipitation from anoxic waters upon encountering a source of dissolved oxygen (Jenne 1976). The precipitation of Mn is less rapid than that of Fe, but isomorphic substitution is much more extensive. The result is that Mn hydroxides are more amorphous than those of Fe.

Over time the initially formed coatings or particles may recrystallize into less reactive micro-crystalline or "aged" crystalline forms. The recrystalization or aging process generally takes long periods of time and can be inhibited by a host of parameters including the coprecipitation of other trace metals, and the sorption of organic complexes and oxyanions. Therefore, these inhibitors of the aging process partially control the nature of the Fe and Mn oxides and hydroxides that are found in any given alluvial deposit.

The most frequently recognized types of Fe oxides and hydroxides in sediments and soils include goethite (alpha $FeOOH$), lepidocrocite (gamma-$FeOOH$), ferrihydrite ($Fe(OH)_3$), hematite (alpha-Fe_2O_3), and maghematite (gamma-Fe_2O_3) (Jenne 1976; Förstner and Wittmann 1979). The type of Mn oxides that may be present is enormous because isomorphic substitutions are extensive and highly variable. In fact, the most common forms are difficult to identify.

Within riverine environments, the formation of Fe and Mn oxides and hydroxides frequently occurs in areas where waters with differing Eh mix. For example, they are frequently formed in zones of groundwater discharge where Fe^{2+} and Mn^{2+} encounter dissolved oxygen and are rapidly oxidized. The pH of the water also has an influence on the precipitation of Fe hydroxides in many contaminated river systems. Perhaps the most commonly cited example is associated with acid mine drainage where Fe oxides can be extensively precipitated within the channel when acidic waters within a tributary mix with waters exhibiting a vastly different pH (Förstner and Wittmann 1979).

2.3.2.3. Organic substrates

2.3.2.3.1.   Humic substances: Organic matter includes all carbon-containing living and nonliving materials. In this section, we are concerned with *humic substances*, defined as nonliving, partially decomposed and visually unidentifiable materials that are characterized by extreme heterogeneity in chemical composition and structure. In fact, humic substances exhibit variations in molecular weights ranging from a few hundred to hundreds of thousands of grams per mole (Jenne 1976; Förstner and Wittmann 1979). Data on the composition of organic matter in riverine environments are limited, but Reuter and Perdue (1977) suggest that 60 to 80% of organic carbon in freshwaters consists of humic substances, either in the dissolved or particulate form. Organic matter found in river waters and sediment is derived from numerous sources, including (1) biota within the channel, (2) erosion of sediment and soil within the watershed, (3) influx of colloids and dissolved substances from zones of groundwater discharge, and (4) deposition of organic substances from the atmosphere. In most cases, these organic constituents are naturally occurring compounds. It is important to recognize, however, that organic substances can be derived from anthropogenic activities (e.g., sewage sludge), and that others do not exist in nature.

Humic substances are often subdivided into humins, humic acids, and fulvic acids on the basis of their molecular weights and, to a lesser extent, their chemical characteristics (Table 2.3). The two most important groups in terms of contaminant mobilization and retention are the humic and fulvic acids. Fulvic acids exhibit lower molecular weights, and are more soluble, than humic acids. As a result, they play a key role in metal transport by fostering their mobility as dissolved organic complexes in river waters. In contrast, the less soluble humic acids tend to occur in greater quantities in channel bed and bank sediment, thereby playing a significant role in the retention of metal contaminants (Jenne 1976).

The precise division between humic and fulvic acids is arbitary and both groups possess a high affinity for trace metals that is due to a large number of functional surface groups that surround the organic solid (Fig. 2.13). Particularly important are the carboxylic and phenolic groups that exhibit a negative surface charge that increases with increasing pH (see discussion on adsorption). Thus, the sorption capacity of humic and fulvic acids, which is generally on the order of 150 to 600 meq metal/100 g of humic substance (Förstner and Wittmann 1979; Schulin et al. 1995) is higher in basic than in acidic waters. The sorption of metals to the organic materials occurs through both adsorption (Schulin et al. 1995) and cation exchange. The strength of the bonds between the metal and the organic material can range from relatively weak (and bioavailable) to strong (so that the metals are unavailable to biota). The relative strength of the bonds is dependent on the character of the organic substrate as well as the nature of the sorbed cation. For

Table 2.3. Types of organic matter in aquatic sediments

1. Synthetic Organic Substances – not known to exist in nature
2. Living Organic Fraction – biomass associated with all living organisms
3. Non-Living Organic Fraction
   a. Macroorganic Matter – nonliving organic matter that is still identifiable
   b. Humus – nonliving component other than macroorganic fraction
      i. Non-humic substances – compounds found as biopolymers or intermediary metabolites in living matter; well-defined chemical structures
      ii. Humic Substances- complex and extremely irregular chemical structures
         • Humins – largely insoluble in aqueous solutions; highest molecular weight
         • Humic Acids – very complex, but soluble under certain conditions; moderate molecular weights
         • Fulvic Acids – occur in dissolved form in aqueous solutions; low molecular weights
         • Yellow Organic Acids – lowest molecular weights

Adapted from Johansson (1977) and Schulin et al. (1995)

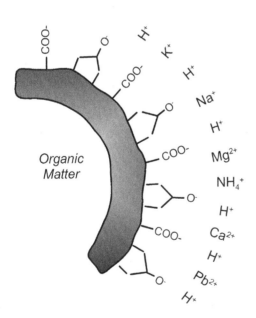

Figure 2.13. Cation adsorption/exchange onto the external surfaces of organic matter. The cation exchange capacity is pH dependent, increasing with a rise in pH and a decline in H^+ activity (From Deutsch 1997)

example, Johansson (1977) has ranked selected divalent cations according to how tightly the metals are associated with humic or fulvic acids. The ranking, which provides insights into the cation's relative bioavailability, is given by:

$$UO^{2+} > Hg^{2+} > Cu^{2+} > Pb^{2+} > Zn^{2+} > Ni^{2+} > Co^{2+}.$$

2.3.2.3.2.   Biofilms: If you have ever waded in a river or lake you have probably experienced a biofilm, although you may not have known what it was called. *Biofilms* are the dense, gelatinous, and slippery coatings which are found on rocks, boulders, and other sediment in rivers, lakes, and wetlands (as well as a host of other surfaces). They are produced by a wide range of microorganisms, including bacteria, algae, fungi, and diatoms. Compositionally, they consist of hydrated exopolysaccharide and polypeptide polymers (Headley et al. 1998) and, thus, they can be considered as part of the organic matter content of the particulate phase.

The analysis of biofilms in aquatic environments has increased significantly in recent years, in part because we have learned that they can have a significant effect on the sorption and desorption of inorganic and organic contaminants. They can, in fact, be particularly important in waste water treatment plants where microbial populations are extensive. These effects are due, in part, to their ability to insulate minerals or other organic substrates from chemical species within the surrounding waters; that is, the ions dissolved within the water must cross the biofilm before they can interact with the substrate upon which microbes are growing. Moreover, biofilms themselves can be highly reactive and may participate directly in the partitioning of trace metals and other contaminants between the dissolved and particulate phase.

The adherence of microbes to solid surfaces is apparently a strategy for microbial survival which provides a means of optimizing their access to nutrients (Costerton et al. 1987; Flemming et al. 1996). Thus, they are ubiquitous features of nearly every aquatic ecosystem, except, perhaps, those which are extremely nutrient deficient. The development of a biofilm begins with the attachment of plantonic forms of microbes to the surface of a solid substance using sticky exopolysaccharide polymers. Exactly which microbes first colonize a surface is a matter of debate, and is likely to vary from place to place. Once attached, the microbes excrete exopolysaccharide (or extracellular) polymer substances (EPS) and become embedded within it. Growth of the biofilm then occurs through a combination of cell division and the recruitment of similar or divergent strains of microorganisms from the surrounding waters (Costerton et al. 1987).

EPS, which comprise about 85 to 95% of the dry weight of a biofilm (Flemming et al. 1996), is particularly reactive, and can selectively collect trace metals from the surrounding solution (Headley et al. 1998). In fact, these polymers are widely used to remove toxic trace metals from wastewaters at sewage treatment facilities. The removal is caused in large part by the extremely high sorption capacity of the EPS, although significant variations in the binding capacity of a biofilm can occur as a function of the specific species of microbe and metal involved

(Flemming et al. 1996). The sorption of a metal by EPS occurs through a variety of mechanisms, including adsorption, cation ion-exchange, and the precipitation and subsequent entrapment of metals as additional polymers are excreted. The total sorption capacity is also known to be pH dependent; the greatest cation binding affinities occur at a pH of approximately eight. This pH dependency is related to the influence of pH on the abundance of negatively charged adsorption sites, particularly those associated with carboxyl and hydroxyl groups.

The cell wall and the cytoplasm of bacteria can also participate in sorption and desorption reactions. Because of its anionic nature, the cell wall is particularly important in some aquatic environments, such as groundwater systems (see, for example, Fein et al. 1997).

Given that biofilms represent living communities, they can produce a microenvironment on a solid surface that is quite different in terms of pH and Eh from that of the surrounding waters. These alterations in microenvironment can either directly or indirectly affect the accumulation of trace metals within the particulate phase. Moreover, bacterial processes can aid in the accumulation of metals within the biofilm. For example, a limited number of bacterial species can oxidize Mn and produce Mn oxide. In other cases, bacteria can produce substances that lead to mineral formation. Perhaps the most commonly cited example is associated with sulfate reducing bacteria that are found in most reducing aquatic environments. These bacteria oxidize organic matter, and in the process use sulfate as an electron acceptor. The net product is hydrogen sulfide (H_2S), which then reacts with metal cations to produce highly insoluble sulfide minerals.

It should be clear that the sorption and desorption processes associated with biofilms are extremely complex. Nonetheless, the increasing perception that they play an extremely important role in controlling metal mobility within riverine environments suggest that more and more attention will be devoted to biofilms in the future.

2.4.   ELEMENTAL SPECIATION

It should now be clear that trace metals are associated with specific reactive phases within alluvial deposits and are incorporated within or bound to particulate matter through a wide variety of mechanisms. With regards to aquatic sediments, this partitioning within the particulate phase is commonly referred to as *speciation*, defined as the molecular form(s) of an element in a mixture of sediment (after Kersten and Förstner 1989). During the past two decades less and less attention has been paid to the total concentrations of trace metals within alluvial sediment, while interest in their speciation has grown dramatically (see, for example, Kramer and Allen 1988; Bartley 1989; Allen et al. 1995). This transition is related to the realization that the mobility of trace metals as dissolved constituents is closely tied to chemical speciation. Moreover, chemical speciation is an important control on the fraction of the total metal content with alluvium that can be accumulated in biota. This follows because the *bioavailable* component depends on the ease with which metals can be stripped from the particulate phase. For example, studies conducted within a wide range of environments have shown that the concentration of metals measured within biota tends not to be correlated with the total concentration of a metal in the sediment. Rather, it is correlated to the proportion of the total concentration of specific reactive phases within the alluvial deposits (Table 2.4).

Elemental speciation within a river system is dependent on host of parameters that control the local physiochemical environment, including the river's biotic characteristics; thus, it varies from river to river, and may in fact change through time or space within any given river system. Attempts to quantify metal speciation in particulate matter have either relied on thermodynamic modeling exercises or various types of chemical and physical laboratory analyses (Tessier and Campbell 1988; Kersten and Förstner 1989). While modeling approaches, such as those based on competitive adsorption, have improved over recent years and hold much promise, they are often plagued by a general lack of data required to generate realistic outcomes. Laboratory analyses, on the other hand, produce operationally defined data that do not necessary provide accurate assessments of the proportion of metals bound to specific minerals (Kersten and Förstner 1989). In spite of these limitations, speciation data undoubtedly provide important insights into the mobility of metal pollutants within a riverine environment, and the potential for specific metals to

Table 2.4. Commonly defined sequential extraction species

Chemical speciesor form	Commonly used reagents for extraction[1]	Relative bioavailability[2]
Dissolved Ions	—	Easily Available
Exchangeable Ions	$CaCl_2$; NH_4OAc; $MgCl_2$; $BaCl_2$; $MgNO_3$; NH_4citrate;	Available
Carbonate Bound	HOAc; commonly at pH 5	Less Available; promoted by chemical alteration
Easily Reducible Substrates	$NH_2OH \cdot HCl$ (pH 2)	Less Available; promoted by chemical alteration
Easily Extractable Organic Materials	$K_4P_2O_7$; NaOCl; NaOCl/DCB; $Na_4P_2O_7$; $SDS/NaHCO_3$	Less Available; promoted by decomposition
Moderately Reducible Oxides	$NH_2OH \cdot HCl/HOAc$; NH_4O_x/HO_x	Available only after chemical alteration
Oxidizable Oxides and Sulfides	H_2O_2/NH_4OAc	Available only after chemical alteration
Ions within Crystalline Structure of Mineral	HF; HNO_3; $HNO_3/HF/HClO_4$ $HF/HClO_4/HCl$	Unavailable unless weathered or decomposed

Adapted from Kersten and Förstner (1989) and Salomons and Förstner (1984)
OAc – Acetate

be bioaccumulated. Thus, it could reasonably be argued that the analysis of metal speciation is an important component of any site assessment program.

Perhaps the most widely used method to assess metal speciation is based on the concept of sequential chemical extractions. Two basic assumptions are inherent in this approach. First, trace metals are associated with discrete phases within the sediment, such as organic matter, Fe and Mn oxides and hydroxides, carbonates, or sulfide minerals (Fig. 2.14). Second, treating the sediment in a stepwise fashion to increasingly more aggressive chemical reagents, or to reagents of a different chemical nature, will produce extracts containing metals from a specific phase, or which are bound to the particles through a specific mechanism. For example, buffered acetic acid will primarily dissolve carbonate minerals and in the process release metals that are only associated with them. Numerous sequential extraction procedures have been proposed during the past three decades, most of which include five to six steps (Table 2.5). The species with which these methods have primarily been concerned include metals that are: (1) exchangeable (via ion-exchange reactions), (2) weakly sorbed to carbonate minerals, (3) associated with easily reducible phases, such as Mn oxides or amphorous Fe oxides, (4) associated with moderately reducible phases, primarily Fe hydroxides, (5) associated with oxidizable organics and sulfides, and (6) the metals contained within residual minerals. Table 2.4 illustrates the potential bioavailability of each of these operationally defined species.

A significant shortcoming of sequential extraction methods is that each step in the process undoubtedly releases metals from chemical components other than those we are seeking to analyze; thus, the individual reactions are not as selective as

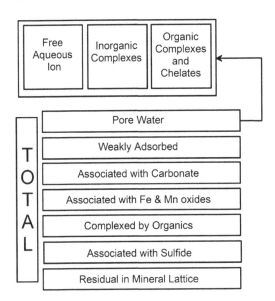

Figure 2.14. Common associations (forms) of metals in sediments (Adapted from Gunn et al. 1988)

Table 2.5. Selected sequential extraction methods

Method	Step 1	Step 2	Step 3	Step 4	Step 5	Ref.
1	HOAc	H_2O_2	$Na_2S_2O_4$	RA	——	1
2	$NH_2OHHCL/$ HNO_3	$NH_2OH \cdot HCL/$ HOAc	$Na_2S_2O_4$	$KClO4/HCl/$ $HNO3$	HF/HNO_3	2
3	NH_4OAc	$NH_2OH \cdot HCL$	H_2O_2/HNO_3	$Na_2S_2O_4/$ HF/HNO_3	——	3
4	$MgCl_2$	NaOAc/HOAc	$NH_2OH \cdot HCL/$ HOAc	H_2O_2/HNO_3	$HF/HCl)_4$	4
5	NaOCl	$NH_2OH \cdot HCL$	$(NH_4)_2C_2O_4/$ $H_2C_2O_4$	$Na_2S_2O_4$	HF/HNO_3	5

Adapted from Tessier and Campbell (1988)
RA – Residual Analysis; OAc – Acetate
1 – Rose and Suhr (1971); 2 – Chao and Theobald (1976); 3 – Engler et al. (1977); 4 – Tessier et al. (1979); 5 – Hoffman and Fletcher (1979)

we would like. Another significant problem is that once the metals are released into solution, they may become sorbed onto other reactive components, thereby altering the defined distribution of the metals within the total sediment package. Despite these errors, numerous studies have shown that sequential extraction data can provide significant insights into the metal speciation within riverine environments.

Speciation within the aqueous phases is also an important control on bioaccumulation. The activity of free metal ions is particularly important (Tessier and Turner 1995). However, the extremely low dissolved concentrations observed in non- or moderately-contaminated rivers makes the direct measurement of specific aqueous species difficult at best. Thus, the activity or concentration of the metal ions is often estimated using modeling routines (e.g., the free ion activity model) that are based on an understanding of the total concentrations of the metals in the solution, and the chemical composition of the water and sediment (Tessier and Turner 1995; Turner 1995).

2.5. CHEMICAL REMOBILIZATION

Up to this point in our discussion, we have primarily been concerned with the physical partitioning of trace metals as a function of grain size and surface area, the reactive phases with which metal contaminants are associated, and the mechanisms responsible for the accumulation of trace metals in sediment. All of these factors are an important component of site assessment because an understanding of the processes that govern metal partitioning in rivers is needed to determine both the mobility and bioavailability of the metals under currently existing conditions. For example, if the majority of the trace metals are loosely adsorbed to mineral surfaces, the potential for adverse impacts on ecosystem or human health are likely to be greater than if all of the metals are tightly bound within the crystalline structure of mineral grains. It is also important to recognize that an understanding of sediment-contaminant relations is essential to predict the potential for future impacts of the pollutant on the aquatic ecosystem.

The importance of predicting future impacts of contaminants on biota is clearly laid out in the concept of a chemical time bomb. Stigliani et al. (1991) defined a *chemical time bomb* as, "a chain of events, resulting in a delayed and sudden occurrence of harmful effects due to the mobilization of chemicals stored in soils and sediments in response to slow alterations to the environment." Förstner (1995) argued that with regards to aquatic ecosystems, two mechanisms could "trigger" a time bomb: (1) the saturation of the sorption sites on the particles, thereby allowing for an increase in the dissolved concentration and bioavailability of the metals, and (2) a fundamental change in the character of the sediment, and presumably the surrounding solutions, that reduces the materials capacity to accumulate toxic materials. Although significant discussion has been devoted to chemical time bombs (see, for example, Salomons and Stigliani 1995), well documented cases of their occurrence are rather limited. Nevertheless, it could be argued that no assessment program is entirely complete without at least a basic understanding of the potential impacts that may result if changes in the primary factors controlling metal mobility are altered.

From our discussion on sorption processes, it should be clear that pH serves as one of the predominant variables controlling the release of metals to river waters, or the pore fluids within the adjacent alluvial deposits. A decrease in pH will have a tendency to release metal *cations*, either through desorption or dissolution, from the sediment to the surrounding fluids. In contrast, an increase in pH will generally lead to sorption of cations and desorption of *anions*, including oxyanions which form with many of the transition metals and metalloids. Alterations in pH within rivers can result from a variety of things, particularly the introduction of acid rain downwind of industrial sites, acid mine drainage, or, as we will see below, the oxidation of sulfide minerals such as pyrite contained within alluvial deposits.

Redox potential, or Eh, is also considered as one of the primary variables in controlling the release of metals to river waters. In fact, it is clear that alterations in Eh can lead to the chemical remobilization of metals that were previously bound to solids. Changes in redox conditions have been attributed to a variety of factors including increased nutrient loading and associated eutrophication (which results in more reducing conditions), the application of sewage sludge to floodplain soils, and deforestation or other changes in land-use. Since our discussion is ultimately focusing on geomorphic processes, it is important to realize that changes in Eh can be initiated by physical changes in fluvial process and form. For example, stream incision often results in a lowering of water tables within the adjacent floodplain, thereby subjecting the alluvial sediment to more oxidizing conditions. Over shorter periods of time, the erosion and resuspension of channel bed sediment can almost instantaneously transfer particles from a reduced environment to the oxygenated water column.

The effects of changing redox conditions on metal solubility are extremely complex, and while the mobility of some metals (e.g., Cr) is directly affected by changes in Eh, others are not over the range of values observed in natural systems (Bourg and Loch 1995). One of the more significant impacts of altering Eh in aquatic environments is related to its influence on the solubility of Fe and Mn oxides and hydroxides; a change from oxidizing to reducing conditions promotes their dissolution. As the Fe and Mn oxides and hydroxides are dissolved, metals which were co-precipitated with the oxides or hydroxides as coating or finely disseminated particles, or that were adsorbed to their surfaces, can be released to the surrounding solution. Whether those metals remain in solution under reducing conditions is dependent in large part on the availability of sulfur and on the formation of metal-bearing diagenetic phases such as metal phosphates. If sufficient quantities of sulfur are available, as is commonly the case for channel bed sediment in low-energy environments or riparian wetlands, the released metals may be removed from solution as sulfide minerals. However, where the sulfur content is low, precipitation of sulfide minerals is prohibited and the anaerobic conditions promote relatively high metal solubility (Bourg and Loch 1995).

Sulfide minerals are themselves affected by changes in Eh. Take, for example, grains of galena (PbS) that have been released from mining operations into an adjacent river where they have been transported and deposited downstream. As long as the sulfide grains remain under reducing conditions, their alteration is likely to be minimal. However, if the grains are exposes to oxygenated waters they may be oxidized, thereby releasing Pb into solution according the following reaction:

$$(18) \qquad PbS + 2H_2CO_3 + 2O_2 \rightarrow Pb^{2+} + HCO_3^- + SO_4^{2-} + 2H^+$$

In some systems, a portion of the Pb^{2+} and SO_4^{2-} may combine to form $PbSO_4$, a precipitate. Nonetheless, as oxidation continues, the concentration of Pb in solution will theoretically increase until its solubility is affected by adsorption (Bourg and Loch 1995) (Fig. 2.15). Equation 18 illustrates that the oxidation of sulfides will also lead to a decrease in pH, unless the system is strongly pH buffered. If pH declines, the solubility of the metal cations will still be dictated largely by adsorption, but the concentration of the cation in solution will occur at a higher value (Fig. 2.15).

Not all rivers possess the same sensitivities to changes in pH and Eh. River waters which flow across carbonates (e.g., limestone or dolomite) are buffered against pH changes. That is, a decrease in pH will result in the dissolution of the carbonate, which will reduce the shift in pH that would otherwise occur in the absence of carbonate rocks. Rocks enriched in aluminum silicates, such as certain clay minerals, can also serve as effective buffers that resist changes in pH. Given the significant role that future changes in pH and Eh can play with regards to remobilizing trace metals, it is important to assess both the pH and Eh buffering capacity of the aquatic systems. Most approaches use important acid consuming or producing reactions in aquatic environments to assess the acid producing and consuming potential of the system, whereas the typical electron-accepting reactions are used to assess the capacity to resist shifts in Eh (see, for example, Bourg and Loch 1995; Förstner 1995).

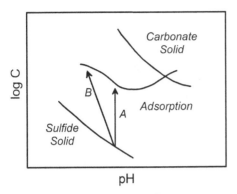

Figure 2.15. Potential pathway for dissolved trace metals as metal sulfides are oxidized. Line A represents a system that is strongly pH buffered. Line B represents an decrease in pH during oxidation (From Bourg and Loch 1995)

In addition to changes in Eh and pH, other environmental alterations can result in the chemical remobilization of trace metals. They include increases in salt concentrations which can result in increased competition for cation sorption sites on reactive particles, an increase in the presence of either natural or synthetic complexing agents, and certain biogeochemical processes (Förstner and Wittmann 1979). All of these factors need to be considered as part of the assessment process if they appear pertinent to the investigated river.

2.6.   SUMMARY

Trace metals and other contaminants are not homogeneously distributed within alluvial deposits, but tend to be concentrated in the fine-grained sediment fraction. The association between fine particles and chemical contaminants is related to the increased surface area per unit mass of the finer particulates, and to an increased abundance of chemically reactive substrates (e.g., clay minerals, organic matter, and Fe and Mn oxides and hydroxides). The accumulation of trace metals on particulate matter occurs through a combination of mechanisms collectively referred to as sorption. Sorption strongly influences the concentrations of trace elements dissolved within the water column, and includes three primary processes: (1) precipitation of material that is compositionally different from that of the underlying particle, (2) incorporation of ions into the crystal structure of the mineral by a set of mechanisms collectively referred to as *absorption*, and (3) the removal of dissolved species from the water, and their accumulation on a particle surface without the formation of a distinct, three-dimensional molecular structure, a process called *adsorption*. Sorption is influenced by a wide range of factors associated with both the substrate (sediment) and the surrounding waters. Thus, in natural systems, the importance of specific sorption processes will change along the river and through time with changes in physiochemical conditions. Nonetheless, the most important controls on dissolved metal concentrations is generally represented by some combination of precipitation/dissolution and adsorption/desorption processes.

The molecular forms that a trace metal assumes within a given media (e.g., water, sediment, soil, etc.) are commonly referred to as its speciation. Speciation is now known to be an important control on both metal mobility and the likelihood that a metal will be accumulated in biota (i.e., its bioavailability). Moreover, an understanding of sediment-contaminant relations not only provides insights into metal mobility under existing physiochemical conditions, but can be used to predict the potential for future impacts of the pollutant on the aquatic ecosystem if those physical and/or chemical conditions change. Alterations in pH and redox potential (Eh) are arguably the two most important parameters affecting the potential release of metals from the particulate to the aqueous phase. A decrease in pH will have a tendency to release metal *cations*, either though desorption or dissolution, from the sediments to the surrounding fluids, whereas an increase in pH will generally lead to sorption of cations and desorption of anions. Increased or decreased sorption can occur over a very narrow pH range. The effects of changing redox conditions on metal solubility are extremely complex, and while the mobility of some metals such as Mn and Fe oxides and hydroxides are directly affected by changes in Eh, others are not for the redox values observed in natural systems. However, in waters which are pH neutral or mildly alkaline, oxygenated or slightly reducing conditions tends to decrease the solubility of many heavy metal cations because they precipitate as oxides, hydroxides or carbonate minerals. Trace metals also exhibit a low solubility in waters characterized by moderately to strongly reducing conditions, but only when enough sulfur is present to allow for the formation of metal sulfides.

2.7. SUGGESTED READINGS

Bourg ACM, Loch JPG (1995) Mobilization of heavy metals as affected by pH and redox conditions. In: Salomons W, Stigliani WM (eds) Biogeodynamics of pollutants in soil and sediments: risk assessment of delayed and non-linear responses. Springer Berlin Heidelberg New York, pp 97–102.

Flemming HC, Schmitt J, Marshall KC (1996) Sorption properties of biofilms. In: Sediments and toxic substances: environmental effects and ecotoxicity, pp 115–157.

Horowitz AJ, Elrick KA (1988) Interpretation of bed sediment trace metal data: methods of dealing with the grain size effect. In: Lichtenberg JJ, Winter JA, Weber CC Fradkin L (eds) Chemical and biological characterization of sludges, sediments, dredge spoils, and drilling muds. American Society for Testing and Materials, pp 114–128.

Kersten M, Förstner U (1989) Speciation of trace elements in sediments. In: Bartley GE (ed) Trace element speciation: analytical methods and problems, CRC Press, Inc, Boca Raton, pp 245–317.

Moore JN, Brook EJ, Johns C (1989) Grain size partitioning of metals in contaminated, coarse-grained river floodplain sediments: Clark Fork River, Montana, U.S.A. Environmental Geology and Water Science 14:107–115.

CHAPTER 3

BASIN PROCESSES

3.1. INTRODUCTION

Rivers, or more precisely, river systems are composed of a single, axial channel that is fed by a number of smaller tributaries. Collectively the axial and tributary channels form an interconnected network of streams that drain a discernible, finite area. This finite area, called a *drainage basin* or *watershed*, is a fundamental unit of the landscape through which solutes, sediment, and contaminated particles are collected, transported, and redeposited by moving waters (Fig. 3.1). Adjacent drainage basins are separated by a drainage divide which mark the zones where surface waters move in opposing directions. As we will see in the following sections, the movement of solutes and particles (contaminated or otherwise) from hillslopes into the drainage network is not solely dependent on the movement of water over the Earth's surface, but involves multiple surface and subsurface flow paths. Exactly which flow path(s) are dominant and how they process the precipitation falling on the watershed will ultimately control the amount of water that reaches the channel and the concentration of dissolved and particulate substances within the water column. It follows, then, that both the hydrologic and geochemical regime of a river is dependent on the physical, chemical, and biological processes that are occurring within the entire watershed. If we were to focus only on the channel, we would be ignoring a significant portion of the equation. This statement may seem intuitively obvious, but its importance is much too often overlooked as many assessments of contaminated rivers focus on the channel and the adjacent floodplain without adequately understanding the links between the river and its basin.

In this chapter, we will examine the movement of water, sediment, and contaminants from hillslopes to the drainage network. Our intent is not simply to summarize the voluminous literature on the subject, but in keeping with our process focus, to document the mechanisms that control the transfer of pollutants from upland areas to the drainage system.

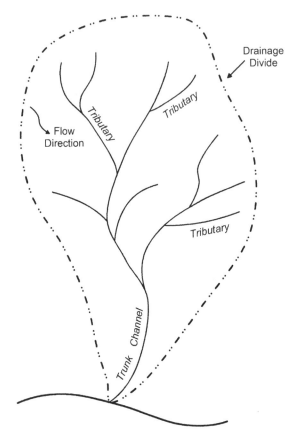

Figure 3.1. Idealized drainage network

3.2.   HILLSLOPE HYDROLOGY

3.2.1.   Direct Runoff Generation

The predominant source of river water is precipitation. Not all of the precipitation falling within a drainage basin makes its way to the ground surface to become runoff; some is trapped by the vegetal cover in a process known as *interception*. Interception is highly variable from location to location because it is dependent on multiple parameters such as the type and density of vegetation, land use, storm duration and intensity, and other hydrometeorological factors. During prolonged storms, the plant cover may become completely saturated at which point a steady-state condition is established in which the same volume of precipitation falling on the vegetation will be passed on to the ground surface. Interception is an important process because it governs the amount of water that may be lost from the system by means of evapotranspiration and, as we will see later, it protects the surface materials from the erosive effects of raindrop impacts.

Upon reaching the ground surface, a portion of the precipitation will be absorbed into the underlying materials in a process referred to as *infiltration*. The rate at which infiltration occurs, called the *infiltration capacity*, is usually measured in mm per hour and is dependent on the complex interaction of three major processes (Ritter et al. 2002): (1) the entry of water into the soil or regolith, (2) the storage of water within pore spaces, and (3) the transmission of water through the material. These three processes are in turn controlled by a host of factors including the thickness of the soil or regolith, vegetation type and abundance, surface slope, and the antecedent moisture conditions (which depends in part on the recent history of precipitation).

The infiltration capacity of most geological materials decreases asymptotically during a rainfall event (Fig. 3.2). This semi-systematic decrease in the infiltration rate is governed by changes in the hydraulic properties of the soil (Knighton 1998). The most important of these changes are related to the compaction of the soil surface by raindrops, the breakage of soil aggregates and the subsequent movement of fine-grained particles into pore spaces, the swelling of clay minerals, and a decrease in the available water storage capacity of the material by the filling of previously dry pores. In most cases, the ultimate infiltration capacity achieved during rainfall is dictated by the rate at which water can be transferred through some relatively impermeable layer. Following an event, the infiltration capacity of the material will recover as the surface dries, aggregates reform, and subsurface pores drain.

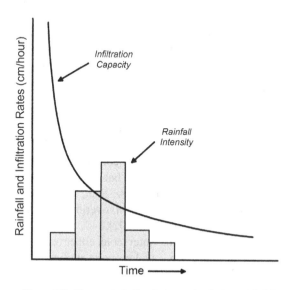

Figure 3.2. Changes in infiltration capacity during a rainfall event. Hortonian overland flow occurs when the rainfall intensity exceeds the infiltration capacity (Modified from Ritter et al. 2002)

Thus, the frequency between rainfall events, which controls the antecedent moisture conditions, can significantly affect infiltration rates.

In a seminal paper published in 1933, Horton argued that runoff could only be generated when rainfall intensity exceeds the infiltration capacity of the surface materials (Fig. 3.2); at all other times the volume of precipitation that reaches the soil infiltrates. It was assumed, then, that infiltration was a primary determinant of overland flow. This type of flow is now referred to as *excess or Hortonian overland flow*, the latter in recognition of Horton's contributions to the field. One of his contributions was to theorize that infiltration largely governed the quantity of water flowing in a channel in direct response to a precipitation event (called *storm* or *direct runoff*). This seemed intuitively correct because water moving over the surface flows at a much faster rate than water that percolates downward to the water table and then moves relatively slowly through the subsurface materials to the channel where it is discharged. Horton's understanding of infiltration was much more sophisticated than this simple model suggests (Beven 2004a–d). Nonetheless, it is his model of excess overland flow, and its influence on direct runoff, which is presented in most textbooks and which formed the basis for engineering hydrology for a period of several decades in the mid-1900s (Beven 2004d). Questions concerning Horton's flow model began to emerge, however, when it was recognized that overland flow rarely occurs in humid regions where the surface materials are covered in a thick blanket of vegetation. Thus, in contrast to arid and semi-arid environments, direct runoff from undisturbed soils in humid climatic regimes could not be attributed entirely to overland flow processes as initially envisioned by Horton (1933).

A primary flaw in Horton's model was the assumption that infiltrating waters could only move downward under the influence of gravity until reaching the water table. We now know that there are a number of routes that a particle of water can follow on its way to a channel, each of which are characterized by differing travel distances and flow velocities (Fig. 3.3). One pathway, referred to as *throughflow* or *interflow*, occurs above the water table, thereby allowing water to take a more direct route to the channel than it otherwise would as groundwater discharge. Throughflow is caused by the development of locally saturated conditions that enables the water to flow laterally. Saturation is generally created when a downward migrating wetting front associated with infiltrating water encounters a zone of low permeability (Dunne and Leopold 1978). The more impermeable unit not only limits the rate of vertical transmission through the geological materials, but diverts the flow downslope, nearly parallel to the ground surface. Throughflow, then, commonly occurs where porous surface materials, such as a soil A horizon, overlie a denser, less permeable unit such a clay-rich B horizon (Ritter et al. 2002).

Some of the infiltrating water will eventually reach the water table, either because it does not encounter, or makes its way through, zones of low permeability. The addition of infiltrating waters to the groundwater system may cause a rise in the water table, particularly in areas adjacent to the channel (Fig. 3.4). This occurs because the depth to the water table near the channel is generally less than it is further

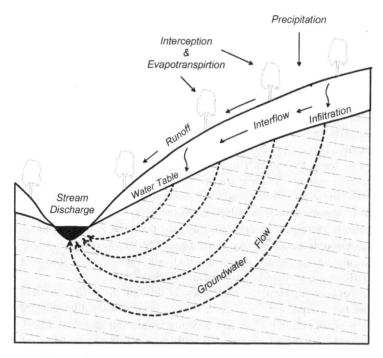

Figure 3.3. Possible hillslope hydrological flow paths (Modified from Ritter et al. 2002)

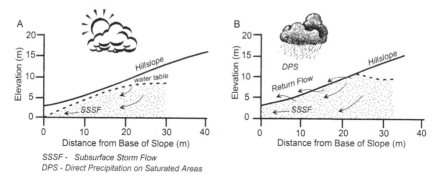

Figure 3.4. Changes in water table conditions adjacent to the channel (Modified from Ritter et al. 2002)

upslope. The infiltrating waters, then, can reach the water table more quickly in these areas, while less water is lost to the filling of pores during transit (Dunne and Leopold 1978). A net effect of rising water tables is a steepened hydraulic gradient which leads to an increase in groundwater flow rates to the river (Fig. 3.4). Flow rates may also increase because the thickness of saturated materials adjacent to the channel, through which the water is moving, has increased, and because the water table may have risen into more permeable horizons (Burt and Pinay 2005). The combined effects of enhanced groundwater flow rates and throughflow is called *subsurface stormflow.*

The rate of flow in the subsurface is dictated in part by whether it occurs as *diffuse flow*, involving the movement of water through a network of interconnected pores, or through linear openings, referred to as *macropores* or *pipes*. In contrast to diffuse flow, the movement of water through macropores and pipes may be rapid and turbulent (Roberge and Plamondon 1987), depending on their size and degree of connectivity. The high rate of flow through macropores and pipes allows subsurface waters to reach the channel quite quickly, in many cases, during the rising limb of the hydrograph. The rapid transit times to the channel also reduces the potential for contaminants to be sorbed onto mineral grains, thereby allowing flows with higher concentrations of dissolved constituents to be discharged to the adjacent river channel. Moreover, macropores have the potential to carry colloids and other particles through the subsurface (Heathwaite1997; Heathwaite and Dils 2000; Lischeid et al. 2002), greatly increasing their significance as a transport pathway, particularly in areas devoid of significant overland flow.

The difference between macropores and pipes is difficult to define in field situations, but the latter represent cavities that have been erosionally enlarged and tend to show a greater degree of interconnection (Anderson and Burt 1990). The formation of both is generally associated with burrows, decaying roots, or other openings created by plants and animals; thus, they are concentrated near the Earth's surface. Quantification of the importance of macropores and pipes to hillslope hydrology has proven difficult. Nonetheless, their occurrence has been widely observed (Holden et al. 2001; Burt and Pinay 2005), particularly in clay-rich soils characterized by shrink-swell processes (Heppell et al. 2000). In some areas, macropores can even dominate the hydrology of the subsurface system, and where laterally continuous may contribute to the buildup of water and pore pressure in the near-stream zone (Burt and Pinay 2005).

During prolonged storms, the accumulation of water immediately adjacent to the channel often leads to a rise in the water table until it intersects the ground surface. Expansion of the saturated zone associated with throughflow may also rise to the surface as water accumulates above a relatively impermeable horizon. When either of these events occurs, *saturation overland flow* is produced (Fig. 3.4). Actually, saturation overland flow consists of two components referred to as *return flow* and the *direct precipitation on saturated areas*. As the name implies, return flow involves the return of infiltrating water to the surface where it flows downslope toward the channel. It is clearly a form of overland flow. However, return flow is distinctly different than Hortonian overland flow in that the water did not originate on the surface, but followed a subsurface path prior to its re-emergence further downslope. Obviously, some precipitation will fall on zones of return flow or saturated throughflow. Most of this precipitation will runoff because saturation of the surface materials reduces infiltration. From a contaminant perspective, saturation overland flow (i.e., return flow plus direct precipitation on saturated areas) is important because it is thought to be a primary contributor to direct runoff, and it is capable of eroding and transporting contaminated particles over the surface to the stream channel.

In light of the above discussion, it should be clear that the area providing runoff to the channel is not constant, but changes seasonally, between storms, and even during a rainfall event. The perception that the area from which runoff occurs varies through time is referred to as the *variable source concept* (Hewlett 1961). Inherent in the concept is the argument that spatial characteristics of the channel network contributing to surface flow will change through time as a function of the local topography, hydrologic characteristics of the soils or regolith, and antecedent moisture conditions (Fig. 3.5) (de Vries 1995). Areas of the watershed with gentle slopes, moderate to poorly drained soils, and within concaved recesses along the valley margins will tend to possess the largest variations in contributing area both during a storm and on a seasonal basis. Topographic expression of the hillslope is particularly important. Hollows, for example, tend to exhibit thicker zones of saturation and higher water tables than adjacent parts of the slope as a result of the convergence of throughflow toward the hollow's center (Anderson and Burt 1978). Moreover, the convergent drainage within the hollow often sustains saturated conditions for longer periods of time, increasing the possibility of saturated overland flow during subsequent rainfall events. On a larger scale, sustained saturation of the upper soil horizons is limited in upstream areas where high-gradient hillslopes merge rapidly with the channel. Farther downstream, however, the potential for soil saturation tends to increase as slope angles decline and floodplains expand.

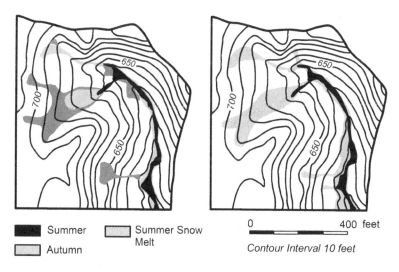

Summer | Summer Snow Melt
Autumn

0 ———— 400 feet
Contour Interval 10 feet

Figure 3.5. Variations in saturated areas which may contribute to surface runoff near Danville, Vermont. (A) Seasonal changes in saturated conditions; (B) Changes in saturated areas resulting from a 46 mm rainfall event. Solid black areas represent saturated areas prior to storm. Lightly shaded areas represent saturated materials following event (From Dunne and Leopold 1978)

The importance of these hydrological contrasts from a pollutant perspective is that they lead to significant differences in the rates of solute, and thus, contaminant transfer to the river channel (Burt and Pinay 2005). More specifically, hotspots tend to correlate with areas of rapid throughflow and saturated overland flow generation.

3.2.2.   Flood Hydrographs

Although some of the water following the various subsurface pathways may be lost to the deep groundwater flow system, most enters a stream channel and becomes direct runoff. Hillslope hydrological processes, then, control how quickly precipitation can enter a channel during a storm, and the total volume of water in the channel at peak flow. Both parameters are critical factors in the assessment of flood hazards and are important considerations in the analysis of sediment and contaminant transport in rivers. Thus, hydrologists have expended significant effort in trying to quantify and model rainfall-runoff processes. The variations in stream flow to precipitation are most often described using a *flood hydrograph* which depicts changes in stream discharge as a function of time (Fig. 3.6).

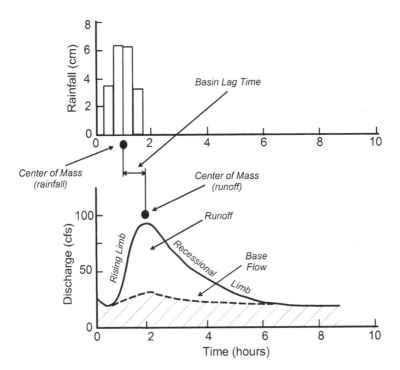

Figure 3.6. Basic components of a flood hydrograph

Discharge is defined as the volume of water that passes a given cross section in the channel during a specified time interval. It is normally presented in units of m^3/sec or ft^3/sec, and can be mathematically expressed as:

(1) $\quad Q = wdv = Av$

where Q is discharge, w is the width of the water surface, d is water depth, v is the velocity at which the water is moving, and A is cross sectional area of flow ($w \times d$). While the width and depth of the flow can be easily measured, determination of the flow velocity for the cross section is problematic. The difficulty arises from the fact that velocity varies across the width and depth of the flow. The net result is that some form of velocity averaging must be applied in order to estimate stream discharge. The most commonly used method involves the partitioning of stream flow into sections of known dimensions (Rantz et al. 1982). The velocity within each section is subsequently measured at a point thought to represent the mean flow rate within that section of the channel (Fig. 3.7). In practice, the channel will often be subdivided into 20 or more distinct sections, and the velocity will be measured several times within each section to ensure a reasonable estimate of the average flow rate. Clearly, then, the direct measurement of discharge can be a time consuming process which is incapable of determining changes in stream flow over relative short time intervals. Fortunately, methods have been developed to indirectly determine discharge over periods as short as a few seconds. The most commonly used technique involves the development of a *rating curve* or *rating table*, both of which relate discharge (measured using the above techniques) to the height of the water surface above or below some datum (Fig. 3.8). Once a rating curve has been developed discharge can be estimated for any flow condition for which the height of the water surface (referred to as *river stage*) has been recorded. During the past decade, numerous sophisticated methods have been developed to measure and record changes in stage for very short time intervals. In nearly all cases, the data is recorded in a digital format, allowing the information to be downloaded on to a lap-top computer in the field, or even transferred in real time from a remote location to an office where the information can then be manipulated.

The discharge shown on flood hydrographs can be separated into direct runoff and base flow using one of several techniques (see McCuen 1998, for a discussion of the methods). Direct runoff, as described earlier, includes the sum of the water derived from the various pathways observed on hillslopes during a precipitation event (e.g., Hortonian overland flow, throughflow, saturated overland flow, etc.). In contrast, base flow is produced by the discharge of groundwater, and can exist within the river when the influences of runoff from a storm are no longer felt within the channel. The maximum or peak flood discharge generally occurs shortly after precipitation ends, and separates the flood hydrograph into two components: the *rising limb* and the *falling* or *recessional limb*. The rising limb represents the influx of water that can be quickly transmitted from the hillslope to the channel; it is therefore characterized by a relatively rapid increase in discharge in comparison to the recessional limb. The recessional limb is primarily controlled by the gradual

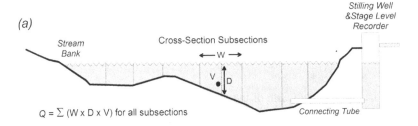

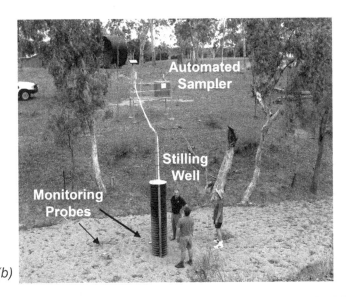

Figure 3.7. (A) Typical design of stream gaging station. Channel cross section is subdivided into small units; discharge is then calculated for each using width, depth, and velocity measurements; (B) Water level gaging station in the Weany Creek Catchment, northeastern Australia (Part A from Ritter et al. 2002)

depletion of water stored in the shallow hillslope flow system. The differences in response between the rising and falling limbs impart an asymmetrical or skewed shape to the flood hydrograph (Fig. 3.6).

An important measure of describing the response of stream flow to a precipitation event is the *basin lag time*. It refers to the time needed for a unit mass of water falling within a basin to be discharged from the basin outlet. It is also composed of two distinct components: the time needed for runoff to reach the channel and the time required for water to be transmitted downstream through the channel to the site of measurement. Quantitatively, basin lag is determined by measuring the time interval between the center of mass of the generated runoff and the center of mass of rainfall (Fig. 3.6). In most cases, shorter basin lag times are associated with higher peak discharges. Moreover, lag times and peak discharges are correlated with basin

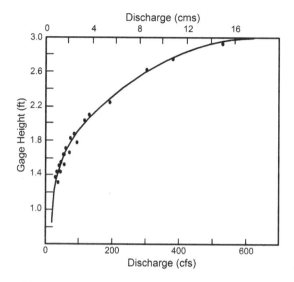

Figure 3.8. Rating curve developed for Rock Creek, located near Red lodge, Montana (From Ritter et al. 2002)

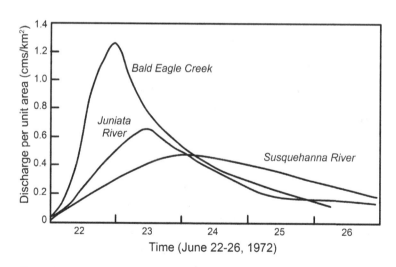

Figure 3.9. Discharge per unit area in the Susquehanna River basin during the 1972 Hurricane Agnes flood. Drainage area progressively increases from Bald Eagle Creek to Susquehanna River. Main stem of river peaks later and discharges water over a longer period as a result of the storage of water on floodplains (From Ritter et al. 2002)

size. Interestingly, however, the peak discharge of an event per unit basin area is inversely related to basin size; small tributary basins tend to generate higher peak flows per unit area than the larger trunk channels (Fig. 3.9). The lower discharge per unit area generated within the master channels is primarily related to their ability to store water on floodplains and within riverine wetlands until after the flood peak has passed.

The peak flood discharge changes as a function of the storm track, duration, intensity, etc. The shape of a flood hydrograph, however, is fairly consistent between events. This has led to the concept of a unit hydrograph in which runoff is examined for the situation in which one inch of precipitation falls uniformly over the basin in a 24-hour period. Theoretically, basins characterized by similar geological materials, topography, and land-uses should possess similar unit hydrographs. It follows, then, that a comparison of unit hydrographs can be used to assess the effects of various physical and biotic attributes of the basin on runoff processes.

3.3. CONTAMINANT TRANSPORT PATHWAYS

3.3.1. The Controlling Factors

The majority of the contaminant load measured in river water can be attributed to the erosion and transport of chemically reactive, fine-grained sediments. We will examine the processes responsible for the generation and delivery of these fine-grained particles to the drainage network later in the chapter. Here we will focus on the movement of dissolved substances from upland areas to the drainage network. Unfortunately, few studies have adequately documented the processes involved in the downslope migration of dissolved trace metals within the hillslope hydrological system. Thus, we will need to rely heavily on investigations pertaining to nutrients, and to a lesser extent, pesticides, which primarily occur as non-point sources of pollution on agricultural lands.

A major thrust of the work on nutrients and pesticides has been the quantification and comparison of contaminant loads associated with direct runoff and base flow (Squillance and Thurman 1992; Ng and Clegg 1997). It might seem obvious that the downslope migration of non-point source pollutants primarily occurs during rainfall-runoff events, and indeed for small- to moderately-sized drainage basins (i.e., basins $< 10^1$ to $10^3 \, km^2$), this is the case. Within tributaries to the Zwester Ohm River in Germany, for example, Müller et al. (2003) demonstrated that more than 70% of the total pesticide load was transported with direct runoff, particularly that contributed by the overland and throughflow components of the hydrological system. However, they found that the temporal distribution of the load was scale dependent; when considering the entire Zwester Ohm River basin, rather than its tributaries, base flow contributed 49% of the total pesticide load in comparison to only 15% by direct runoff. The observed trend, which seems to hold for many other sites as well, is significant because it suggests that the relative contributions of contaminants from direct runoff and base flow will change and may even be reversed in the downstream direction. The position of a site's monitoring station within the watershed, then, becomes an important consideration when interpreting the major sources of dissolved constituents within the stream channel.

If we assume that contaminant migration on hillslopes primarily occurs during rainfall-runoff events (which as we mentioned is likely for small- to moderate-sized watersheds), a significant question that arises is what controls the downslope movement of dissolved constituents to the stream waters during a storm event? Logic would tell us that part of the answer lies in the characteristics of the storm itself, including the form, duration and intensity of precipitation. Two additional hydrologic factors of importance are the discharge capacity of the various hydrological pathways, and the frequency with which those pathways are utilized (Heathwaite and Dils 2000). Other factors of importance include the amount, concentration, and chemical form of contaminants in the soil, their spatial distribution across the landscape, and the nature of the underlying geological materials, particularly with regards to their hydrological and geochemical (sorption) properties. The residence time of the water on the hillslope will also be important because it controls the amount of time that the water has to react with the surrounding geological substrates.

There is widespread agreement that the above parameters are of most importance in determining the rate and quantity at which contaminants can be transported off of the upland areas. Nonetheless, quantifying the sources and pathways through which pollutants move through the hillslope hydrological system has proven to be an extremely complicated process which we are just beginning to understand (Müller et al. 2003). The difficulty arises in part from the number of distinct flow paths which the constituents can follow, and the dynamic nature of the hillslope hydrological system in terms of both time and space (Heathwaite and Dils 2000). To illustrate the complexity involved, let us look at the movement of a contaminant at or near the ground surface in an area of initially unsaturated soils. During a short, low intensity rainfall event, the dissolved contaminants are likely to be transported downward toward the water table under the influence of gravity. The downward migration, however, may be partitioned between water which is moving slowly through the soil matrix and flow confined to macropores and pipes which moves at considerably higher rates. As the rainfall continues, temporary saturation above the water table may lead to the generation of throughflow and the lateral movement of the contaminant through the subsurface. Eventually, the contaminant may not only move downward toward the water table, or laterally as throughflow, but may reemerge downslope as return flow as the water table intersects the ground surface. In addition, at one or more instances during the event, the rainfall intensity may exceed the infiltration capacity of the soil, at which point Hortonian overland flow will become an important pathway along which contaminants are moved from the hillslope to the adjacent stream channel. Now consider the difficulties one faces when predicting the downslope transport of contaminants to the channel for the entire watershed in response to a series of quickly moving thunderstorms, each of which encompass only part of the basin, and which may activate one or more of the possible transport pathways.

One of the most significant problems encountered is the inability to adequately instrument and monitor the discharge and chemistry of each of the potential flow paths. In the absence of direct measurements, numerous studies have attempted to use environmental isotopes (e.g., ^{18}O, ^{2}H, and ^{3}H) to separate flood hydrographs into the proportion of the water supplied by precipitation over the watershed during the event (generally referred to as *new* or *event* water) and that fraction of the total which was stored within the basin prior to the event (often called *old* or *pre-event* water). The underlying basis of this approach is that the source, pathway, and residence time of the water in a basin will strongly influence water chemistry. Water that moves slowly through the basin will be in contact with the underlying geological materials longer than that which exits the basin quickly; thus, the isotopic composition of the slower moving, pre-event waters will have had a greater opportunity to be altered by such processes as adsorption-desorption, dissolution, and cation exchange (Whitehead et al. 1986; Buttle 1994). The resulting differences in the isotopic composition of the slower- and faster-moving waters can then be used in mixing models to determine the relative proportion of event and pre-event water in the channel and the primary pathway(s) through which it traveled.

There are several pathways through which new (event) water may reach the channel. They include Hortonian overland flow, runoff of precipitation from saturated areas, and flow via macropores and pipes that discharge waters directly into the channel (Fig. 3.10). In contrast, a primary mechanism through which pre-event waters enter the channel is referred to as translatory flow (Hewlett and Hibbert 1967). Translatory flow is the displacement of old (pre-event) water stored in soils and sediments between runoff events by new (event) water, and its subsequent release to the stream channel. The process is particularly important in the near-stream zone where rises in the water table are likely to occur. In addition, the

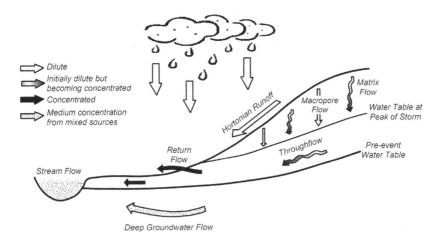

Figure 3.10. Diagram showing possible transport pathways of solutes within the hillslope hydrological system (Adapted from Burt 1986; Buttle 1994; Burt and Pinay 2005)

flushing of water and solutes from pores in newly saturated areas, such as in zones of throughflow, rapidly imparts an old (pre-event) signature to the water. These waters are then quickly transferred downslope to the channel, although not as rapidly as would be the case for Hortonian overland flow or flow through macropores. Saturated overland flow is likely to exhibit chemical traits somewhere between that of the true event and pre-event waters because it represents a mixture of the two. Buttle (1994), for example, argues that in many situations only minor contributions of event water is required to create return flow as water tables tend to be close to the ground surface immediately adjacent to the river. Thus, the emerging return flow will initially exhibit a pre-event signature. However, as it moves rapidly toward the channel, it will be mixed with waters from direct precipitation, thereby altering the composition of the saturated overland flow which ultimately enters the channel (Buttle 1994).

Most studies using hydrograph separation methods have found that direct runoff is predominantly composed of pre-event waters (Buttle 1994). This was a rather unexpected finding as it seemingly contradicted many physical hydrological studies which had shown that the runoff response of hillslope hydrological systems was rapid, and should therefore be dominated by event water. Part of the discrepancy between the geochemical and physical analyses could be explained by taking a more detailed look at the mechanisms that were capable of rapidly delivering pre-event water to the channel during a precipitation event. These processes, which were reviewed by Buttle (1994), include (1) translatory flow, (2) increases in water table elevation, saturated thickness, and hydraulic gradients in the near channel zone, thereby enhancing the discharge of groundwater to river, (3) saturation-overland flow, (4) macropore flow, (5) kinematic waves, and (6) the release of water from surface storage. However, not all of the discrepancy in the results developed using the two approaches can be explained by the movement of pre-event water to the channel during storms. This has led to the belief that while predictions based on isotopic hydrograph separations are generally correct and may provide valuable insights into system hydrology, the method is unlikely to yield results that provide the level of detail associated with the direct observation of hydrological processes (Anderson and Burt 1982; Burt and Pinay 2005).

The shortcomings associated with the isotopic separation methods are related to three factors: (1) violations of the fundamental underlying assumptions, including the invariant (constant) geochemical composition of the various water sources, (2) the possibility that a particular isotopic response is related to a number of different process (often called equifinality), and (3) a decoupling of the hillslope geochemical and hydrological system from the channel system as a result of processes operating in the near-stream zone. With regards to the latter, it is now clear that the near-stream zone has the potential to act as either a barrier or conduit to the movement of water from the adjacent hillslope (Burt and Pinay 2005). Whether or not they serve as a barrier is dependent on numerous factors, and varies in both time and space. Floodplains, for example, typically possess high water tables as a result of upwelling groundwater, inflow from hillslopes, recharge during floods and through the bank from the stream channel, and from rainfall (Burt and Pinay 2005). When water tables are high, hydraulic gradients are likely to be toward the channel, and the hillslope hydrological system is linked to the river. The chemistry of the river water will reflect water from the upland areas. However, during the late summer, or periods of prolonged drought, the water table may decline within the floodplain, reversing the hydraulic gradients. Flow is then from the river to the floodplain, thereby breaking the hydrologic connectivity between the hillslope and channel systems. As rainfall increases with a change in season, the connectivity is likely to be re-established. Such changes in connectivity are likely to be more pronounced downstream where floodplains are most extensive and least pronounce in upstream areas where the drainage network is inset into the adjacent hillslopes.

Geochemical decoupling can also occur, even when the hillslope and channel are hydrologically connected. The degree of geochemical decoupling depends on the chemical evolution of hillslope waters within the floodplain and other deposits adjacent to the channel; if it is significant, the chemical nature of the water from the hillslope can be "reset" prior to entering the river as direct runoff. The extent to which this chemical evolution will occur is dependent on the degree to which surface and groundwater is mixed prior to entering the channel and the residence time of the water in the near-stream zone. Again, neither of these factors will be temporally or spatially constant.

3.3.2.   Transport via Hortonian Overland Flow

In light of the preceding section, it may appear that Hortonian overland flow is not important in the transport of solutes to the drainage network. While the solute content of Hortonian overland flow is likely to be low in comparison to subsurface waters (Burt and Pinay 2005), it is extremely important to the delivery of sediment-borne contaminants, as will be described later in the chapter. In addition, it is a mistake to think that it does not provide solutes to the channel. In fact, the transfer of dissolved constituents from contaminated soils to overland flow (and their subsequent downslope transport) has received considerable attention in recent years (Parr et al. 1987; Wallach et al. 1988; Wallach and Shabtai 1992; Wallach et al. 2001).

The transfer of solutes from soil to runoff is dependent on the interactions of a wide range of factors such as the vertical distribution of dissolved pollutants within the surface materials, rainfall intensities and duration, infiltration rates, slope, runoff rate and depth, and antecedent moisture conditions. Wallach et al. (2001), for example, used data from computer simulations to argue that the transfer of constituents to overland flow is dependent on the time required to saturate the soil prior to the onset of runoff. It is therefore a function of the rainfall intensity and antecedent moisture conditions, among a variety of other factors that influence infiltration. Conceptually, the model shows that the potential transfer of chemicals from the surface materials to runoff is reduced by the downward displacement of dissolved constituents within the pore spaces by infiltrating waters. The more time that is available to displace a dissolved constituent downward, the less likely it is to contaminate overland flow once it begins. After overland flow has begun, the concentration of a contaminant at the base of the slope is primarily dependent on the time during which the runoff was in contact with the soil surface (Wallach et al. 2001). Thus, concentrations increased for lower rates of rainfall and slope, and higher values of surface roughness.

3.3.3. Mapping Spatial Variations in Metal Sources

The remediation of a contaminated river not only requires a sound understanding of the primary sources of metals to the channel (as described in the previous section), but knowledge of their distribution throughout the entire watershed. The developed understanding of metal sources must include those which enter via both surface and subsurface flow routes. There is a tendency, however, for the assessment of rivers to focus on contaminated sediments that form the channel perimeter or on the influx of metals via surface water routes. The various subsurface routes, which may be difficult if not impossible to visibly identify, are often overlooked (or at least, poorly quantified). For example, a review of the Remedial Invesitigation/Feasiblity Study conducted on the mine contaminated Coeur d'Alene River basin in Idaho found that the groundwater sources of dissolved metals (primarily Zn) were not adequately characterized within the watershed in spite of the fact that groundwater served as the primary source of dissolved metals to the channel (NRC 2005). As a result, it was impossible to determine if the removal of contaminated sediment proposed as part of the remediation program would lead to the desired effects.

Recent studies have shown that a synoptic sampling approach that can be used to map both surface and subsurface sources along a channel at the watershed scale (Kimball 1997; Kimball et al. 2002). The analysis is often conducted as part of a three step process. First, discharge within the trunk channel as well as within tributaries and other sources of inflow are quantified. In doing so, the channel is subdivided into discrete reaches which may be bound upstream and downstream by points of known surface (or subsurface) inflow (e.g., Kimball et al. 2002). In addition, tracer-dilution methods may be used to estimate discharge (Fig. 3.11). The advantage of using tracer-dilution methods is that the discharge measurements include waters that are both within the channel and which are flowing through the hyporheic zone (i.e., the channel bed), the latter of which may be significant in coarse grained channels. Second, spatial variations in water chemistry are documented during a synoptic sampling campaign which is conducted at the same time and location that the discharge measurements are performed. Synoptic sampling refers to the collection of water samples from a large number of locations during a short-period of time (usually a few hours to days). As such, the data provide a "snapshot" of the changes in water chemistry along the channel at a given

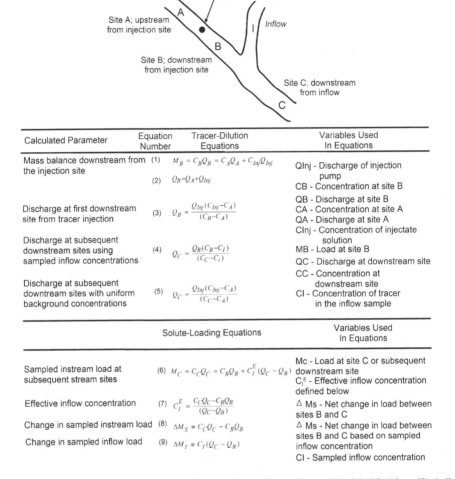

Figure 3.11. Equations used in tracer-dilution and synoptic sampling studies (Modified from Kimball et al. 2002)

instant in time (Kimball 1997). Third, the discharge and elemental concentration data are combined to calculate loads within each reach of the channel and the identified inflow points (Fig. 3.11). The discharge and load calculations can then be used (as described by Kimball et al. 2002) to (1) determine the significances of each inflow source (which were previously identified in the field), (2) estimate, on the basis of a hydrologic balance, the location and significance of unsampled, diffuse sources of metals from groundwaters, and (3) assess trace metal attenuation along the river.

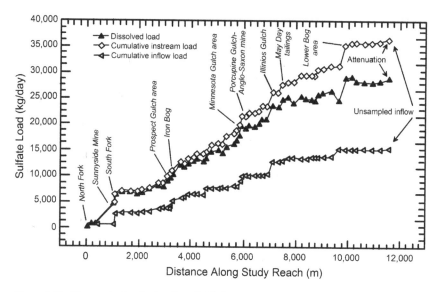

Figure 3.12. Changes in dissolved sulfate (SO_4^{2-}) load along Cement Creek, Colorado (From Kimball et al. 2002)

Figure 3.12 provides an illustration of the method as applied to Cement Creek, Colorado by Kimball et al. (2002) for sulfate (SO_4^{2-}). The graph is based on data from 58 stream reaches along the trunk channel and 45 inflow points. The cumulative instream load was determined using equation 8, whereas the cumulative inflow load was based on equation 9. The cumulative instream SO_4^{2-} load, which assumes no loss from the system, was 36,300 kg/day (Fig. 3.12). Some SO_4^{2-} was lost (attenuated) from the system as illustrated by the difference between the cumulative in-stream load and the load actually measured at a point along the channel. If there had been no attenuation (loss), the two trend lines would be the same. Perhaps more importantly, the cumulative inflow load (which indicates the total load derived from known sources) is considerably below that of the total instream load. The difference, in this case, is primarily due to the influx of SO_4^{2-} from diffuse, subsurface sources. The most significant inflow from these diffuse sources is located where the in-stream and inflow cumulative plots diverge on Fig. 3.12. Along Cement Creek such zones of divergence tend to correspond to areas of fractured bedrock and presumably enhanced groundwater flow (Kimball et al. 2002).

It is important to recognize that the sources of delineated SO_4^{2-} may differ from those determined for other constituents. Thus, the analysis is constituent specific. In addition, synoptic sampling and discharge measurements are usually conducted during base flow conditions. Low flow conditions enhance the subsurface geochemical signal and tend to be easier and safer to sample. However, the significance of the metal sources is likely to change as a function of the sampled flow conditions.

3.4.   HILLSLOPE EROSION

In Chapter 2 data were presented to demonstrate the close relationship between fine-grained sediment and the concentration of various inorganic and organic contaminants. The nearly universal acceptance of this relationship has lead to the assumption that the primary pathway for most hydrophobic, non-point source pollutants to enter and affect river waters is through the erosion of contaminated particles. It follows, then, that any attempt to assess contaminant loadings to a river must not only document the distribution of those constituents within the upland soils, but the magnitude and rate at which contaminated particles are moved off the hillslope and exported from the basin. In the following sections, we will examine the methods used to measure and/or predict the rates at which sediment and sediment-associated trace metals are eroded from upland areas and delivered to the river channel. First, however, we need to review the mechanics involved with the erosion and downslope transport of hillslope sediment.

3.4.1.   Basic Mechanics

Erosion involves two distinct processes: (1) the detachment of particles from the underlying geological materials, and (2) the transport of the detached grains downslope (Knighton 1998). From a mechanical perspective, erosion occurs when the driving forces initiating detachment and particle transport exceed the material's potential to resist erosion (a property called its erodibility).

Detachment can be thought of as a process that increases the potential for movement of substrate particles by the breaking of chemical and physical bonds that tend to hold them in place. The detachment process is driven in large part by the impact of raindrops on the soil surface. Actually, raindrop impacts have several erosion-inducing effects on the soil, only one of which is detachment. Others include soil compaction and the sealing of the soil surface by the movement of fine grained particles from broken aggregates into pore spaces. The combined effects of these two processes are important because they can greatly decrease infiltration, thereby increasing the potential for soil loss by Hortonian overland flow. Raindrops are also involved in the local transport of sediment on the hillslope as particles are ejected from the splash crater and redeposited in a 360° arc. Sediment movement, then, occurs both up- and downslope, but because the particles move farther downhill than uphill, raindrops produce a net flux of material down the hillslope (Dunne and Leopold 1978).

The ability of raindrops to detach and transport sediment is typically expressed in terms of the kinetic energy which they attain during their fall. A drop's kinetic energy is a function of its mass and velocity and can be considerable, depending on drop size, wind conditions, etc. For example, during intense precipitation events, drops can attain a diameter of 6 mm and reach a terminal velocity of 9 m/sec. The impact of such drops on a soil surface can eject particles that are 10 mm in diameter into the air and set even larger particles into motion (Ritter et al. 2002).

In addition to its kinetic energy, the amount of sediment that is eroded by raindrop impacts depends on other interrelated factors, including the gradient of the slope, the nature of the geological materials in terms of texture, aggregate size and cement, and the antecedent moisture conditions. As you might expect, raindrop erosion is particularly significant in areas devoid of thick vegetation which can intercept and dissipate the kinetic energy of the drops.

Earlier it was stated that Hortonian overland flow was produced when rainfall intensity exceeds the infiltration capacity of the soil. Originally overland flow was envisioned as a thin sheet of water of uniform thickness which flowed downhill; thus, it was commonly referred to as *sheet flow*. In reality, overland flow (or sheet flow) is rarely of uniform thickness, but is characterized by zones of deeper and faster moving water that are interspersed with threads of shallower and slower moving water. Sheet flow is, nonetheless, of limited depth, and it is generally incapable of detaching large quantities of particles from soil or other geological materials unless slopes are particularly steep. The primary role of overland flow is the downhill transport of particles previously detached by raindrop impacts and other processes.

Horton (1945) recognized that erosion was not uniformly distributed over the length of a hillslope but semi-systematically varied from top to bottom. The variations in erosion are primarily dependent on changes in the transport capacity of the runoff rather than changes in particle detachment. Near the drainage divide, very little transport occurs as the space available to generate overland flow is limited. Farther downhill, increases in water depth and surface slope enhance sediment transport rates. Near the base of the slope, the transport capacity declines, commonly producing zones of deposition.

In many areas devoid of significant vegetation, sheet flow tends to produce small, subparallel microchannels, called *rills*, at a surprisingly uniform distance from the drainage divide. Rills are usually only a few centimeters in width and depth and are generally obliterated between rainstorm events, or even during an event, by soil heaving and other processes. Nonetheless, the formation of rills is important because they temporally increase hillslope transport. In addition, their periodic destruction and reformation insures that erosion rates averaged over longer periods of time are fairly constant across the slope. The net result is a fairly uniform rate of surface lowering.

Saturated overland flow can also lead to particle detachment and transport. However, the dense vegetation with which it is typically associated tends to limit the total amount of erosion that can occur. In most cases, there is a tendency for erosion by saturated overland flow to be concentrated in localized zones of relatively thin vegetation or within concaved portions of the slope where there is a convergence of groundwater flow (Knighton 1998).

3.4.2.   Measurement of Erosion Rates

One can argue that the best method to determine the rates of hillslope erosion is to directly measure them over some specified period of time. Direct measurement has, in fact, become an essential tool in understanding the rates of soil loss, and a wide variety of methods have been employed to measure hillslope erosion during the past several decades. The most frequently used techniques include video surveillance, repeated photography, the quantitative comparison of surface topography at specific time intervals, the measurement of overland flow and soil loss at the outlet of small runoff troughs, and the installation and monitoring of erosion pins (Fig. 3.13) (Imeson 1974; Peart and Walling 1986). Inherent in these techniques is the assumption that the collected data are representative of the erosion rates of a specific type of soil, land cover, slope, etc. Nonetheless, interpretations of the data are plagued by the fact that the measurements encompass only a small part of the landscape,

Figure 3.13. Erosion pins (A) and runoff flume (B) used to measure hillslope erosion in the Weany Creek basin, northeastern Australia

and the extrapolation of the documented rates of soil loss to other areas is often questionable. Moreover, long periods of monitoring (> 5–10 years) are required to insure that the data are representative of the rainfall conditions (intensity, duration, frequency) that characterize the area (although monitoring times may be reduced by using some form of rainfall simulation).

Another concern that frequently needs to be addressed has been called resolution mismatch by Lawler (2005). *Resolution mismatch* refers to the case in which the temporal resolution (time span) of measured rates of erosion or deposition does not coincide with the temporal resolution of data collected to describe the magnitude and variations in the driving forces (e.g., precipitation intensity, pore pressures, and soil temperature). This mismatch makes it difficult to interpret system dynamics that may be important to adequately describe the erosional and depositional processes. For example, the topography beneath small stationary grids is often surveyed multiple times during a year and compared to determine the *net* loss of soil from a hillslope. The surveys do not, however, provide us with any information on the movement of material on the slope between or during individual precipitation events (Lawler 2005). Therefore, it cannot be determined how rainfall intensity, for instance, influences particle motion.

In order to enhance the short-term temporal linkages between data sets, a number of innovative methods have been developed to measure the magnitude and rates of erosion. Examples include the use of high frequency terrestrial photogrammetry (Lane et al. 1993, 1998) and Photo-Electric Erosion Pins (Lawler 1991, 2005), both of which allow for the quasi-continuous measurement of hillslope processes. Advances in the collection and processing of remotely sensed data are also making it possible to assess rates of upland erosion, particularly that associated with gullies, over large areas.

3.4.3. Prediction of Erosion Rates

It is probably fair to say that in most situations those responsible for the assessment of contaminated sites (or watersheds) do not have the time or resources to directly measure the rates of upland erosion. In the absence of directly measured rates, some means must then be undertaken to estimate the current and future rates of soil loss from the contaminated site. The utilized approaches generally fall into three categories. First, erosion rates directly measured from other locations may be extrapolated to the site of interest, provided that the biophysical properties (topography, geology, vegetation, etc.) are similar between sites. Second, various hillslope erosion models can be applied to estimate erosion rates for various periods of time. Third, geochemical and isotopic methods can be applied to indirectly estimate the past rates and patterns of erosion, and the deciphered rates can be used as a guide to the future. The three approaches are not mutually exclusive. In fact, it is common to apply more than one approach to a site in order to obtain a more complete understanding of soil (and contaminant) loss from upland areas.

3.4.3.1.   Erosion and transport models

The approaches used to model the erosion and transport of sediment and contaminants vary widely, so much so that at the present time there is no universally accepted classification system of the utilized models (Table 3.1). Letcher et al. (1999) argue, however, that they broadly fall into three categories: empirical/metric models, conceptual models, and physics-based models. Empirical models make no attempt to describe the physical processes occurring within the watershed, but use empirically derived relationships between sediment/contaminant generation and basin characteristics in one area to predict that which is occurring in another. The data requirements for these models are generally less than for other types, but because they are not founded on the physical processes that are involved with sediment generation and transport, their ability to predict the effects of changing basin characteristics can be limited.

Perhaps the empirical model that is most widely used to assess hillslope erosion is the Universal Soil Loss Equation (USLE) (Wischmeier and Smith 1965). The USLE was developed using more than 10,000 plot years of erosion data from the U.S., and calculates the average annual soil loss according to the following expression:

$$(2) \qquad A = RKLSCP$$

where A is soil loss in tons, R is the rainfall erosivity index, K is the soil erodibility index, L is the hillslope-length factor, S is the hillslope-gradient factor, C is the cropping-management factor, and P is the erosion-control practice factor. The USLE was originally intended as an agricultural management tool to determine the correct combination of land use and management practices to control erosion at a specific site. Its use, however, has been extended well beyond its original intent and has been applied in a wide range of countries. Moreover, interest in the USLE has led to numerous modifications in its code (e.g., the USLE-M; Kinnell 2000) as well as the development of a revised version (the RUSLE) which was published in 1997 (Renard et al. 1997). The factors used in the Revised Universal Soil Loss Equation (RUSLE) are the same as those contained within the original, but the specifics of how the various factors are determined have changed to improve upon the estimates of soil loss.

The USLE and RUSLE assume that the modelled hillslope possesses uniform characteristics and, thus, its application using a given set of parameter values is spatially limited. This limitation may be partially overcome by partitioning the basin into small parcels, each defined by a different set of nearly homogenous conditions, and then using the USLE to estimate hillslope erosion for each parcel. An additional shortcoming of the USLE is that it does not take into account sediment which may be redeposited at the base of or on the slope prior to entering a channel, or the storage of sediment on floodplains or wetlands located along the channel when it is applied to small catchments. As a result, rates of hillslope erosion estimated by the USLE cannot be compared to sediment loads or basin sediment yields measured at a gaging station.

Table 3.1. Selected erosion/sediment transport models

Model	Type	Scale	Output	Comment
AGNPS Agricultural Non-Point Source Model Young et al. (1989)	Conceptual	Basin	Runoff volume, peak flow, eroded and delivered sediment; N, P, COD concentrations in runoff and sediment	Output is site specific; spatially distributed and event based; can be integrated with GIS
ANSWERS Areal Nonpoint Source Watershed Environment Response Simulation Beasley and Huggins (1982)	Physical	Basin	Sediment yield; nutrient loads in water and sediment; runoff	Spatially distributed, event model; can be integrated with GIS; erosion estimated for both overland flow and raindrop impacts
CREAMS Chemical Runoff and Erosion from Agricultural Management Systems Knisel (1980)	Empirical	Field size 40–400 ha	Erosion, deposition, transport (slope to 2nd-order channels)	Event oriented; basin assumed to consist of uniform soils, topography and land use
EPIC Erosion-Productivity Impact Calculator Sharpley and Williams (1990)	Conceptual	Field < 100 ha	Nutrients, sediments, runoff, pesticides, plant growth	Event based model; can be integrated with GIS; climatic parameters, soils and management are assumed to be uniform; erosion based on USLE.

(*Continued*)

Table 3.1. (Continued)

Model	Type	Scale	Output	Comment
HSPF Hydrologic Simulation Program-Fortran Donigian et al. (1984)	Conceptual	Basin	Runoff, flow rate, sediment load, nutrient concentration, water quality	Event based, spatially distributed model; can be integrated with GIS; Relies on calibration with field data.
SWMM Storm Water Management Model Huber and Dickinson (1988)	Physical	Basin	Assessment of urban runoff, prediction of flows and pollutants	Event and non-event, spatially integrated model; primary for urban areas with point sources.
SEDNET & Annex (for nutrients) Sediment River Network Model Prosser et al. (2001)	Conceptual	Regional or Basin	Erosion, sediment and nutrient loads	Non-event based, spatially distributed model; can be integrated with GIS; Erosion determined using RUSLE.

Modified from Letcher et al. (1999)

In contrast to the empirical approach, conceptual models view basins as a series of interconnected compartments and pathways through which sediment and contaminants pass prior to export from the basin. Basic physical processes of sediment production and runoff are typically described, but the level of their description varies widely (Letcher et al. 1999).

An example of a conceptual model that is used throughout Australia and many other countries is the *Sediment River Network Model* (SedNet; Prosser et al. 2001). SedNet utilizes a sediment budgeting approach and spatially distributed data, meaning that data are collected and applied systematically for specific units or parcels of the basin. The specific unit in this case is a river link, defined as a reach of a river between adjacent stream junctions (Fig. 3.14). Each of these river links possesses an internal drainage area which delivers sediment and contaminants directly to it. Thus, for each link, a mean annual mass balance (expressed in units of t/yr) can be calculated using the expression (McKergow et al. 2005):

$$(3) \qquad Y_i = T_i + H_i + G_i + B_i - D_i$$

where Y_i is the mean annual sediment yield for the link, T_i is the upstream tributary input, H_i is the amount of hillslope erosion, G_i is the amount of sediment generated by gully erosion, B_i is the magnitude of bank erosion, and D_i is the net quantity of sediment deposition or storage on the river bed and floodplain. Estimates of hillslope erosion for each river link are calculated using the RUSLE, although the numbers are modified to account for the proportion of sediment that is deposited before reaching the channel. The model ultimately generates an estimate of the total annual sediment load at the mouth of the basin and an understanding of the spatial variations in sediment generation within the watershed. The latter allows for the identification of the primary sources of sediment within the catchment that can then be targeted for remediation. Slope length and gradient data used in the RUSLE can

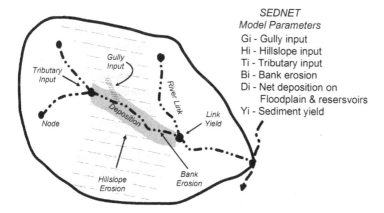

Figure 3.14. Conceptual sediment budget used in SedNet for a river link and its associated subcatchment (light shading) and floodplain (dark shading) (From McKergow et al. 2005)

be obtained from digital elevation models, whereas land cover data can be entered and managed from a geographic information system (GIS). Thus, large areas can be efficiently addressed (McKergow et al. 2005).

Physically based models, as the name implies, are based on a detailed description of the physical processes driving sediment and contaminant generation to estimate constituent exports from the watershed. Theoretically, because physics-based models are driven by an understanding of the operating processes, their ability to predict the effects of changing basin characteristics on sediment and contaminant export is quite good. However, these models tend to possess large data requirements, much of which is often lacking. Moreover, they are computationally intensive and, thus, can be quite expensive to run.

It is important to recognize that the different types of models are not mutually exclusive. Some are even described as hybrids because they contain significant components of more than one class of model. Each type of model possesses strengths and weaknesses, and the modeling approach that is chosen for a particular assessment will depend on a host of factors, including the availability of data, the size of the watershed or study area, the available computational resources, and, most importantly, the question that is being asked.

Since most of the erosion models were originally developed for agricultural purposes, they have been most frequently used to estimate losses of nutrients (N and P) from a given area. It is probably fair to say, however, that most erosion and sediment transport models can be modified to estimate the losses of trace metals from diffuse sources. Vink and Peters (2003), for example, attempted to quantify the fluxes of heavy metals in the Elbe basin from 1985 to 1999 using a GIS based model called METALPOL. The total flux of metals to the river was modeled as a function of industrial emissions, wastewater treatment plants, runoff from urban areas, groundwater emissions, and the erosion of contaminated soils (Fig. 3.15). The flux associated with eroded sediments was determined by modifying a model developed by Behrendt and Opitz (1999) to estimate nutrient emissions via erosion to rivers and streams. The modified model linked soil loss, determined using the USLE, to the heavy metal content of the soil, an enrichment factor, and a sediment delivery ratio to account for sediment deposition and storage (Fig. 3.15).

One of the advantages of using a modeling approach is that once it is developed for an area, parameters within the model can be altered to determine their net effect on sediment and contaminant transport rates. For example, it may be highly profitable to examine the change in sediment or contaminant exports from a basin if the land use is changed in a particular way. A word of caution is warranted, however, when using modeling data for predictive purposes (e.g., the future rates of contaminant loadings from the erosion of hillslopes). Models undoubtedly simplify the complex nature of natural systems and, thus, one cannot precisely understand how a system will respond to a change in the external controlling factors. This has led some to argue that their use for true prediction is impossible (Pilkey and Thieler 1996; Libbey et al. 1998). Wilcock and Iverson (2003) point out, however, that modeling exercises play a crucial role in organizing our understanding of complex

(a) $L_{i,T} = [E_i + E_{wtp} + E_{urban} + E_{gw} + E_{fast} + E_{erosion} + E_{atm} - E_{retention}]$

$L_{i,T}$ Total load of consitutent at monitoring station i over time T

E_i Annual avearge industrial emissions to the surface water

E_{wtp} Annual average emissions from wastewater treatment plants to the surface water

E_{urban} Annual average emissions from urban areas to the surface water

E_{gw} Annual average groundwater emissions to the surface water

E_{fast} Annual average emissions room fast runoff (surface runoff and interflow) to the surface water

$E_{erosion}$ Annual average emissions from erosion to the surface water

E_{atm} Annual average emissions from direct atmospheric deposition to surface water

$E_{retention}$ Annual average retention of material in surface water upstream

(b)
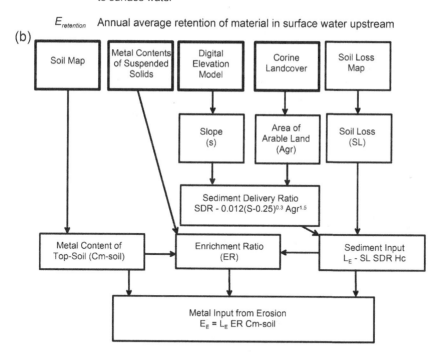

Figure 3.15. (A) Primary input parameters used to calculate constituent loads in METALPOL. (B) schematic of the process used to estimate heavy metal inputs associated with erosion (From Vink and Peters 2003)

systems, and therefore, provide a foundation for making management decisions. Nonetheless, it is clear that a host of factors can introduce uncertainty into model outputs, and the generated data need to be interpreted in light of their underlying assumptions and the quality and resolution of the utilized data.

3.4.3.2.   *Assessment by short-lived radionuclides*

The potential use of ^{137}Cs to indirectly determine average rates of soil loss was recognized as early as the 1960s (Menzel 1960; Rogowski and Tamura 1965). Since the initial hypothesis was proposed, the concept has been intensively tested and is now a widely accepted methodology for determining average rates of both erosion and deposition over approximately the past four decades. In fact, Ritchie and Ritchie (2005) identified 3,184 papers related to the transport and fate of Cs in the environment, a significant number of which pertain to the use of ^{137}Cs as a tool for analyzing erosional and depositional processes.

Cesium-137 is produced by nuclear fission, and its presence in the environment is due to its release from nuclear reactors and, more importantly, nuclear bomb tests (Ritchie and McHenry 1990). Although the first weapons tests were carried out in 1945, the release of ^{137}Cs on a global scale began in 1952 with the initiation of thermonuclear weapons testing (Perkins and Thomas 1980). These thermonuclear explosions injected ^{137}Cs into the troposphere where it circulated globally before falling back toward the Earth's surface. The rate of fallout and, thus, its distribution, is closely linked to the amount and intensity of precipitation (Longmore 1982).

The amount of ^{137}Cs in soils in many areas of the northern hemisphere reached detectable concentrations in 1954 (Ritchie and McHenry 1990). Measurable concentrations in the southern hemisphere did not occur until 1958 and are generally much lower than in the northern hemisphere as a result of differences in the extent of thermonuclear testing between the two regions. In both hemispheres, concentrations began to decrease following the 1963 Test Ban Treaty. By the mid-1970s, fallout had declined below detectable levels in the southern hemisphere, whereas in the northern hemisphere it could no longer be measured after 1983/1984 (Ritchie and McHenry 1990).

The cycling of ^{137}Cs in the near surface environment has been reviewed by Ritchie and McHenry (1990) and is illustrated in Fig. 3.16. In most environments, atmospheric ^{137}Cs is deposited directly on soils (and other geological materials), upon water bodies, and on vegetation. That portion which is deposited on vegetation can either be absorbed or adsorpbed. Adsorped Cs tends to be washed off during later precipitation events and is incorporated in to the surrounding soils. The absorped Cs remains within the plants until they die, after which it is incorporated into the soil as the plant materials decay. In either case, the atmospherically deposited ^{137}Cs ends up in the soil where it is strongly bound to fine-grained particles in the upper 20–30 cm of the soil profile (Fig. 3.17). Downward diffusion or leaching of Cs is thought to be negligible. It follows, then, that the amount of ^{137}Cs within an undisturbed soil profile is equal to the ^{137}Cs directly deposited from the atmosphere, that which was washed off of vegetation, and that which is associated with decaying plant materials. Because all

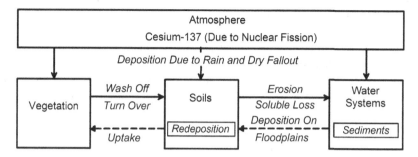

Figure 3.16. Conceptual diagram of ^{137}Cs cycling in the near surface environment. (From Ritchie and McHenry 1990)

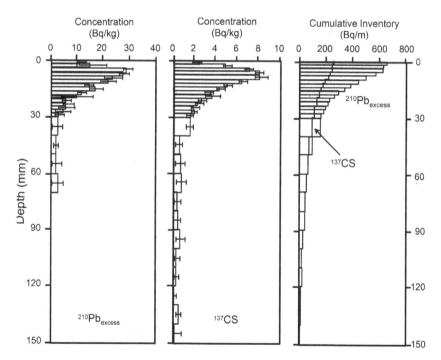

Figure 3.17. Radionuclide depth profiles in an undisturbed site within the St. Helens State Forest, Australia. (A) ^{210}Pb$_{ex}$ concentration, (B) ^{137}Cs concentration (C) Cumulative inventories (Modified from Wallbrink and Murray 1996)

three sources of ^{137}Cs were originally derived from the atmosphere, the ^{137}Cs content of an undisturbed soil profile is a close representation of the total atmospheric ^{137}Cs flux to the Earth's surface (provided that the uptake of Cs by plants from the soil is minimal, which is almost always the case). The ^{137}Cs content of the soil may, however, be altered by the erosion or deposition of particles to which the ^{137}Cs is bound. Thus, it is theoretically possible to determine the net rates of soil loss or gains to a site by determining changes in the ^{137}Cs content of the soil through time.

Two primary methods are used to determine rates of erosion and deposition at a site on the basis of temporal changes in the ^{137}Cs content of the soil (Walling and Quine 1990). One approach is to measure the ^{137}Cs content of the soil at two distinctly different times (de Jong and Kachanoski 1988). Erosion is indicated by a loss of ^{137}Cs, whereas deposition results in an increase in the ^{137}Cs inventory. The primary advantage of this technique is that the time period over which the measurements are made is precisely known, allowing for a more accurate assessment of the average erosion rates. The method, however, is rarely used because measurement intervals of 10 years or more is normally required to obtain analytically detectible differences in the ^{137}Cs content of the soil (Walling and Quine 1990). The second approach is to determine the ^{137}Cs content of the soil at an undisturbed (reference) site where erosion and deposition has been minimal. Rates of erosion or deposition can then be determined for other sites by comparing their ^{137}Cs inventories to that measured for the undisturbed area. Losses in ^{137}Cs indicate net erosion over the time interval extending from the onset of atmospheric fallout to the time of sampling.

Cesium-137 concentrations typically decrease with depth in undisturbed areas (Fig. 3.17). Thus, a significant question that must be considered is how to convert the net loss of ^{137}Cs into an absolute quantity of eroded sediment. Several methods have been developed to tackle this problem (e.g., Kachanoski 1987; Fredericks and Perrens 1988; Elliot et al. 1990; Ritchie and McHenry 1990; and Zhang et al. 1990), and have been reviewed by Walling and Quine (1990). Most of these techniques (except that presented by Zhang et al. 1990) were developed for cultivated fields where the vertical distribution of ^{137}Cs in the eroded materials could be assumed to be uniformly distributed.

A number of other short-lived radionuclides have been applied to estimate erosion rates utilizing the approaches developed for ^{137}Cs. The two most common nuclides are ^{210}Pb and ^{7}Be. Lead-210 differs from ^{137}Cs in that it is naturally produced as part of the ^{238}U decay series. Its immediate parent along the decay series is ^{222}Rn, a gas emitted from soils and sediment to the atmosphere. Most ^{210}Pb that is created by the decay of ^{222}Rn is subsequently returned to the Earth's surface within a few days where it may become incorporated and strongly attached to soil particles (Dickin 1997). Not all of the Pb, however, is derived from the atmosphere. Some, referred to as supported ^{210}Pb, is produced in the soils by the decay of ^{226}Ra. The amount of supported ^{210}Pb activity can be determined and, thus, removed from the total to determine the total excess ^{210}Pb$_{ex}$ in the soil which is derived from atmospheric deposition. As was the case of ^{137}Cs, most ^{210}Pb is found in the upper soil horizons allowing total inventories to be determined by analyzing the near surface materials (Fig. 3.17).

The half-life of ^{210}Pb is 22.26 years. In most cases, the ^{210}Pb activity is such that it can continue to be measured in soils for a period of about four to five half-lives, or about 100 years. Thus, the calculated rates of soil erosion represent average values over a period of approximately the past century.

Beryllium-7 is a naturally produced radionuclide that is created from the cosmic bombardment of atmospheric N and O. The half-life of ^{7}Be is extremely short – about 53 days. Thus, it can be used in a way similar to ^{210}Pb to determine average erosional or depositional rates, but the timeframe considered is considerably shorter (about 1 year).

A significant advantage of using short-lived radionuclides to determine erosion rates is that the required data can be generated rather quickly and may involve as little as one trip to the site. Moreover, use of more than one radionuclide may allow erosional rates to be determined for a variety of differing timeframes. However, the analysis of radionuclides can be quite expensive, and the generated results are likely to contain more uncertainties than those associated with the direct measurement of hillslope erosion rates. For example, two assumptions inherent in the comparison of ^{137}Cs, ^{210}Pb, or ^{7}Be contents between undisturbed and disturbed sites is that the atmospheric fallout of the nuclides is uniform over the area, and that this distribution was not modified before the radionuclides were bound to sedimentary particles. However, microclimatic variations across the landscape have been shown to result in non-uniform rates of atmospheric deposition. In the case of ^{137}Cs, variations in the initial (pre-disturbed) ^{137}Cs content may reach as much as 40% (Sutherland 1994; Wallbrink et al. 1994; Wallbrink and Murray 1996). Although the effects of these variations may be reduced by using ratios of the short-lived radionuclides, such as ^{210}Pb$_{ex}$ to ^{137}Cs (Wallbrink and Murray 1996), the derived results should be used with a high degree of caution.

3.5. SUMMARY

The downslope transport of pollutants from upland areas to the drainage network primarily occurs during rainfall events and is largely governed by the discharge capacity of the various hydrological pathways, the frequency with which those pathways are utilized, and the time with which the water is in contact with the basin's geological materials (Heathwaite and Dils 2000). Contaminant migration is also dependent on a number of other parameters such as the amount, concentration and chemical form of contaminants in the soil, their spatial distribution across the landscape, the nature of the underlying geological materials, particularly with regards to their hydrological and geochemical (sorption) properties.

The majority of the contaminant load measured in direct runoff can generally be attributed to the erosion and downslope transport of chemically reactive, fine-grained particles by either Hortonian or saturated overland flow. The best way to determine the rates of upland erosion is through the direct measurement of soil loss. In the absence of direct measurements, erosion rates determined for other locations may be extrapolated to the site of interest, provided that the topography, geology, hydrology, and vegetation of the sites are similar. Alternatively, potential erosion rates may be assessed using various types of erosion models, or through the application of short-lived radionuclides, the latter of which provide estimates of the average rates and patterns of erosion over a period of several decades.

Hillslope hydrological processes ultimately control the amount of water that reaches the channel and the concentration of dissolved and particulate materials within the water column. Thus, the hydrologic and geochemical characterization of river channels, and the water within them, should never be divorced from an analysis of the basin as a whole.

3.6.   SUGGESTED READINGS

Burt TP, Pinay G (2005) Linking hydrology and biogeochemistry in complex landscapes. Progress in Physical Geology 29:297–316.

Heathwaite AL, Dils RM (2000) Characterizing phosphorus loss in surface and subsurface hydrological pathways. The Science of the Total Environment 251/252:523–538.

Lawler DM (1991) The importance of high-resolution monitoring in erosion and deposition dynamics studies: examples from estuarine and fluvial systems. Geomorphology 64:1–23.

Ritchie JC, McHenry JR (1990) Application of radioactive fallout Caesium-137 for measuring soil erosion and sediment accumulation rates and patterns: a review. Journal of Environmental Quality 19:215–233.

Wallach R, Grigorin G, Rivlin J (2001) A comprehensive mathematical model for transport of soil-dissolved chemicals by overland flow. Journal of Hydrology 247:85–99.

Walling DE, Quine TA (1990) Calibration of Caesium-137 measurements to provide quantitative erosion rate data. Land Degradation and Rehabilitation 2:161–175.

CHAPTER 4

THE WATER COLUMN – CONCENTRATION AND LOAD

4.1.   INTRODUCTION

Site characterization and assessment involves the collection of data to address a wide range of questions pertaining to the potential risk that the contaminant(s) pose to human health and the environment. Three such questions related to water chemistry serve as examples: (1) does the quality of the river water violate regulatory guidelines, (2) what is the potential for accumulation of trace metals in biota, particularly aquatic biota, and (3) will the rate of downstream transport lead to negative environmental consequences in wetlands, lakes, or estuaries beyond the current area of contamination? A sound answer to all three questions requires an understanding of the dissolved and particulate trace metal concentrations within the water column. It should come as no surprise, then, that characterizing the concentration of a given constituent in river waters is an important component of almost every site assessment program. Unfortunately, the measurement of trace metal concentrations within a river is a difficult process because they continually fluctuate with changing flow conditions. Concentration also depends on a host of other parameters, such as the source of the metal and the hydrological pathways responsible for its delivery to the channel. Thus, even when discharge is static, trace metal concentrations can change (Horowitz et al. 1990, 2001). To make matters worse, changes in concentration through time are not necessarily consistent from one trace metal to another (Nagano et al. 2003; Nagorski et al. 2003), and the trends observed for a given constituent in one part of the basin may differ from that determined in another part of the basin (Brooks et al. 2001). Therefore, to characterize any given site, assessment programs must not only document the nature of the temporal and spatial trends in concentration, but also understand why such variations occur.

Here, we will examine commonly encountered, temporal trends in contaminant concentration within both the dissolved and particulate loads at a cross section. Particular attention is given to the processes that are responsible for creating the observed changes in concentration through time and to the influence of fluctuating flow conditions on trace metal concentrations and loads. To simplify the analysis, discussions of the dissolved and particulate load have been separated into separate sections.

103

It should be recognized, however, that in natural systems the dissolved and particulate load are interrelated and changes in concentration in one may affect those in the other.

4.2.   TEMPORAL VARIATIONS IN CONCENTRATION

4.2.1.   Dissolved Constituents

4.2.1.1.   Concentration-discharge relations

The dissolved load in rivers has been the focus of intense investigation since the late 1960s. Historically, dissolved constituents were defined on an operational basis as any substance that passes through a $0.45 \mu m$ filter. More recent studies have found that a portion of the constituents that pass through these filters may not be dissolved, but are attached to colloidal materials. Nonetheless, most investigations, including those typically conducted for site assessments, define dissolved constituents according to its historical definition and we will do so here as well.

The concentration of dissolved substances in river waters is widely recognized to fluctuate in response to multiple parameters, the most important of which include changes in discharge, the source of the solutes and their proximity to the river, redox and sorption/desorption processes (both within and external to the channel), and alterations in the dominant hydrological flow paths during flood events. In many cases the most significant of these controlling factors is discharge. Thus, variations in trace metal concentrations can often be expressed as:

(1) $C = aQ^b$

where C is the dissolved concentration of the metal of interest, Q is discharge, and a and b are constants. For a large number of river systems, trace metal concentrations tend to decrease with increasing flow magnitudes (Fig. 4.1). For example, in a global survey of 370 rivers by Walling and Webb (1986), 97% had b values less than 0, the majority falling between 0 and -0.4. This inverse relationship between concentration and discharge is generally attributed to dilution of solutes in base flow, derived primarily from groundwater, by diffuse inputs of meteoric waters during runoff events. The relatively high concentration of trace metals in groundwater results from its longer contact time with rocks or unconsolidated sediment prior to emerging within the channel during periods of low flow.

Negative trends in trace metal concentrations with discharge have also been found in areas affected by point sources of contamination (House and Warwick 1998; Gundersen and Steinnes 2001). In this case, high contaminant concentrations which enter the channel at a specific site, and which are concentrated during low-flow conditions, are diluted by relatively clean waters derived from upstream areas during floods. This process has been commonly recognized along stream segments located downstream of waste water treatment plants, industrial sources, and seepages from mine sites, among others.

Not all rivers exhibit decreases in dissolved elemental concentrations during floods. In fact, it is now recognized that concentrations may decrease, increase,

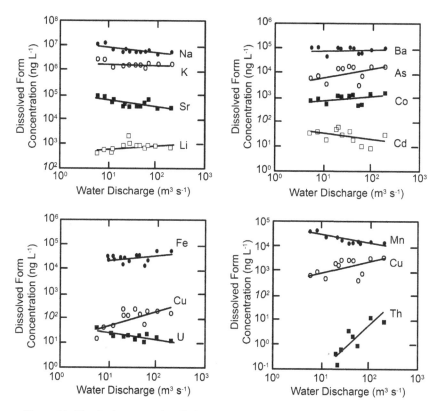

Figure 4.1. Dissolved concentration-discharge relationship for selected elements in the Kuji River, Japan. Regression lines possess positive and negative slopes indicating that concentrations can both increase or decrease with rising water levels, depending on the element (From Nagano et al. 2003)

or follow no systematic trend with rising discharge conditions (Sherrell and Ross 1999; Brooks et al. 2001), depending on the complex interaction between the controlling factors. Moreover, the relationships between concentration and discharge are element specific, and it is not uncommon to observe direct relationships (positive b values) for one suite of elements and negative relationships for another (Fig. 4.1). In general trace metal concentrations vary more through time than do the concentrations of the major ions. The larger variations observed for trace metals is presumably due to their higher sensitivity to changes in pH, redox, temperature, and various complexing agents within the water column (see, for example, Nagorski et al. 2003).

An excellent example of the complexities observed in dissolved trace metal concentrations has been provided by Olías et al. (2004) for the Odiel River of southwestern Spain. The Odiel River, which drains the Iberian Pyrite Belt, is highly contaminated by mining wastes, and river waters are characterized by low pH conditions (typically less than three). During the summer months, when rainfall is extremely limited, Fe hydroxysulfates are precipitated in mine dumps, tailings

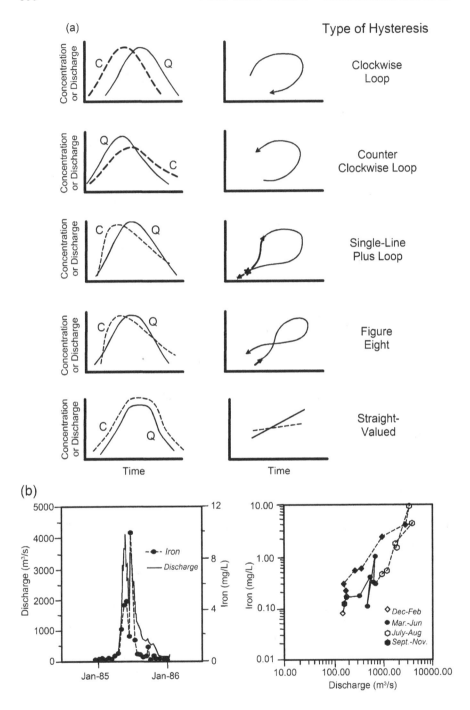

piles, and alluvial deposits along the river. In addition, the intensity of sulfide mineral weathering increases as a result of the hot, dry conditions, releasing selected trace metals to the environment. The net effect of these two processes is the development of relatively high dissolved concentrations in the water column for a number of metals (e.g., Zn, Mn, Cd, and Pb). With the onset of autumn rains, metal concentrations increase further as Fe hydroxysulfates are dissolved and the products of weathering are flushed from the upper soil horizons. However, with continued rainfall, and an increase in discharge, metal concentrations decline as a result of dilution and minor increases in pH, which affects metal solubility. The depletion of solutes from within the soil horizons also plays a role.

In contrast to the analyzed metals, arsenic concentrations reach maximum values in winter and minimum values in summer as it is strongly sorbed to and/or coprecipitated with the Fe hydroxysulfates. Thus, not only do the trends in elemental concentrations change through time as water levels rise, but the trends differ between the various elements that were analyzed.

4.2.1.2. Hysteresis

The data presented above from Olías et al. (2004) were collected for specific runoff events, and the timing of collection with respect to the rising or falling limb of the hydrograph was known. In most cases, however, concentration-discharge relationships, such as those described by Eq. 1, are derived from multiple flow events that have occurred over a period of months to years. Thus, the observed trends, where they are found to exist, represent generalized relationships which are characterized by significant variability. A large portion of this variability is associated with a phenomenon known as hysteresis, in which the concentrations of an element for a given discharge differ between the rising and falling limb of the hydrograph (Fig. 4.2). Hysteresis occurs because the timing of maximum elemental concentration does not precisely correlate with the timing of peak flow. It can also be related to the mixing of waters from different sources during a runoff event, each of which have a distinct solute chemistry (Hooper et al. 1990; Evans and Davies 1998).

The most commonly observed type of hysteresis forms a clockwise loop where the constituent concentrations increase faster than discharge, creating higher values on the rising limb of the hydrograph than on the falling limb (Fig. 4.2). The development of clockwise hysteretic loops is primarily attributed to the flushing of solutes from soils during the onset of a runoff event, followed by the influx of relatively dilute meteoric waters after solutes within the soil reservoirs have been depleted (House and Warwick 1998). The accumulation of solutes between events is generally associated with particle weathering (e.g., sulfide oxidation).

Figure 4.2. (a) Types of hysteresis recognized by Williams (1989). Curves developed for suspended sediments, but similar types have been recognized for dissolved solutes (figure modified from Williams 1989); (b) time series plot and hysteretic loop for iron for the Skeena River showing typical types of variations associated with "real" data (From Bhangu and Whitfield 1997)

It is, therefore, commonly observed along contaminated rivers, particularly those affected by mining operations. For example, along the Odiel River studied by Olías et al. (2004) the flushing of solutes by autumn rains, followed by the subsequent affects of dilution, produces a clockwise hysteretic loop if examined for a single year. This type of hysteretic process is often referred to as the *first flush* phenomenon.

Counter-clockwise loops have also been observed in some environments, but with much less frequency. Where counter-clockwise loops have been described, they are most often attributed to the initial dilution of river waters by runoff possessing low solute concentrations, followed by an increased influx of solutes associated with the arrival of throughflow (House and Warwick 1998). This particular behavior is more likely to be observed for relatively mobile solutes, such as nitrate, than for strongly reactive trace metals which are sorbed to particulate matter (House and Warwick 1998). Nonetheless, it demonstrates the importance of hydrologic flow paths in controlling the generalized relationships between solute concentrations and discharge, and the direction of the hysteretic loops.

A number of attempts have been made to model hysteresis within rivers of a given region in order to gain an understanding of the controls on the rotational direction and overall shape of the loop (Bazemore et al. 1994; DeWalle and Pionke 1994; Elsenbeer et al. 1995). These efforts have led to the argument that hysteretic loops may be used to determine the primary source of a contaminant measured within the channel (McDiffett et al. 1989; Evans and Davies 1998; House and Warwick 1998). Most models are based either on a two or three component system; thus, the sources which are defined are rather vague. Nonetheless, the methods can provide useful information regarding the chemical dynamics of a watershed.

The modeling of two component systems typically considers pre-event and event sources of water. Pre-event water is assumed to be primarily derived from groundwater discharge and from point sources of contamination. Both sources of water tend to possess relatively high levels of trace metals which are delivered to the channel between floods. Event water refers to the influx of water from various hillslope hydrological pathways operating during rainfall-runoff events. Its chemical signature is generally assumed to be that of precipitation, although it is recognized that its composition may be altered during its migration to the channel. Under these constraints, the concentration of solutes at any point along a river is dependent on (1) the influx of constituents from point sources, such as waste water treatment plants (pre-event water), (2) water from diffuse or non-point sources (event water), and (3) groundwater discharge which generates the existing base flow (pre-event water) (see, for example, House and Warwick 1998).

Three component models differ from two component modeling techniques in that the event water is subdivided into that derived from soil and that derived from some combination of channel interception and/or overland flow. Evan and Davies (1998), for example, applied a three component model to streams within the northern Appalachian Plateau of Pennsylvania in which the concentration of a solute within the channel was described by a simple mass balance equation:

(2) $C_t Q_t = C_g Q_g + C_s Q_s + C_{sr} Q_{sr}$

where C is concentration, Q is discharge, and the subscripts *t*, *g*, *s*, and *sr*, refer to total stream water, groundwater, soil water, and surface (runoff) water, respectively. They then examined the nature of the hysteretic loops that could be developed by varying the relative solute concentrations of each of the sources, while assuming that the dominant supply of water to the channel followed a specific pattern during any given runoff event. The progressive change in flow types, which had been observed in the area, was (1) base flow, (2) surface runoff and channel interception, (3) shallow subsurface flow carrying soil water, and (4) base flow. The variations in relative solute concentrations, combined with the hydrologic assumptions, yield six possible hysteretic loops (Fig. 4.3a–f). Two additional trends could be produced when the concentrations within two of the components were equal (Fig. 4.3g, h). The significance of the generated plots is that data collected from any location

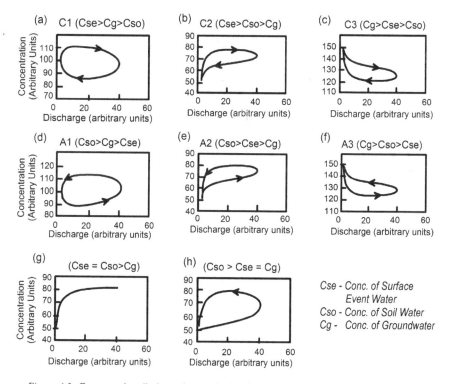

Figure 4.3. Concentration-discharge hysteresis developed using a three component model. Model based on change in discharge dominance which follows the progression: base flow (G), surface runoff and channel interception (SE), shallow subsurface flow carrying soil water (SO), and base flow. Relative source contributions shown at top of graphs (Adapted from Evans and Davies 1998)

within a channel can be compared to the various forms of hysteresis to determine the relative concentrations of the solutes that exist within each of the three defined sources. Thus, the primary source of a contaminant can be identified through the collection of a relatively simply data set (Evans and Davies 1998).

Hysteresis has been well documented for both individual flood events as well as for seasonal variations in runoff, such as those associated with spring snowmelt (Bhangu and Whitfield 1997). Not all rivers, however, exhibit hysteresis. Linear or curvilinear relationships between concentration and discharge have also been observed (Fig. 4.2). These linear trends occur when the various constituent sources (e.g., soil water, groundwater, or overland flow) possess indistinguishable geochemical signatures, or when sources regulating the geochemistry of the water provide similar quantities of the constituents during both the falling and rising limb of the hydrograph (Nagorski et al. 2003).

4.2.1.3.   Diurnal (Diel) variations

Variations in the concentration of trace metals associated with flood events have been relatively well documented over time periods ranging from a few hours to several months (e.g., those associated with spring runoff; Fig. 4.4). Few studies,

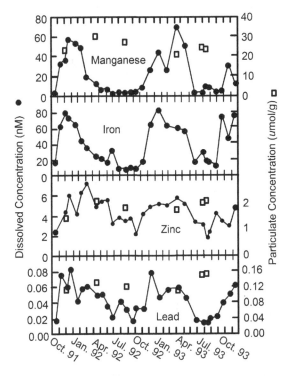

Figure 4.4. Time series plot showing seasonal variations in dissolved and particulate concentrations for selected elements in the Mississippi River immediately upstream of Baton Rouge, Louisiana (From Shiller 1997)

however, have attempted to document variations in trace metal concentrations for a 24-hour period during which there are no floods. Nonetheless, the available data clearly demonstrate that solute concentrations can systematically fluctuate on a diurnal basis (Fig. 4.5) (Wieder 1994; Brick and Moore 1996; Sullivan et al. 1998). The timing and processes controlling these daily changes depend on the physiochemical and biological conditions that exist as well as the constituent of interest.

In rivers where there is a distinct daily fluctuation in discharge, such as those receiving snowmelt or which head at the foot of a glacier, both the dissolved and particulate concentrations can vary as function of changing hydrologic conditions. The timing of the maximum and minimum concentrations observed during any 24-hour period will typically vary throughout the year (Sullivan et al. 1998), and depends on whether concentrations directly or indirectly correlate to discharge, as described earlier. Minor, diurnal variations in metal concentrations within the channel may also be caused by daily fluctuations in the exchange of solutes between the surface- and subsurface flow systems. For example, a reduction in the flux of water and solutes from the channel bed sediment to the water column is known to occur as a result of enhanced evapotranspiration during daylight hours and has been suggested as a potential cause of diurnal variations in Mn and Zn concentrations within the upper Clark Fork River (Brick and Moore 1996).

Not all of the observed short-term variations in concentration are associated with changes in system hydrology. Diurnal fluctuations observed for Mn along some contaminated rivers have been attributed to the reductive dissolution of Mn oxides (see, for example, Brick and Moore 1996). Although the dissolution of Mn oxides can be driven by several processes (Stone 1987; Meyers and Nealson 1988), the one which is of most importance on a daily timescale is the reduction brought on by a localized decrease in oxygen as daylight ceases and photosynthesis is replaced by respiration as the dominant biotic process.

Iron cycling is also influenced by redox processes. Of most importance to the diurnal cycling of Fe is the reduction of ferric iron, Fe(III) to the more soluble ferrous form, Fe(II) by a photochemical process during the day (Fig. 4.6). Sullivan et al. (1998), for example, found that maximum Fe concentrations, which occurred at about mid-day, correlated with UV radiation, the wavelength which is primarily responsible for the photoreduction of Fe hydroxide (Fig. 4.6). They also found that Fe(II) was reoxidized at night, thereby reducing Fe concentrations during the hours of darkness.

As discussed in Chapter 2, Fe and Mn oxides and hydroxides are important scavengers of trace metals. Their dissolution, then, can result in an increase in other dissolved trace metal concentrations in river waters. Increases in trace metal concentrations are also likely to be due to the desorption of metal cations from the oxides/hydroxides in response to daily variations in pH that result from diurnal fluctuations in microbial and mactrophytic respiration, among host of other factors.

The daily variations in solute concentrations that have been observed in riverine environments tend to be much less than those documented during individual floods

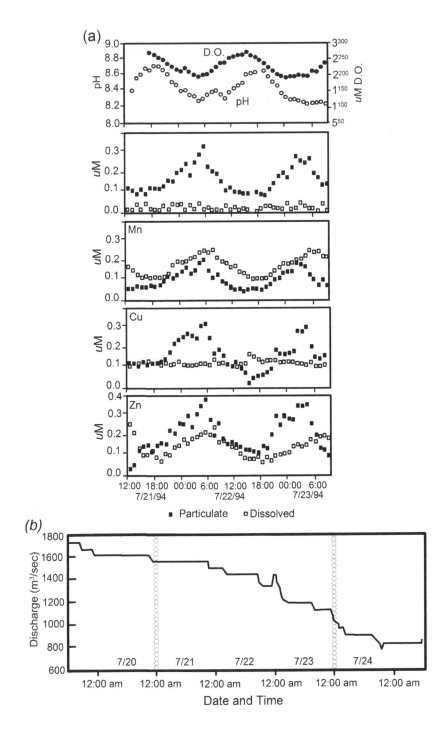

or over an entire season. Nonetheless, they can be significant. In the case of the upper Clark Fork River, Montana, Brick and Moore (1996) found that dissolved concentrations of Mn and Zn, and acid-soluble (particulate) concentrations of Al, Fe, Mn, Cu, and Zn varied on a diurnal cycle. The total (acid-soluble) concentrations of Cu were particularly interesting in that they were well below the EPA aquatic life chronic toxicity criteria during the day when flows were minimal. However, at night Cu concentrations exceeded the criteria (Fig. 4.7). Without having sampled the river throughout an entire 24-hour period, it would never have been recognized that the EPA criteria were exceeded during low flow conditions. Moreover, because many of the processes driving diurnal fluctuations in dissolved and particulate metal concentrations lead to elevated levels at night, daytime sampling may underestimate both the impacts to aquatic organisms, and the flux of metals to and through the river.

4.2.2. Particulate and Particulate-Borne Contaminant Concentrations

The clarity of a mountain stream is due, in part, to a lack of particulate matter. While these clear, cool waters are something to be enjoyed and preserved, they create a significant problem for the investigator charged with determining the particulate load for a contaminant because it is extremely difficult to collect adequate quantities of material for analysis. Even in turbid rivers, the collection of suspended sediment during low flow conditions can be problematic. As a result, the traditional method of determining particulate trace element concentrations (TC) does not involve the direct analysis of suspended sediment. Rather two samples (one filtered and the other unfiltered) are collected and analyzed separately. Particulate trace metal concentrations are then determined by subtracting the concentration of the dissolved fraction from the concentration measured in the unfiltered sample (i.e., water plus sediment; TC)(Office of Water Data Coordination 1982). While the technique simplifies the process of determining trace metal concentrations in the suspended load, the indirect nature of the analysis can lead to biased results. The bias is often related to the analysis of small quantities of suspended sediment within the collected sample which are not representative of the suspended sediment within the entire cross section (Horowitz 1995; Horowitz et al. 2001). In other words, the measured concentrations do not accurately portray the bulk trace metal content of the suspended sediment in the river at the time of sample collection.

Figure 4.5. (a) Daily time series of selected parameters within the upper Clark Fork River, Montana; (b) U.S. Geological Survey discharge measurements for the upper Clark Fork. Dissolved metal concentrations increased by two to three fold at night in response to changes in pH and dissolved oxygen. Fluctuations of solute concentrations are thought to be due to redox reactions or changes in the influx of water from the channel bed (hyporheic zone). Particulate concentrations are related to increased suspended sediment, which may be due to the enhance biotic activity of benthic organisms at night (Adapted from Brick and Moore 1996)

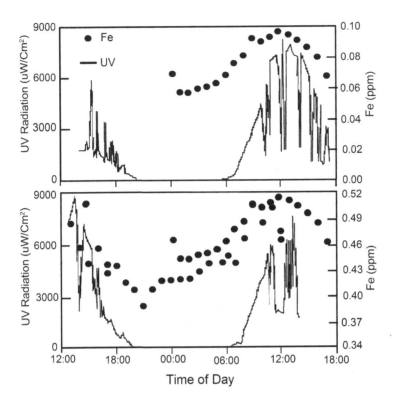

Figure 4.6. Time series of iron concentration and UV radiation in (a) July and (b) mid-August measured along Peru Creek, Colorado. Increased Fe concentrations are likely due to redox reactions associated with increased UV radiation during daylight hours (From Sullivan et al. 1998)

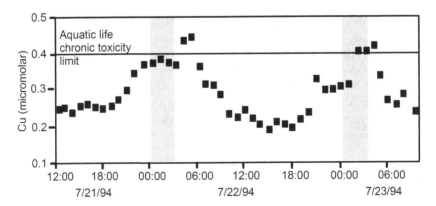

Figure 4.7. Copper concentrations measured on the upper Clark Fork River, Montana. The USEPA aquatic life chronic toxicity limit is exceeded at night, when samples are not generally collected. Such variations could lead to underestimations of the ecological effects of metals in contaminated rivers (From Brick and Moore 1996)

An alternative approach that is widely used by the academic community involves the filtering of small, bulk water samples to separate the particulates from the fluid (Horowitz et al. 2001). Either the sediment or the sediment and filter combination are then analyzed independently of the "clean" water to determine the particulate and dissolved fractions, respectively. This procedure is likely to generate more accurate results, although it is time consuming and expensive. A third, closely related approach that is now being frequently used is to collect large volumes (on the order of 10 to 100 L) of unfiltered water in the field, and then to separate the particulate fraction for analysis using a continuous centrifuge (see, for example, Horowitz et al. 2001). This method also allows for the direct analysis of the suspended particles. However, it is possible that changes in trace metal partitioning can occur between the water and sediment by sortion and desorption processes during sample shipment to the laboratory.

Acceptance of a single method of determining trace metal concentrations of suspended particles appears to be a long-time off. Thus, the approach that is utilized for an assessment will likely depend on the lab and sampling facilities that are available, the purpose of the analysis, and the existing regulatory requirements. For many regulatory purposes the traditional technique of indirectly determining suspended sediment-trace metal concentrations remains the accepted method.

The suspended sediment-borne trace element concentration in river waters (TEC_{susp}, expressed in mg/l) is a function the trace metal content of the suspended particles (TEC_p, $\mu g/g$) and the suspended sediment concentration within the water column (TSS, mg/l). It can be calculated using the following formula, which accounts for differences in the units of measurement (Horowitz et al. 2001):

$$(3) \qquad TEC_{susp} = ((TEC_p)(TSS))/1,000$$

Equation 3 shows that variations in TEC_{susp} associated with changing flow conditions will be driven in part by the way in which TSS is affected by changes in discharge (see, for example, Nagano et al. 2003). The relationship between discharge and TSS in most rivers is similar to that expressed for dissolved constituents in which,

$$(4) \qquad TSS = aQ^b$$

where Q is discharge and a and b are constants. The values of the exponent b generally range between one and two (Walling and Webb 1992), indicating that the increase in sediment to the river during periods of rising discharge is greater than the influx of water to the channel. There is, however, considerable variability in the established relationships. For example, suspended sediment concentrations measured within the Creedy River at a discharge of $10 \, m^3/s$ ranged from approximately 1 mg/l to 1,000 mg/l (Fig. 4.8). The observed variability is related to a number of factors, including flood magnitude. Consider, for instance, that an observation of $10 \, m^3/s$ might be at the peak of one storm, midway on the rising limb of another, or most of the way down the falling limb of a third.

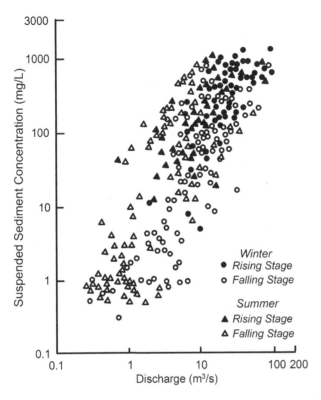

Figure 4.8. Suspended-sediment concentration-discharge relationship for the Creedy River, Devon, U.K. Note differences in relationship for different seasons and stage (Modified from Knighton 1998; data after Walling and Webb 1987)

Another important source of variability in suspended sediment concentration is the source of the suspended particles. Data collected during flood events have shown that channel bed aggradation can occur during periods of increasing suspended sediment concentration. Since more sediment is added to the channel bed during aggradation than is removed, the suspended particles could not have been derived from the erosion of the channel floor (Leopold and Maddock 1953). These observations have led to the widely held assumption that suspended sediment is primarily derived from (1) the erosion of the channel banks, which is dependent on the ability of the flow to abrade the bank materials, and (2) the erosion of hillslope soils by overland flow, rilling, or gullying processes (Knighton 1998; Ritter et al. 2002). Hillslope erosion is commonly found to be the dominant source (see, for example, McKergow et al. 2005). It follows, then, that suspended sediment concentrations are usually controlled more by the supply of sediment from the hillslope than by the magnitude of the flow within the river.

The supply of sediment from upland areas to the channel is affected in no small way by the availability of easily eroded particles within hillslope sediment reservoirs. A case in point is the seasonal variation shown in Fig. 4.8 which primarily results from a greater availability of erodible materials during the summer months. Changes in sediment availability can also cause hysteresis effects during seasonal runoff or short-term precipitation events. Clockwise hysteresis, for example, occurs where suspended sediment concentrations reach a maximum before peak flow is attained. It is generally attributed to the input of easily eroded sediment to the channel during the early stages of a runoff event, followed by a decline in sediment influx as the easily eroded materials are depleted from the upland areas (Williams 1989; Knighton 1998). Hysteresis, in turn, produces much of the scatter in the concentration-discharge relationship (as it did for the dissolved load) because different values of TSS occur for a given discharge, depending on whether it is associated with the rising or falling limb of the hydrograph.

In addition to the linear relationships between sediment concentration and discharge, Williams (1989) identified five possible types of hysteresis (Fig. 4.2). These types of hysteresis are similar to those identified for dissolved constituents, but the causative factors are different. In the case of the suspended particles, exactly which type of hysteresis is formed, if any, depends on the interaction of a large number of factors in addition to sediment availability. These factors include the intensity and areal distribution of precipitation within the basin, the amount and rate of runoff, the amount of sediment stored along the channel, and differences in downstream flow rates between the sediment and the water (Williams 1989). Clockwise hysteresis is probably the most common type, but is known in larger basins to change downstream to counter-clockwise loops. The downstream change in the type of hysteresis is presumably related to the proximity of the reach to sediment sources and to differences in the rate at which water and suspended sediments move through the channel (Knighton 1998). In headwater areas, hillslope processes can deliver sediment to the channel relatively quickly, allowing suspended sediment concentrations to peak during the early stages of an event. However, peak flow conditions may occur downstream before maximum suspended sediment concentrations are realized because flood waves move considerably faster than the suspended particles. The shift in the timing of peak discharge relative to maximum suspended sediment concentrations produces a reversal in the rotational direction of the hysteretic loop.

A common purpose for understanding the relationship between TSS and discharge is that it provides insights into the changes in the total suspended, sediment-borne trace metal concentrations in the river water (TEC_{susp}) that are likely to occur during a flood event. In other words, variations TSS are seen as a surrogate for changes in the TEC_{susp} for times or locations where geochemical data are unavailable. Moreover, because particulate trace metal concentrations are usually much greater than the dissolved concentrations, TSS can often be used as a surrogate for the total (unfiltered) trace metal concentration (TC) in river water. It is important to remember, however, that TEC_{susp} is not only dependent on TSS, but on the trace metal content of the suspended particles (TEC_p). Thus, while TSS typically increases with discharge, TEC_p may increase, decrease, or show no distinct relationship with

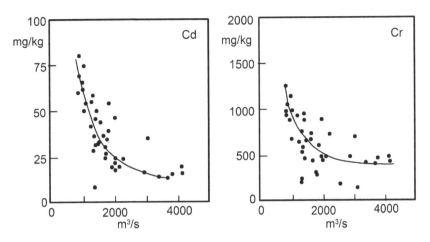

Figure 4.9. Change in Cd and Cr concentration with increasing discharge in the suspended sediments of the Rhine River. The inverse relationship is generally attributed to dilution of contaminated particles by "clean" sediment (Adapted from Salomons and Eysink 1981)

increasing discharge. The diversity of trends is driven largely by the effects of dilution on particulate metal concentrations by the influx of "clean" sediment from upstream areas and tributaries. Where dilution is significant, concentrations will likely decrease with increasing discharge (Fig. 4.9). The point, here, is that the use of TSS as a surrogate for TEC_{susp} or TC for times in which there are no data should be performed with caution, and only when the existing data suggest that they are strongly correlated to discharge.

4.3.   SEDIMENT AND CONTAMINANT LOADS

4.3.1.   Load Estimation

The *load* that a river carries refers to the mass of material that passes a given point in the channel in a given period of time. It is expressed in units of mass (e.g., kg or tons) because the timeframe is implied from the context of how it is used. Load differs from loading rate (or flux) which refers to the instantaneous rate at which load (sediment) moves past a given point or cross-section in the channel as expressed in units of mass per unit time (e.g., g/sec or tons/day).

Although there are exceptions, most assessments of contaminated rivers will spend considerable time and effort to document the river's load, in addition to simply monitoring concentrations as described above, as it is considered an essential ingredient of ecological risk assessments. One of the many uses of load, for example, is to determine the amount of the contaminant, in our case trace metals, that are transported offsite and may therefore affect downstream wetlands, lakes, and estuarine environments located beyond the borders of the polluted area. Load estimations are also used with considerable frequency to document trends in water quality, especially with regards to deciphering whether water quality targets are being met.

In order to calculate the *total* annual load of a contaminant one must estimate and combine the flux of the pollutant as a dissolved constituent and as a constituent bound to sediments moving in suspension and along the channel bed. For most gravel-bed rivers, however, the contaminant load associated with particles moving along the channel floor (i.e., bedload) is rarely ever considered because of the difficulties in measuring bedload transport and its non-reactive nature (Horowitz et al. 2001). Even for sand-bedded rivers, the importance of bedload may be minimal as sediment-borne trace metal transport tends to be dominated by the movement of fine-grained particles (Fig. 4.10) (Gibbs (1977).

Load, or more precisely loading rates, cannot be directly measured, but must be calculated as the product of discharge and concentration. Because both discharge and concentration are continually changing, the flux also changes. If it were possible to collect an infinite number of discharge and concentration measurements from a perennial stream, we could form a continuous plot of the alterations in flux through time (commonly referred to as a time series) (Fig. 4.11). The area under the curve, such as that shown in Fig. 4.11, represents the load of that constituent for the time period over which the measurements were collected (Richards 2001). In the case of the particulate load, TEC_{susp} is used to perform load calculations.

Two problems emerge when we begin to develop a plot of flux versus time using real data. First, we can never collect enough samples to form a perfectly continuous series. Second, the number of measurements that are collected will likely differ between discharge and concentration. The USGS, for example, commonly collects and stores

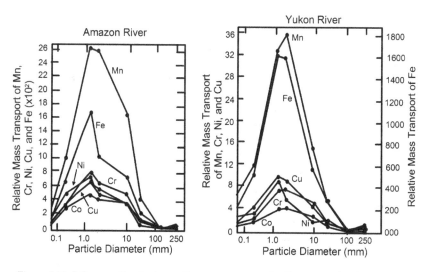

Figure 4.10. Influence of grain size on the mass transport of sediment associated trace metals within the Amazon and Yukon Rivers. Note that trace metal transport is dominated by fine particles (Modified from Gibbs 1977)

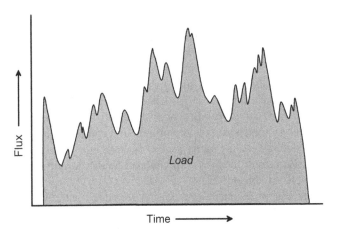

Figure 4.11. Changes in sediment/contaminant flux (or loading rate) through time. Load is equal to the area under the curve (From Richards 2001)

discharge data for large rivers on an hourly basis; on smaller rivers where runoff is rapid data may be obtained over shorter intervals, such as every 15 minutes (Richards 2001). Such detailed measurements of concentration are seldom if ever obtained on an annual basis because it is extremely expensive and time consuming. The issue, then, becomes one of knowing how to combine our discrete measurements of discharge and concentration (which may differ in number) in order to most effectively portray the true load of constituents in the river through time.

A wide range of approaches have been developed to address the above difficulties in estimating load (Table 4.1). In general, these techniques fall into two broad categories: interpolation procedures (include averaging and ratio estimators) and extrapolation methods. Interpolation procedures assume that discharge and concentration data collected at an instant in time can be used to estimate the load during periods between samples for which no data exist (Walling and Webb 1981; Degens and Donohue 2002). Extrapolation methods are based on the development of mathematical functions (e.g., regression equations) that extrapolate a limited number of concentration data over the entire period of measurement.

Given the variety of approaches that are available to calculate load, one is almost certainly faced with the question of which method is most appropriate for the river they are studying. There is no easy answer to this question because analyses comparing the accuracy and precision of the various techniques have clearly demonstrated that no one method is ideally suited for all rivers (Preston et al. 1989). In fact, some approaches may work better than others for different constituents within the same river. Fortunately, there are several excellent guidance documents that can be consulted for information on the advantages and disadvantages of the most frequently used methods, where they are likely to be applicable, and how and why they should be linked to an effective sampling program (Letcher et al. 1999; Richards 2001; Degens and Donohue 2002). There are, however, a few general

Table 4.1. Selected methods used to estimate constituent loads

Techniques	Method	Reference
Averaging Techniques		
$Load = K(\sum_{i=1}^{n} \frac{c_i}{n})(\sum_{i=1}^{n} \frac{q_i}{n})$	1	Walling and Webb (1981)
$Load = K\left(\sum_{i=1}^{n} \frac{c_i q_i}{n}\right)$	2	Walling and Webb (1981)
$Load = K\bar{q}_r \left(\sum_{i=1}^{n} \frac{c_i}{n}\right)$	3	Walling and Webb (1981)
$Load = \dfrac{K\sum_{i=1}^{n}(c_i q_i)}{\sum_{i=1}^{n} q_i}\,\bar{q}_r$	4	Walling and Webb (1981)
$Load = K\sum_{i=1}^{n}\left(c_i \bar{q}_{pi}\right)$	5	Walling and Webb (1981)
$Load = K\sum_{m=1}^{12}(\bar{c}_m \bar{q}_m)$	6	Walling and Webb (1981)
Ratio Estimators		
$Load_a = \bar{l}i\left(\frac{\bar{q}_a}{\bar{q}_i}\right)\left(\dfrac{1+(\frac{1}{n}-\frac{1}{N})\frac{S_{lq}}{\bar{l}_i \bar{q}_i}}{1+(\frac{1}{n}-\frac{1}{N})\frac{S_{qq}}{\bar{q}_i^2}}\right)$	7 (Beale)	Richards (2001)
$Load_a = \frac{\bar{l}_i}{\bar{q}_i} Q_t$	8	Cochran (1977)
General Extrapolation Equation		
$\ln c_i = \beta_0 + \beta_1 \ln q_i$	9	Degens and Donohue (2002)

K = conversion factor to take account of period of record; c_i = concentration of individual samples; c_m = mean monthly concentration; q_i = discharge at time of sampling; q_m = mean monthly discharge; q_r = mean discharge for period of record; q_{pi} = mean discharge for interval between samples; q_a = mean discharge for year; Q_t = total discharge for period of record; n = number of samples; N = number of expected samples based on sample period and interval; S_{lq} = covariance between load and flow; S_{qq} = variance of flow; l_i = mean load based on individual samples; l_a = mean load for year; bar over symbol represents mean value

comments which consistently emerge from these and other documents that are worth noting here for selecting an appropriate method for a particular setting, constituent or problem. First, for fixed-interval sampling (where samples are collected at a uniform time increment), the quality of the load estimates and the resolution of the determined variations, increase as the time interval between sample collection decreases and the number of samples collected and analyzed increases (Fig. 4.12). Second, load estimates using averaging methods, which are the easiest to use, may be highly precise where there is an abundance of data. Thus, they may be particularly well suited in some situations to determine temporal trends in water quality. Third, regression (extrapolation) approaches allow for the calculation of loads for periods or flow conditions for which concentration data are lacking (e.g., the 100-year flood). They tend to perform well when the statistical relationships between discharge and concentration are well-defined, linear, and constant throughout the

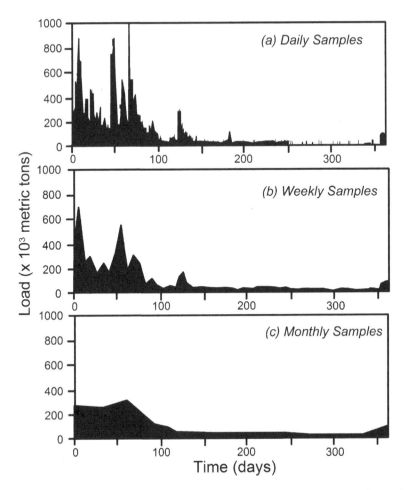

Figure 4.12. Pollutant loads determined for the Sanduscky River, Ohio, between October 1, 1984 and
October 1, 1985. Loads based on (a) daily samples, (b) weekly samples, and (c) monthly samples.
Frequent sampling increases the reliability of the total load estimates
(From Richards 2001)

sampling period. They should not be used where a strong relationship between
discharge and concentration does not exist. Fourth, ratio estimators which include
a correction factor for bias (e.g., the Beale ratio estimator shown in Table 4.1) tend
to out perform regression and averaging techniques, particularly when used with
data that is stratified according to specific flow conditions (e.g., periods of low-
and high-flow). They should strongly be considered for sites where the appropriate data are available. Finally, load estimates based on samples collected for
specific flow conditions which are then analyzed independently and combined
almost always leads to an improvement in precision. This is not necessarily true

when data stratification is performed after sample collection (Richards 2001). Thus, if post-sampling stratification is contemplated, then the load calculations should be performed both with and without the use of the flow stratified data and the more precise calculations should to be utilized.

4.3.2. The Effective Transporting Discharge

Sediment and contaminant loads nearly always increase with discharge. This relationship holds true even when concentrations of the constituents decrease with rising water levels because the increased volume of water passing a given point in the channel per unit time more than compensates for the decline in concentration. One may assume, then, that most of the load is transported during flood events. The question that arises, however, is whether most of the sediment and contaminant load is transported by extremely large floods that occur infrequently (e.g., every 100 years), or by relatively small floods that do not transport as much material, but which occur more often. It turns out that the answer to this question is rather important because it provides valuable insights into the conditions which are most responsible for contaminant dispersal as well as the interpretation of sediment and contaminant load data collected as part of the assessment program.

The first comprehensive examination of this question was provided by Wolman and Miller (1960) in a seminal paper on geomorphic work. They argued that the flow conditions responsible for transporting the most sediment over a period of time could be identified by (1) subdividing the range of flow conditions observed at the site into specific discharge classes, (2) multiplying the rate of sediment transport for a given discharge range (applied stress) by the frequency with which that discharge range occurs, and (3) examining the differences in the flows transporting ability for each of the discharge categories. The product of this process is graphically shown in Fig. 4.13, where the peak in curve C represents the effective transporting discharge, defined here as the magnitude and frequency of the flow which transports the majority of

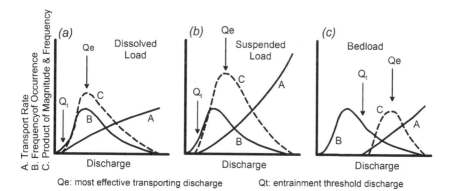

Figure 4.13. Most effective transporting discharge for (a) dissolved, (b) suspended, and (c) bedload (From Knighton 1998)

a constituent through a reach over periods of several decades. When the approach was performed using actual data Wolman and Miller found that about 90% of the suspended sediment load was transported by relatively moderate floods that recurred once every 5 to 10 years on average. While megafloods transported more sediment than the smaller events, they occurred so infrequently that their contribution to the total load exported from the watershed over a period of years was minimal.

The original curves presented by Wolman and Miller were based on data for suspended sediments. More recent studies have shown that different curves will be produced for constituents moving in solution or along the channel bed (Fig. 4.13a, c). In general, the recurrence interval (frequency) of the most effective transporting discharge will progressively decrease from the dissolved load, through the suspended load, and to the bedload (Knighton 1998), reflecting the energy (applied stress) required for transport (Fig. 4.13). It has also become clear that the recurrence interval of the effective discharge varies between rivers more than was originally thought. Much of the observed variation is dependent on differences in the frequency distribution of flood events within the basins (Nash 1994; Knighton 1998). For example, rivers in small watersheds and/or located in semi-arid and arid climatic regimes generally possess a relatively high proportion of large floods. These larger events tend to transport a greater part of the load, as curve B in Fig. 4.13 is shifted to the right. Interbasin differences in the effective discharge are also related to the threshold for particle motion. In rivers characterized by large particles or erosionally resistant bed and bank materials, a higher applied stress is required to initiate transport, thereby increasing the magnitude (and reducing the frequency) of the effective discharge (Fig. 4.13).

Table 4.2. Effective Transporting Discharges

Load Type-Reference	River	Basin Area (km^2)	Qe, %[1]
1. **Dissolved Load**			
Webb & Walling (1982)	Creedy, UK	262	~20
2. **Suspended Load**			
Webb & Walling (1982)	Creedy, UK	262	~2
Ashmore & Day (1988)	Saskatchewan basin (21 sites)	$10 \rightarrow 10,000$	Generally 1–10
Brahmaputra	–	18	
Thorne et al. (1993) Biedenharn & Thorne (1994)	Lower Mississippi	–	13
3. **Bedload**			
Andrews (1980)	Yampa River (15 Sites)	52–660	0.4–3

Modified from Knighton (1998)

[1] Duration of Effective Discharge (% time of occurrence)

It is important to recognize that the frequency distribution of large floods and the threshold for particle motion tend to be interrelated. Rivers in smaller watersheds, for instance, are not only characterized by a higher proportion of large events, but tend to be composed of larger particles. Thus, the controlling parameters can work in concert to increase the magnitude of the effective transporting discharge.

While considerable variability in the effective transporting discharge occurs between rivers, and even between different reaches of a given river, the existing data demonstrate that the basic tenants of the Wolman and Miller model – that most of the sediment transported through a reach is done so by moderate floods – appears to hold true for a majority of the rivers investigated. Nonetheless, the duration at which these flow conditions occur is generally quite low; thus, sediments, and presumably contaminants, are primarily transported to downstream reaches by flows that occur less than about 10 to 20% of the time (Table 4.2). These findings stress the need to sample high magnitude, low frequency events to accurately assess contaminant loads. Moreover, the data indicate that short-term monitoring programs, during which time high magnitude flow events did not occur and, thus, were never sampled, may not provide an accurate indicator of the average long-term export of contaminants from the system.

4.4.   SUMMARY

Most site assessments will need to develop a thorough understanding of the temporal changes in trace metal concentrations within the water column and the processes which govern the variations in concentration through time. The existing data demonstrate that the relationship between metal concentration and discharge is related to a suite of parameters including the source of the contaminants and their proximity to the river, redox and sorption/desorption processes within and external to the channel, and alterations in the dominant hydrological pathways through which contaminants migrate to the drainage network during flood events. Sediment-borne trace metal concentrations are also influenced by the transport of easily erodible sediment from the hillslope to the channel. The concentration-discharge relationship for both the dissolved and the particulate load can often be described by an exponential equation, but considerable variability can exist around the best fit line. A large portion of this variability is associated with a phenomenon known as hysteresis, in which the concentrations of a constituent (e.g., suspended sediment, trace metals, etc.) for a given discharge differ between the rising and falling limb of the hydrograph. The causes of hysteresis differ between the dissolved and the particulate load, but for both the nature of the hysteretic loops can provide insights into contaminant sources and the transport mechanisms to the drainage network.

In order to calculate the *total* annual load of a contaminant exported from a polluted site one must estimate and combine the flux of the pollutant as both a dissolved and sediment-bound constituent. Estimation techniques fall into two broad categories: interpolation procedures and extrapolation methods. Interpolation procedures include averaging estimators and ratio estimators; both assume that discharge and concentration data collected at an instant in time can be used to estimate the load for periods between which there are no data. Extrapolation (regression) approaches allow for the calculation of loads for periods or flow conditions for which concentration data are lacking (e.g., the 100-year flood). No one method is ideally suited for all rivers and for all purposes. Nonetheless, in many cases, ratio estimators tend to out perform regression based methods, both of which yield better results than averaging estimators.

4.5.   SUGGESTED READINGS

Brink CM, Moore JN (1996) Diel variations of trace metals in the Upper Clark Fork River, Montana. Environmental Science and Technology 30:1953–1960.

Brooks PD, McKnight DM, Bencala KE (2001) Annual maxima in Zn concentrations during spring snowmelt in streams impacted by mine drainage. Environmental Geology 40:1447–1454.

Evans C, Davies TD (1998) Causes of concentration/discharge hysteresis and its potential as a tool for analysis of episode hydrochemistry. Water Resources Research 34:129–137.

Richards RP (2001) Estimation of pollutant loads in rivers and streams: a guidance document for NPS programs. Report to the US Environmental Protection Agency, Region VIII, Grant No. X998397-01-0.

Sherrell RM, Ross JM (1999) Temporal variability of trace metals in New Jersey Pinelands streams: relationships to discharge and pH. Geochimica et Cosmochimica Acta 63:3321–3336.

Shiller AM (1997) Dissolved trace elements in the Mississippi River: seasonal, interannual, and decadal variability. Geochimica et Cosmochimica Acta 61:4321–4330.

CHAPTER 5

THE CHANNEL BED – CONTAMINANT TRANSPORT AND STORAGE

5.1.   INTRODUCTION

In the previous chapter we were concerned with trace metals occurring either as part of the dissolved load, or attached to suspended particles within the water column. Rarely are those constituents flushed completely out of the system during a single flood; rather they are episodically moved downstream through the processes of erosion, transport, and redeposition. Even dissolved constituents are likely to be sorbed onto particles or exchanged for other substances within alluvial deposits. The result of this episodic transport is that downstream patterns in concentration observed for suspended sediments is generally similar to that defined for sediments within the channel bed (although differences in the magnitude of contamination may occur as a result of variations in particle size and other factors) (Owens et al. 2001). While this statement may seem of little importance, it implies that bed sediments can be used to investigate river health as well as contaminant sources and transport dynamics in river basins. The ability to do so greatly simplifies many aspects of our analyses as bed materials are easier to collect and can be obtained at most times throughout the year, except, perhaps during periods of extreme flood. Moreover, bed sediment typically exhibits less variations in concentration through time, eliminating the need to collect samples during multiple, infrequent runoff events. Sediment-borne trace metals are not, however, distributed uniformly over the channel floor, but are partitioned into discrete depositional zones by a host of erosional and depositional processes. Thus, to effectively interpret and use geochemical data from the channel bed materials, we must understand how and why trace metals occur where they do.

   In this chapter, we will examine the processes responsible for the concentration and dispersal of sediment-borne trace metals, and the factors that control their spatial distribution within the channel bed. We begin by examining the basic mechanics involved in sediment transport, including an analysis of open channel flow. We will then examine how these basic processes produce spatial variations in sediment-borne trace metal concentrations, both at the reach scale and at the scale of the entire watershed.

127

5.2.   SEDIMENT TRANSPORT

5.2.1.   Modes of Transport

Our understanding of sediment transport processes is hampered by the difficulties of obtaining measurements in natural channels. Rivers are characterized by ever changing discharge and velocity and often are bounded by erodible, inhomogeneous bed and bank materials. In addition, river water is usually opaque, thereby making visual examination of the channel bed impossible. In order to overcome these difficulties, investigators have supplemented field investigations with laboratory studies that allow the transport of sediment to be observed in flumes where slope and discharge can be changed, and other selected variables can be held constant. These laboratory studies have been instrumental in deciphering the mechanisms responsible for initiating particle motion, and for predicting the relationships between sediment transport rates and discharge. Nonetheless, our ability to predict the size and quantity of sediment (and thus, contaminants) that a river can carry under a given set of flow conditions remains less than perfect.

The movement of sediment in a channel varies with both time and space. It is a function of the size of the sediment that is being transported as well as the energy that is available to perform mechanical work. In the previous chapter, we examined the downstream flux of fine-grained particles transported within the water column as part of the suspended load. These suspended particles have little, if any, contact with the channel bed. They are held in suspension by short, but intense, upward deviations in the flow as a result of turbulent eddies which are generated along the channel margins. The more frequent and intense the turbulence, the more material, and the larger the grains, that can be held within the water column (Mount 1995). We generally think of the suspended load as consisting of only fine grained sediment. However, during extreme floods in steep, gravel bed-rivers, particles a few tens of centimeters in diameter have been observed to move in true suspension.

Suspended sediment typically moves at a rate that is slightly slower than that of the water, and may travel significant distances downstream before coming to rest. In contrast, *bedload*, which is transported close to the channel bottom by rolling, sliding, or bouncing (saltating), is unlikely to be moved great distances before it is deposited and stored, either temporarily or semi-permanently, within the channel. It is important to recognize that sediment which moves as suspended load during one flood event, may move as bedload during another, depending on the hydraulic conditions to which the particles are subjected.

The ability of rivers to transport sediment is described in terms of competence and capacity. *Competence* refers to the size of the largest particle that a river can carry under a given set of hydraulic conditions. *Capacity* is not concerned with the size of the material, but is defined as the maximum amount of sediment that a river can potentially transport. Note that capacity represents a theoretical maximum. As such, it is almost always larger than the actual sediment load being carried by the river.

It should come as no surprise that as discharge increases the amount of sediment being transported also will increase. The relationship, however, is much more complicated than one might think. For example, the correlation between the suspended load and discharge is generally quite poor because rivers can typically transport more fine-grained sediment than is supplied from the watershed; the load is controlled by supply rather than discharge. In contrast, the amount of material available for transport as bedload generally exceeds that which can be carried by the river. The correlation between discharge and bedload transport rates, then, is much better than for suspended load, but it is still less than perfect. In the case of bedload, the correlation is significantly influenced by complexities associated with particle entrainment (discussed below). Another important problem is that bedload transport is difficult to measure because the movement of particles along the channel bed is primarily associated with floods when flow depths and velocities are elevated. Working in these conditions can be difficult and dangerous. In addition, where bedload transport rates have been accurately measured along a heavily instrumented river, they are found to vary across the width of the channel bed and through time (Leopold and Emmett 1977; Hoey 1992; Carling et al. 1998). These variations are at least partly related to the movement of large-scale bedforms along the channel floor. The periodic movement of a bedform through a cross section hinders the use of hand held instruments to measure bedload transport rates because they can only sample a single location for a short period of time.

The difficulties of directly measuring bedload has led to the use of mathematical models to estimate a river's load for given a set of channel, sediment, and flow conditions (Meyer-Peter and Muller 1948; Einstein 1950; Bagnold 1980; Parker et al. 1982). The accuracy of the estimated transport rates can be difficult to assess because of a general lack of direct measurements with which to verify the model outputs. Nonetheless, it is generally accepted that these transport models only provide estimates of bedload transport (perhaps within 50 to 100% of the actual value). Gomez and Church (1989), for example, examined ten transport formulas and found that none of them were entirely adequate for predicting bedload transport in coarse-grained (gravel-bed) rivers.

5.2.2. Channelized Flow

The ability of a river to transport sediment or sediment-borne trace metals is governed by the balance between driving and resisting forces. The principal driving force is gravity (g), which is oriented vertically to the Earth's surface and equal to $9.81 \, m/s^2$. The component of this gravitational force that acts to move the available mass of water in the channel downslope is equal to $g \sin \beta$, where β is the average slope of the channel (Leopold et al. 1964). The resisting forces are more difficult to quantify and originate from a variety of sources. Attempts in recent years to identify these sources, and their relative importance, has met with only limited success. Nonetheless, the total resistance to flow can be broadly categorized into three principal components referred to as free surface resistance, channel resistance, and boundary resistance (Bathurst 1993). *Free surface resistance* is caused by disruptions in the water surface by waves or abrupt changes in gradient. *Channel resistance* represents the loss of energy as water moves over and around undulations in the channel bed and banks, or is distorted by changes in the channel's cross-sectional or planimetric configuration. The final component, *boundary resistance*, is produced by the movement of water over individual clasts or microtopographic features (e.g., dunes and ripples). Resistance produced by dunes and ripples is referred to as *form drag*, whereas resistance due to individual particles is called *grain roughness*. Other factors such as vegetation, suspended sediments, and internal viscous forces within the water also affect the resistance to flow.

Hydraulic engineers have been concerned with quantifying the relationships between the driving and resisting forces inherent in channelized flow for centuries. The net result of these studies has been the formulation of a number of resistance equations presented in Table 5.1. The Manning equation is perhaps the most widely used of these in the U.S., although the Darcy-Weisbach equation is growing in popularity within the academic community. The Manning equation was developed in 1889 by Manning in an attempt to systematize the available data into a useful form, and is expressed as:

(1) $V = (R^{2/3} S^{1/2})/n$

where V is velocity, R is the hydraulic radius (i.e., the cross sectional area of flow divided by its wetted perimeter), and S is slope. The total resistance in the channel is defined by n, called the Manning roughness coefficient.

It is important to recognize that Manning's roughness coefficient, as well as the other resistance factors, cannot be directly determined. Rather, it must be indirectly defined by measuring hydraulic parameters such as flow velocity, depth, and slope. For example, values of n are commonly determined by rearranging Eq. 1 and solving for n at a site in which the velocity, slope and hydraulic radius have been measured.

A common problem encountered in hydrological investigations is the need to calculate flow velocities for a range of conditions where roughness values have

Table 5.1. Selected flow resistance equations

Name	Equation	Unit notes
Manning equation	$V = k(R^{2/3}\ S^{1/2})/n$	$k = 1$ (SI units)
		$k = 1.49$ (imperial units)
		$k = 4.54$ (cms)
Chezy equation	$V = C\sqrt{RS}$	
Darcy-Weisback equation	$ff = (8gRS)/V^2$	

From Knighton (1998)
Symbols: C, n, ff – resistance coefficients; V – mean velocity; R – hydraulic radius; S – slope; g – gravitational constant

not been determined. Roughness values must therefore be estimated. This task is frequently undertaken by visually comparing the river or river segment being studied to other rivers of known roughness using some form of visual guide (e.g., Barnes 1968). Difficulties in using these visual guides arise because channel roughness varies as a function of flow depth and depends on factors, such as suspended sediment concentration, that cannot be easily accounted for using visual techniques. Needless to say, the method leaves a lot to be desired and estimated n values can lead to significant errors in the generated data. In fact, roughness estimates are one of the most important sources of uncertainty associated with numerical attempts at predicting sediment and contaminant transport rates.

The resistance equations presented in Table 5.1 show that the total resistance to flow is closely related to flow velocity. The velocity with which the water is traveling can vary vertically, horizontally, and downstream within the channel as well as through time at any given point. The variations in velocity that occur have led to various classifications of flow in open channels. For example, flow is frequently classified according to whether it is uniform or steady. Under *uniform flow* conditions, velocity is constant with respect to position in the channel; *steady flow* refers to the situation in which velocity is constant through time at a single point. Even a causal inspection of moving water in a river will reveal that the flow is inevitably non-uniform and unsteady. Flow can also be classified according to whether it is laminar or turbulent. In laminar flow, layers of water slide smoothly past one another without disrupting the path of the adjacent layers. Under this type of flow regime, the magnitude of the resistance is primarily governed by the molecular viscosity of the fluid, where viscosity is dependent on such variables as temperature. In contrast, turbulent flow is characterized by packets of chaotically mixing waters that transmit shear stresses across layer boundaries in the form of eddies. Eddies, or more correctly, eddy viscosity, greatly increases the resistance to flow and the dissipation of energy.

The transition between laminar and turbulent flow depends on both the flow depth and its velocity. The conditions at which the transition occurs can be predicted by the Reynolds number (Re), in which:

$$(2) \qquad Re = VR(\rho/\mu)$$

where V is the mean velocity, R is the hydraulic radius, ρ is the density, and μ is the molecular viscosity of the water. In wide, shallow channels, R can be closely approximated by mean depth and, thus, they are frequently interchanged for one another. When values of Re are less than 500 flow is laminar; in contrast, values greater than 2,000 are indicative of turbulent flow. Flows possessing Re numbers between 500 and 2,000, called transitional flow, can exhibit characteristics of both laminar and turbulent flow conditions.

Flow in natural channels is turbulent, with the possible exception of a very thin layer of quasi-laminar flow along the channel bed, referred to as the *laminar sublayer*. Most turbulence is generated along the channel perimeter, causing the resistance to flow to increase, and the velocity to decrease, as the water-sediment interface is approached. The maximum flow velocity, then, is generally observed immediately below the water surface near the center of the channel. However, the precise position of the maximum flow rates will vary with cross sectional shape as well as the channel's planimetric configuration (Fig. 5.1).

5.2.3.   Entrainment

A significant problem associated with the prediction of sediment transport is determining how the properties of flow in open channels combine to produce particle entrainment. *Entrainment* refers to all of the processes involved in initiating motion of a particle from a state of rest. It is clearly dependent on the erosive power of the flow. However, describing the relationship between the largest particle that can be entrained and the flow's erosive power has proven difficult, because: (1) individual particles within the channel bed are acted upon by multiple forces, each of which is best described by a different parameter of flow, (2) flow velocities, particularly during floods, are neither constant, nor easily measured, and (3) the size, shape, and packing arrangements of particles within natural channels are highly variable,

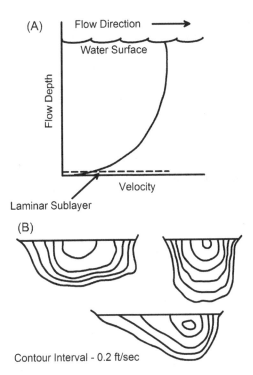

Figure 5.1. (A) Variations in flow velocity as a function of water depth. The laminar sublayer may be absent or discontinuous in coarse-grained rivers; (B) Typically observed variations in velocity through a cross section (Modified from Wolman 1955; Ritter et al. 2002)

and these parameters can cause divergent responses in particle motion to the same flow conditions (Ritter et al. 2002).

Historically, two methods have been utilized to predict a river's competence: critical bed velocity and critical shear stress. The adjective "critical" refers to the bed velocity or shear stress at the precise time at which the particle begins to move. Conceptually, the two approaches are quite different. Bed velocity refers to the impact of water on the exposed portions of the grain, and its strength is related to the momentum of the water (mass × velocity). Tractive force or boundary shear stress (τ) is associated with the downslope component of the fluids weight exerted on a particle in the channel bed. In the form of an equation, it is expressed as:

(3) $\tau = \gamma RS$

where γ is the specific weight of the fluid, R is hydraulic radius, and S is slope. In wide shallow channels such as those which typically transport gravel, water depth (D) is a close approximation of the hydraulic radius; thus, tractive force is proportional to the product of depth and slope.

Hjulström (1939) presented a set of curves that illustrates the relationships between velocity, particle size and process (Fig. 5.2). The dark, upper line on Fig. 5.2 represents the mean velocity above which a particle of a given size will be entrained. The lower dashed line indicates the approximate velocity at which the particle will be deposited. Hjulström's curves, then, illustrate that higher velocities

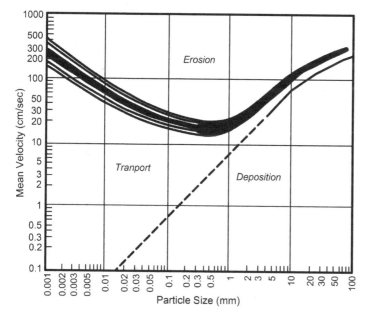

Figure 5.2. Mean velocity at which particles of various size are eroded, transported, and deposited (From Hjulström 1939, In Trask P (ed.) Recent Marine Sediments, American Association of Petroleum Geologists, used with permission)

are required to initiate the movement of a particle from a state of rest than to keep it in motion after it has been eroded. More importantly, the erosional threshold between grain size and velocity (upper set of lines) is not characterized by a linear trend in which continually increasing velocities are required to entrain larger and larger particles. Rather sand-sized particles between approximately 0.25 and 2 mm are the easiest to erode, and higher velocities are required to entrain particles that are both larger and finer than this size range. The higher velocities needed to entrain fine-grained (silt- and clay-size) particles can be attributed largely to the material's cohesive properties which bind the particles together. In fact, these fine-grained sediments tend to erode as aggregates rather than as individual particles (Knighton 1998), adding even more complexity to the problem of entrainment.

It is important to note that the Hjulström diagram is based on mean velocity and not critical bed velocities for the simple fact that the rate at which the water is moving immediately adjacent to the channel floor is extremely difficult to measure in high-energy rivers. In contrast, the depth and slope of the water in a channel are easily measured, making critical shear stress a more attractive parameter.

In the case of critical shear stress, the threshold for particle entrainment has been most notably illustrated by the Shields' diagram in which dimensionless critical shear stress (θ) is plotted against the grain Reynolds number. The grain Reynold's number is expressed as D_i/δ_o, where D_i is grain diameter and δ_o is the thickness of the laminar sublayer (Fig. 5.3). It essentially describes the extent to which an individual particle projects above the laminar sublayer into the zone of turbulent flow. Because D_i/δ_o is related to particle size, Fig. 5.3 is similar to the Hjulström diagram in that it depicts the force that is required to entrain a particle of a given size at the time of erosion. The Shields' diagram shows that the threshold of erosion reaches a minimum of approximately 0.03, which corresponds to particles in the size range of approximately 0.2 to 0.7 mm. The dimensionless critical shear stress required to entrain particles smaller than 0.2 mm increases because the particles reside entirely within the laminar sublayer where they are less likely to be affected

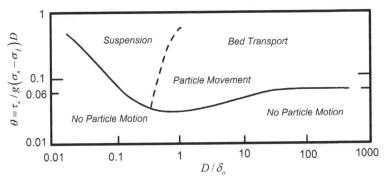

Figure 5.3. Shields' diagram for entrainment of bed particles. D is grain diameter, τ_c is critical shear stress, σ_s is sediment density, σ_f is fluid density, and δ_o is thickness of laminar sublayer (From Ritter et al 2002)

by turbulence. For particles larger than about 0.7 mm, which extend above the laminar sublayer, dimensionless critical shear stress (θ) increases before reaching a constant value, which has been shown to range from 0.03 to 0.06 (Komar 1989). If this constant value of dimensionless critical shear stress is substituted into its mathematical expression, and rearrange to solve for τ_c we see that:

(4) $\qquad \theta = \tau_c/[g(\rho_s - \rho_f)D_i]$

(5) $\qquad \tau_c = \theta g(\rho_s - \rho_f)D_i$

where ρ_s is the density of the particle, ρ_f is the density of the fluid, g is the gravitational constant, and θ is a constant of dimensionless shear stress (which falls within the range of 0.03–0.06).

Equation 5, and similarly developed mathematical expressions for critical shear stress, provide a powerful tool for assessing the flow conditions required for entrainment because the shear stress required to move a particle of a given size can be easily determined by assuming an appropriate value of θ.

A disadvantage of using either critical shear stress or critical bed velocity is that both approaches ignore the forces of lift on the particle. Lift is an upward directed force that can be created by the development of turbulent eddies on the downstream side of a particle, or by the establishment of a pressure gradient as a result of differences in flow velocities over and around a clast. In either case, lift has the potential to reduce the velocity or shear stress that is theoretically required to entrain a clast of a given size (Baker and Ritter 1975).

A slightly different approach that has been used to quantify sediment transport is to estimate a flow's stream power. Stream power is mathematically defined as:

(6) $\qquad \omega = \gamma QS$

where ω is stream power, γ is the specific weight of the fluid, Q is discharge, and S is slope. If we consider the stream power that is available to transport sediment per unit width of the channel, then Eq. 6 becomes:

(7) $\qquad \omega/w = \gamma QS/w = \gamma(wdv)S/w = \gamma vdS = \tau v$

where w, d, v represent the width, depth, and velocity of the flow, respectively. Equation 7 illustrates that stream power is a parameter that includes measures of both shear stress and velocity. As a result, it has been widely used to describe the sediment transport capacity of rivers.

Studies of particle entrainment in coarse-grained rivers have demonstrated that coarser clasts tend to be more mobile, and finer clasts less mobile, than would be predicted for sediment of uniform size, shape, and packing arrangements (Parker et al. 1982; Paola and Seal 1995). The importance of these characteristics on entrainment seems to rest with their influence on the exposure of individual particles to the overlying flow. For example, it is now clear that finer particles can be partially hidden by larger clasts, thereby increasing the force required for entrainment of the smaller particles. Entrainment may also be complicated by the development of stratigraphy within the channel bed in which coarse-grained sediments overlie and bury finer grained sediment, a process that accentuates hiding effects (Paola and Seal 1995). These observations have led to a concept known as the *equal mobility hypothesis.* This hypothesis suggests that the effects of particle hiding and sediment layering are so significant that all of the clasts in the channel bed will begin to move at the same shear stress (Parker et al. 1982; Andrews 1983; Andrews and Erman 1986). In other words, equal mobility implies that the size of the particles that are in motion within a channel does not change with increasing discharge in gravel-bed rivers; rather all of the clasts will begin to move at the same time upon reaching some critical threshold of shear stress. Actually, equal mobility in the strictest sense of the concept is unlikely to occur in natural channels except, perhaps, under extreme flood conditions (Komar and Shih 1992). Nonetheless, there is little question that entrainment processes are more significantly influenced by the size, shape, and packing of the bed sediment than originally envisioned.

5.3. PROCESSES OF CONTAMINANT DISPERSAL

If it were somehow possible to completely understand the mechanisms controlling the dispersal of contaminated particles, we could precisely determine the geographical patterns of trace metal concentrations within the river and its associated landforms, making site assessments considerably easier and less costly. Unfortunately, the dispersal of sediment-borne trace metals is dictated by a host of parameters that interact in complex ways to produce what are often confusing spatial trends in trace metal concentrations. Five distinct, but interrelated, processes are commonly cited as the predominant controls on contaminant transport rates and dispersal patterns (Lewin and Macklin 1987; Macklin 1996). They include (1) hydraulic sorting, (2) sediment storage and exchange with the floodplain, (3) dilution associated with the mixing of contaminated and uncontaminated sediment, (4) biological uptake, and (5) geochemical remobilization or abstractions. As we will see, the importance of each process varies between rivers as well as between reaches of a given river, making the geochemical patterns in trace metal concentrations all the more difficult to decipher.

5.3.1. Hydraulic Sorting

Hydraulic sorting involves the partition of particles into discrete zones along the channel according to their size, density, and shape. The sorting process actually involves several different mechanisms, the importance of which may vary with both time and space. The sorting mechanisms include: (1) selective entrainment as discussed earlier, (2) differential transport, in which smaller particles move farther than larger clasts following entrainment (Fig. 5.4), and (3) selective deposition

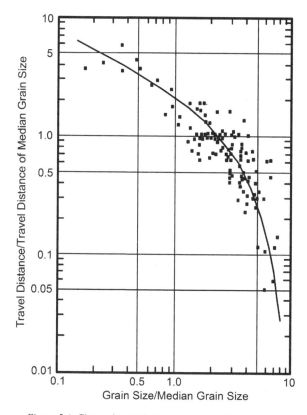

Figure 5.4. Change in travel distance as a function of grain size (After Hassan and Church 1992)

produced by differences in the settling velocity of the particles (Knighton 1998). The latter depends on the density and viscosity of water as well as the size, shape, and density of the sediment.

Later in the chapter we will see that hydraulic sorting can play a significant role in distributing trace metals within distinct morphological features in the channel, such as point bars and riffles. On a larger (basin) scale, it is primarily of significance where trace metals are associated with hydraulically heavier particles, such as metal enriched sulfides. In this case, the denser particles can be concentrated closer to the point source, and move more slowly downstream, than metals attached to hydraulically lighter particles which are transported entirely in suspension. For example, hydraulic sorting of sulfide minerals along the Rio Abaróa downstream of an impoundment containing mining and milling wastes is illustrated in Fig. 5.5. Here, hydraulic sorting produced a downstream decrease in both grain size of the sediment and the abundance of sulfide grains, thereby producing a minor downriver decline in metal concentrations.

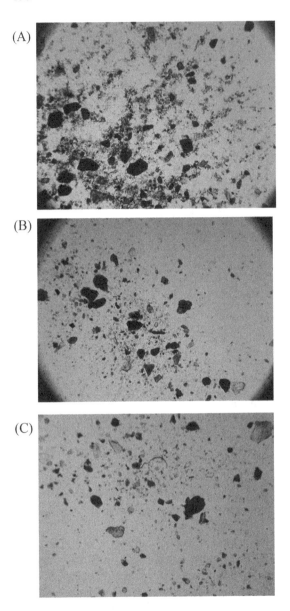

Figure 5.5. Photo micrographs of mill tailings at the Abaróa Mine, southern Bolivia (A) and downstream
channel bed sediments (B, C). Black grains are sulfide minerals (From Villarroel et al. 2006)

5.3.2.   Dilution and Exchange with the Floodplain

A second important process influencing geographical patterns in trace metal concentration is the dilution of contaminated sediment by non-contaminated, or less

contaminated, sediment. The latter materials can be derived from tributaries or the erosion of the channel banks, floodplain, or some other geological feature. Dilution is one of the most commonly cited causes of decreasing downstream concentrations from a point source. It can occur gradually along the channel, or abruptly. Abrupt dilution is illustrated in Fig. 5.6 for the Rio Tupiza of southern Bolivia. Here Pb, Zn, and Sb concentrations decline upon mixing with the relatively clean sediments from the Rio San Juan del Oro. Dilution has less to do with the actual transport of contaminated particles than it does with altering the composition of the sediment load prior to its deposition.

Dilution can occur through three distinct processes. To illustrate their differences, let us assume that a sample of sediment from the channel bed of a river is contaminated with Pb and that all of the Pb is associated with silt- and clay-sized particles. One way in which this hypothetical sample can be diluted is by adding sediment of the same mineralogical composition and grain size, but which has a lower Pb concentration. In contrast, the material can also be diluted by adding sediment in which the concentration of Pb in the silt and clay is exactly the same as that in our contaminated sample, but the dilutant contains less fine-grained material (and, thus, possesses a lower bulk Pb concentration). Dilution, then, occurs as a result of a change in the grain size distribution of the original sediment. This type of dilution will be discussed later with regards to the use of grain size and compositional correction factors. While differences in the mechanisms of dilution are subtle, they can have significant effects on data interpretation. Take, for instance, the influx of sediment from a tributary in which the concentration of Pb in silt and clay sized particles is the same as that observed along the axial channel. Trace metal concentrations would appear unchanged downstream of the tributary if only the

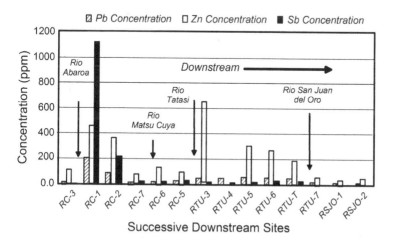

Figure 5.6. Concentration of Pb, Zn, and Sb in channel bed sediments of the Rio Chilco-Rio Tupiza drainage system, southern Bolivia. Antimony mines occur within the Rio Abaróa and Rio Matsu Cuya basins; Polymetallic tin deposits are mined in the Rio Tatasi (From Villarroel et al. 2006)

< 63 μm sediment fraction was analyzed. However, concentrations downstream of the tributary could increase, decrease, or remain the same if we analyzed the bulk sample. Exactly which result will occur depends on the percentage of sand in the tributary and axial channel sediment, and the effects of dilution associated with their mixing.

In addition to the processes of dilution described above, dilution can also occur by dispersing a finite amount of contaminated sediment over a larger area. This form of dilution is particularly important following the removal of a contaminant source(s) by some form of remediation.

5.3.3.   Sediment Storage and Exchange Mechanisms

Sediment storage and exchange with floodplain and channel bed deposits is often a significant control on contaminant transport rates and dispersal patterns. Storage can be thought of as the reverse of dilution. Rather than adding non-contaminated particles to the mixture, contaminated particles are removed from the load being transported by deposition upon the channel bed or floodplain, thereby decreasing their overall abundance.

The storage of sediment within the channel bed is closely linked to the reworking of bed sediments during flood events. Along many rivers the scour of the channel bed occurs during rising water levels as flow velocities and tractive force increase. The channel is subsequently filled during the waning stages of the flood (Fig. 5.7). During the scouring and filling episode contaminated particles are incorporated

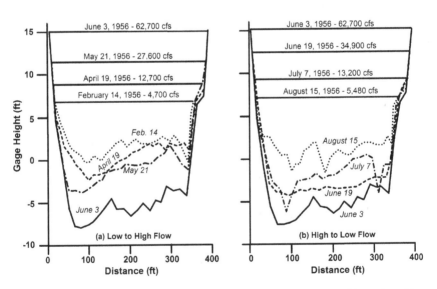

Figure 5.7. Scour and fill during the passage of a flood along the Colorado River at Lees Ferry, Arizona. (A) low to high flow; (B) high to low flow (From Leopold et al. 1964)

into the bed material where they are stored until they are remobilized by a subsequent high water event. It follows, then, that the amount of contaminants stored within the channel will depend, in part, on the degree to which the channel bed is eroded and refilled during floods, and the frequency of scouring and filling events. The significance of scour and fill tends to be particularly pronounced within ephemeral channels of semi-arid or arid environments. It is much more subdued within perennial rivers found in humid regions, and in coarse grained mountain streams which are armored by the winnowing of finer sediments during low flow (Leopold et al. 1964). Even where scour and fill is minimal, however, fine-grained, contaminated sediment can be deposited and stored between larger, relatively immobile clasts, or infiltrate into the channel bed materials.

Very few studies have attempted to quantify contaminant storage within the channel. The existing studies on the topic have shown that in-channel storage of both fine-grained sediment and trace metals is generally small. For example, in several studies conducted in the U.K., the storage of silt- and clay-sized particles to which trace metals are most likely attached was found to be less than about 10% of total suspended sediment load (Walling et al. 1998; Owens et al. 1999). Similarly, the storage of Pb and Zn in the Rivers Aire and Swale was less than about 3% of the total annual load (Walling et al. 2003). These data led Walling et al. (2003) to suggest that channel storage is of limited importance in regulating the downstream transport of contaminants through many river systems. It should be recognized, however, that the above data are for relatively stable rivers. In systems where aggradation has (or is) occurring, channel storage may be of more importance (Brookstrom et al. 2001, 2004; Villarroel et al. 2006). A case in point was provided by Brookstrom et al. (2001) who found that 51% of the Pb within the Coeur d'Alene River valley was associated with riverbed sediment. The main stem of the Coeur d'Alene was severely impacted by mine tailings and the storage of Pb within the channel was primarily driven by nearly 3 m of historic aggradation.

In most rivers, the amount of sediment and sediment-borne trace metals stored within the channel is significantly less than that found within the floodplain where more than 40–50% of the total annual load may be deposited (Marron 1992; Macklin 1996; Owens et al. 1999; Walling et al. 2003). The residence time of the contaminants also differs substantially between channel and floodplain deposits. Although poorly constrained, residence times for fine-grained channel sediments is generally less than 5 years (Walling et al. 2003); in contrast, contaminants may be stored for decades, centuries, or even millennia in floodplain environments (Macklin 1996; Miller 1997; Coulthard and Macklin 2003).

Contaminant storage downstream of a point source generally leads to downstream declines in metal concentrations. However, storage is a time dependent phenomenon and while it may initially remove contaminated particles from the river's transported load, its affects on contaminated particles may decrease or even become reversed in later years. This follows because sediment is not only deposited on floodplains, but eroded from them. Thus, once significant quantities of metal-enriched sediments have been deposited on (and stored within) a floodplain, the floodplain itself can become a contaminant source to the river during bank erosion. When this occurs, longitudinal patterns in metal concentrations may more closely reflect the exchange of particles to and from the floodplain rather than their distribution from the original point sources. An example of the importance of storage processes on contaminant concentrations is illustrated by the Carson River system of west-central Nevada. During the late 1800s significant quantities of Hg-enriched sediments were deposited along the valley floor downstream of mill processing facilities. Although there is significant variation in the data, Hg concentrations within the valley fill generally decline downstream from the mills (Fig. 5.8a). However, Hg concentrations within the modern channel bed deposits do not decrease downstream from the mills, but quasi-systematically increase for approximately 25–30 km before subsequently declining (Fig. 5.8b). Miller et al. (1998) argued that the observed longitudinal trend resulted from the progressive erosion of dense Hg–Au and Hg–Ag amalgam grains stored within the valley fill, and their incorporation into the channel bed materials. Subsequent declines downstream were presumably related to dilution caused by the influx of clean sediment from a tributary, and the erosion of uncontaminated, pre-mining bank deposits which were more extensively exposed downstream.

5.3.4.   Geochemical Processes and Biological Uptake

Contaminant dispersal patterns are not only controlled by physical processes, but also by biogeochemical processes including biological uptake. While the accumulation of trace metals by biota must theoretically influence sediment-borne trace metal concentrations, in most systems its effects are negligible. Other biogeochemical processes, however, can play a significant role. These trends are perhaps most evident in rivers affected by atypical Eh and pH conditions, such as systems impacted by acid mine drainage.

In rivers with typical Eh and pH conditions, sediment-borne trace metal concentrations can be significantly influenced by spatial variations in geochemical processes, even in rivers devoid of acid drainage. Hudson-Edwards et al. (1996), for example, found that in addition to hydrodynamic dispersal processes (i.e., dilution, hydraulic sorting, and storage), downstream declines in sediment-borne metal concentrations in the River Tyne were related to geochemical changes in sediment mineralogy which occurred during particulate transport. More specifically, thermodynamically unstable minerals, such as galena and sphalerite, oxidized to form secondary minerals (primarily Fe and Mn hydroxides) which contained proportionately less trace metals than the original mineral grains. Clearly, the chemical form (speciation) of the trace metals changed along the channel as well.

Evans and Davies (1994) also demonstrated that geochemical processes can affect the metal content of the sediment as well as their chemical form. In their study, Mn coatings within the River Ystwyth in mid-Wales adsorbed significant quantities of Pb within bedrock confined channel reaches. In contrast, most of the Pb along meandering reaches, bound by alluvial sediments, was associated with Fe-oxides

(a)

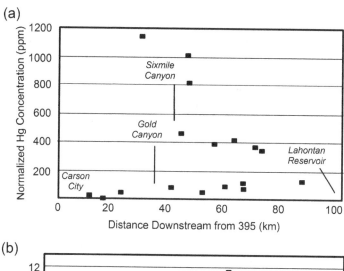

(b)

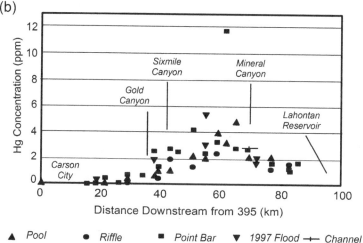

Figure 5.8. (A) Downstream changes in Hg concentration within valley fill located along the
Carson River, Nevada. Elevated values are related to the release of Hg from ore processing mills
which utilized Hg amalgamation methods; (B) downstream changes in Hg concentrations with the
channel bed sediments. Note differences in concentration at a site between point bars, pools, and
riffles, and the similarity in concentration between pre- and post-1997 flood deposits
(Modified from Miller et al. 1998)

(and was therefore more tightly bound to the sediments). The observed differences
in chemical speciation was attributed to the development of turbulent conditions
within bedrock chutes that changed the oxygen status of the water enough to
induce the precipitation of Mn(IV) onto the surface of sand sized particles. Thus,
oxyhydroxide precipitation zones were created that were directly attributable to
channel morphology and which impacted the chemical speciation of Pb along
the river.

5.4.   DOWNSTREAM PATTERNS

Every contaminated river exhibits its own unique downstream trend in elemental concentrations. Nonetheless, some recurring patterns are found to exist at the basin scale. In industrialized areas metal concentrations in rivers tend to increase semi-systematically downstream as they flow from relatively undeveloped headwater areas to more developed parts of the watershed, characterized by broad floodplains. Within the Aire-Calder River system in the U.K., for example, Pb, Cr, Cu, and other contaminants increase downstream as a result of urban and industrialized inputs to the mid- and lower reaches of the watershed (Owens et al. 2001; Walling et al. 2003) (Fig. 5.9).

Another frequently recurring pattern is associated with mining operations in mountainous terrains. In this situation, mining and milling debris, primarily from historic operations, is flushed into headwater tributaries and largely transported downstream until reaching the mountain front where the waste materials are deposited on agricultural lands. Metal concentrations tend to decrease downstream with distance from the mines, but deposition and storage of the contaminated debris may increase several fold upon exiting the mountain front.

Clearly, the above examples show that geographical patterns in concentration reflect points or zones of contaminant influx. Therefore, documenting the spatial patterns for a given metal allows us to identify potential sources of contamination even when multiple sources exist. The potential to identify contaminant

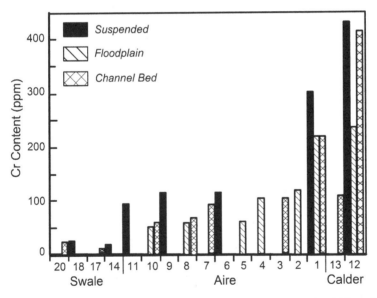

Figure 5.9. Downstream changes in Cr concentration in suspended, floodplain, and channel bed sediment of the Rivers Swale, Aire, and Calder. Numbers represent consecutive downstream sampling sites. Only the < 63 μm fraction of floodplain and channel bed sediment was analyzed (From Owens et al. 2001)

sources is demonstrated in Fig. 5.6 for the Rio Chilco-Rio Tupiza drainage system of southern Bolivia. Increases in Pb, Zn, and Sb concentrations increase immediately downstream of tributaries affected by mining operations indicating that waste products from the mines are entering the axial drainage system. Moreover, the downstream patterns provide insights into the relative quantities of metal influx from each source, and differences in the composition of the mine wastes. The Rio Abaróa, contaminated by an Sb mine, primarily provides Sb to the Rio Chilco. In contrast, the Rio Tatasi delivers large quantities of Pb and Zn to the axial drainage, and only limited quantities of Sb (Fig. 5.6).

Along most rivers, dispersal processes combine to produce a downstream decrease in metal concentrations from a point or zone of trace metal influx (Fig. 5.10). The downstream decay in metal concentrations can vary significantly from one point source to another, even along the same river. This too is well illustrated by the Rio Chilco-Rio Tupiza system. Between the Rio Abaróa and the Rio Chilcobija concentrations rapidly decrease downstream, whereas trace metal content below the Rio Tatasi decays gradually until reaching the Rio San Juan del Oro where concentrations abruptly decline (Fig. 5.6). Moreover, the downstream declines in Sb differ from those of Pb indicating that the trends are metal specific.

Considerable attention has been given in recent years to documenting and modeling such trends in sediment-borne trace metal concentrations downstream of point sources. The first quantitative descriptions were provided by Wolfenden and Lewin (1977) using exponential decay and regression analyses. Since their seminal

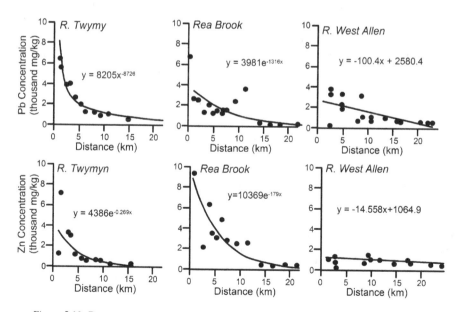

Figure 5.10. Downstream trends in Pb and Zn concentrations for three rivers in the U.K. Lines represent best-fit regression models (Modified from Lewin and Macklin 1987)

work was published, a number of other investigators have used the regression approach, applying linear, logarithmic, polynomial, power, and exponential decay models to a wide range of river systems (Lewin and Macklin 1987; Marcus 1987; Taylor and Kesterton 2002). Part of the interest in modeling these downstream trends has been the assumption that differences identified between specific trace metals for a given river provide clues regarding the transport rates and dispersal mechanisms. Comparison of the regression models, for example, has been used to identify downstream patterns in trace metal dilution, differential movement of metals by hydraulic sorting, and the relative importance of floodplain deposits as a source of sediment-borne contaminants (Macklin and Dowsett 1989; Macklin 1996). Care must be taken, however, in using regression equations to determine physical dispersal mechanisms because as we saw earlier in our discussions, dispersal is not always dominated by the physical transport and deposition of particles, but by geochemical processes (Macklin and Dowsett 1989; Taylor and Kesterton 2002). Where geochemistry prevails, the regression models may provide a basic understanding of metal concentrations between sampling sites, but they could potentially lead to erroneous conclusions as to why those concentrations exist.

Another less frequently used approach is to model the downstream dispersion of sediment-borne trace metals by assuming that concentration at any given point along the channel is related solely to the mixing of contaminated and non-contaminated sediments. The concepts for this approach were derived by exploration geochemists who recognized that downstream changes in trace metal concentrations were primarily driven by mixing of sediment from a geochemical anomaly with un-enriched tributary sediment. In the case of exploration, the area and grade of the anomaly can be predicted using readily available data including elemental content within axial channel sediments, basin area, and the background concentration of the element measured in tributaries. The model, however, can be reversed to predict the concentration of an element within the axial channel, provided that data are available concerning the size and grade of the upstream anomaly or contaminant point source (Helgen and Moore 1996). The revised formula is expressed as:

$$(8) \qquad [C_a]_{km} = C_b \frac{(A_t - A_o)}{A_t} + C_o \frac{A_0}{A_t}$$

where $[C_a]_{km}$ is the concentration of the metal in the axial channel at a given distance from the point source, C_b is the average background concentration in the basin, A_t is the total basin area upstream of the sampled site, A_o is the area of the anomaly or point source, and C_o is the concentration of the metal in the surface anomaly. As written the model is most easily applied to natural geochemical anomalies. However, it has been modified and used by Helgen and Moore (1996) to describe downstream trends in concentration where the additional influx of trace metals occurs as a result of mining activity. They also demonstrated that a comparison of metal dispersion prior to and following mining activity can be used to characterize the degree of contamination along the channel and to set remediation targets.

The mathematical formula used by Helgen and Moore (1996) does not allow for variable rates of erosion and sediment influx from tributary channels. Marcus (1987), however, utilized a different form of mixing model to account for varying rates of sediment influx to the axial channel from tributary sources. In this case, downstream trends in Cu concentration along Queens Creek, Arizona were predicted on the basis of:

$$(9) \qquad C_r = \left| \frac{(C_m)(X_m)}{(X_m)(X_t)} + \frac{(C_t)(X_t)}{(X_m)(X_t)} \right|$$

where C_r is the concentration downstream of the confluence along the axial channel, C_m is the concentration of the metal upstream of the confluence, C_t is the concentration of the metal within the tributary, and X_m and X_t are the basin areas or basin sediment yields of the axial channel and tributary, respectively. For Queens Creek, the mixing model performed better when compared against actual data than did developed regression models. The primary advantage of this particular mixing model is that it more accurately simulates the effects of sediment input on trace metal concentrations from tributaries characterized by variable sediment yields and trace metal contents.

Mixing models in general, however, are based on several simplifying assumptions that can limit their application in some rivers. The assumptions are that: (1) the downstream trends in concentration are the sole product of dilution resulting from the steady-state mixing with tributary sediment, (2) the terrain is characterized by uniform rates of erosion, either throughout a tributary catchment or over the entire basin, (3) there is only one source of metal-enriched sediment within the basin, and (4) the trace metals behave conservatively within the channel (i.e., they are not affected by geochemical processes). Perhaps the most common violations of the assumptions for contaminated rivers involves the influx of contaminated sediment stored in floodplain deposits during bank erosion, and the effects of processes other than dilution on trace metal dispersal (e.g., associated with hydraulic sorting or downstream variations in sediment storage).

Regardless of whether regression and mixing models are applied to the river, their fit to the actual data is often poor. For example, Macklin and his colleagues have intensively studied the River Tyne in the U.K. They found that at the basin scale, downstream trends in concentration could be adequately described using simple regression models, but when viewed at a finer scale, systematic variations in the data appeared. These variations, in part, could be attributed to increases or decreases in concentration immediately downstream of tributaries. More frequently, however, spatial changes in concentration formed a quasi-systematic wave like pattern that could not be attributed to tributary junctions. Macklin and Lewin (1989) argue that the underlying control on the observed pattern was the organization of the valley floor into alternating "transport" and "sedimentation" zones. The transport zones tended to be steeper and narrower than average, allowing sediment entering the reach to move more quickly through it with only limited storage. In contrast, the sedimentation zones possessed wider valley floors and shallower gradients promoting deposition. Thus, it was along these reaches that most of the Pb and Zn from historic mining operations were deposited and where metal concentrations were the highest (Macklin and Dowsett 1989; Macklin 1996). Sedimentation zones have been recognized along many other river systems as well, such as the Rio Pilcomayo where deposition is closely linked to valley width (Fig. 5.11).

Data from the River Tyne also suggest that sediment-borne trace metals may move through the system as a series of sediment pulses, or slugs, complicating the pattern predicted by standard regression or mixing models (Nicholas et al. 1995; Macklin 1996). Sediment slugs are injections of clastic sediment into the channel as a result of both natural and human activities that exceed the transport ability of the river (Hoey 1992; Bartley and Rutherfurd 2005). The injected sediments result in cycles of aggradation and degradation which propagate downstream. Slugs, then, are associated with periods of disequilibrium that range from minor disturbances to basin-scale perturbations in sediment supply that may lead to a total readjustment of the valley floor (Table 5.2) (Gilbert 1917; Nicholas et al. 1995). With regards to metal transport, slugs are important because aggradation commonly increases sediment and contaminant storage and reduces their rate of downstream transport, while the subsequent period of degradation is likely to remobilize sediment associated trace metals, thereby increasing downstream transport rates.

5.4.1.   Impoundment Failures

The transport processes described thus far have been related to "normal" rainfall-runoff events. However, waste disposal impoundments (ponds) associated with mining and milling operations are located along a large number of rivers, particularly in mountainous terrains. While the nature of these facilities is highly variable, the potential exists for the confining earthen structures to fail, catastrophically releasing

Figure 5.11. Transport and sedimentation zones along the Rio Pilcomayo, Bolivia

Table 5.2. Classification of sediment slugs

Slug size	Dominant controls	Impact on fluvial system
Macroslug	Fluvial process-form interactions	Minor channel change
Megaslug	Local sediment supply & valley-floor configuration	Major channel change
Superslug	Basin-scale sediment supply	Major valley-floor adjustment

From Nicholas et al. (1995)

metal enriched sediment into an adjacent stream channel. In fact, during the past 40 years, more than 75 major failures have released trace metals and other contaminants into riverine environments (Table 5.3). This equates to an average of nearly two major tailings dam failures per year, not including those in secluded regions, which are seldom reported.

Few studies have documented the resulting downstream trends in metal concentrations immediately after a failure (and before reworking by subsequent flood events). Graf (1990), however, found that the flood wave resulting from the 1979 Church Rock uranium tailings spill did not produce a systematic downstream trend

Table 5.3. Selected tailings dam failures

Mine (location – year)	Volume of material released – m^3
Aznalcóllar (Spain - 1998)	1,300,000[a]
Harmony, Merries (South Africa – 1994)	600,000[b]
Buffalo Creek (USA-1972)	500,000[b]
Sgurigrad (Bulgaria – 1996)	220,000[b]
Aberfan (UK – 1966)	162,000[b]
Mike Horse (USA – 1975)	150,000[b]
Bilbao (Spain – 1969)	115,000[b]
Baia Mare (Romania - 2000)	100,000[c]
Ages (USA – 1981)	96,000[d]
Huelva (Spain – 1998)	50,000[b]
Stancil, Perryville (USA – 1989)	39,000[b]
Dean Mica (USA – 1974)	38,000[b]
Arcturus (Zimbabwe – 1978)	30,000[b]
Maggie Pie (UK – 1970)	15,000[b]
Borsa (Romania – 2000)	8,000[e]

Modified from MRF (2002)

[a] Metal enriched tailings

[b] Waste

[c] Effluent & tailings

[d] Coal refuse

[e] Cyanide contaminated effluent & tailings

in [230]Th concentrations within an entrenched arroyo system in New Mexico. Instead, concentrations were inversely correlated to unit stream power and the length of time that shear stress exceeded critical values during the passage of the flood wave (Fig. 5.12). In other words, sediment enriched in [230]Th was preferentially deposited within low energy environments. Along reaches where unit stream power was relatively high, most of the contaminated sediment was transported downstream. As was the case for the River Tyne described above, the concentration of sediment-borne contaminants was closely linked to the existing morphology of the channel and valley system.

In contrast to morphologically controlled cases, the 1998 Aznalcóllar Mine spill in Spain produced a highly sediment-laden flow ($\sim 660\,g/L$ in solid weight) that resulted in a semi-systematic decrease in thickness of deposited tailings on the floodplain, despite the fact that significant differences in channel form and slope were present (Gallart et al. 1999) (Fig. 5.13). Crevasses splays (described in more detail in Chapter 6) were, however, preferentially formed downstream of narrow reaches and channel bends (Gallart et al. 1999). Splays and other overbank deposits produced during subsequent floods between January and May, 1999 also exhibited a general downstream decrease in metal concentrations (Hudson-Edwards et al. 2003). Differences in the downstream transport of sediment-borne trace metals during dam failures such as at Church Rock and Aznalcóllar are clearly an area requiring additional study. Nonetheless, diffferences are related to the regional physiography, the morphology of the channel and valley floor below the impoundment, the rheology of the flow, the timing of the failure relative to rainfall-runoff events, and the amount and size of the waste material that is released, among others factors.

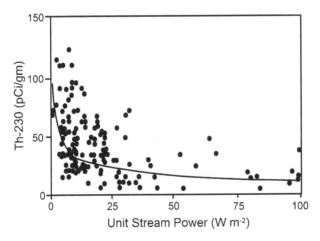

Figure 5.12. Change in [230]Th concentration with stream power, Puerco River, New Mexico
(From Graf 1990)

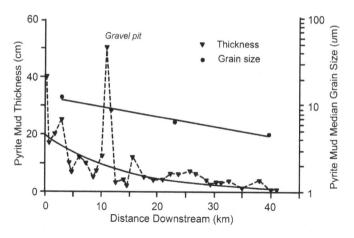

Figure 5.13. Change in the thickness and grain size of pyrite enriched mud deposits associated with the Aznalcóllar Mine tailings spill (From Gallart et al. 1999)

5.5. DEPOSITION AND STORAGE ALONG A REACH

Up to this point we have been concerned with the dispersal of sediment-borne trace metals over long reaches of the channel, often encompassing 10 to 100s of kilometers in length. However, basin scale trends in trace metal concentrations are the product of entrainment, transport, and deposition of particles at much smaller scales. The interaction of these geomorphic processes over the length of a reach (measured in 10's of meters) is of significant importance to site assessments because it can result in the concentration of trace metals within morphological features (e.g., bars, pools, and riffles) possessing distinctive sedimentological characteristics. In other words, the selective nature of specific particles sizes and densities by these morphological units is an extremely important, but often overlooked, consideration because it produces significant spatial variations in trace metal concentrations (Ladd et al. 1998; Taylor and Kesterton 2002). Failure to recognize these trends can lead to either over- or underestimation of the average (or maximum) concentrations that exist. In addition, sampling of channel bed sediments without regard to the depositional feature from which the materials were derived can confuse larger (basin) scale geographical patterns, thereby resulting in erroneous conclusions.

The morphological features found on the channel floor can be subdivided into small transitory deposits and larger (alluvial bar) deposits (Table 5.4). The transitory deposits are intimately tied to local flow conditions and fluctuations in those conditions through time. Larger, more persistent features, which represent our primary concern, are more closely related to discharge, sediment supply, and macroscale channel processes than local fluid hydraulics (Knighton 1998). These larger features are both created by and influence the mean flow configuration through the reach, and, therefore, have specific associations with a given channel pattern

Table 5.4. Characteristics of channel deposits

Scale	Characteristic
Transitory deposits	Bedload temporarily at rest
Micro-forms	Coherent structures such as ripples with λ ranging from 10^{-2} to 10^{0} m
Meso-forms	Features with λ from 10^{0} to 10^{2} m; includes dunes, pebble clusters and transverse ribs
Alluvial bars	Formed by lag deposition of coarse-grained sediment
Macro-forms	Structures with λ from 10^{1} to 10^{3} m such as riffles, point bars, alternate bars, and mid-channel bars
Mega-forms	Structures with $\lambda > 10^{3}$ m such as sedimentation zones

Adapted from Knighton (1998), Church and Jones (1982) and Hoey (1992)

(e.g., point bars, pools, and riffles with meandering streams, alternate bars within straight channels, etc.) (Knighton 1998).

Historically, channel patterns were classified as straight, meandering and braided (Leopold and Wolman 1957). Most geomorphologists now recognize that this classification scheme is insufficient to describe the wide range of channel patterns that exist in nature. Schumm (1981), for example, combined the traditional scheme with the predominant type of load transported by the river to develop a classification system with 13 different patterns (Fig. 5.14). Others have subdivided rivers into those with single channels (straight and meandering) and those with multiple channels (e.g., braided and anabranching) (Knighton 1998). No matter which classification system is used, it is clear that the boundaries are indistinct. In fact, many rivers exhibit a single channel pattern during low flow conditions, but during floods possess a distinctly braided configuration. Thus, a river may acquire morphological features characteristic of both meandering and braided systems. Nonetheless, river classification is a useful process in that it aids in the description of the formative processes associated with the various planimetric configurations, and illustrates the general trends in channel pattern with changes in hydrologic and sedimentologic regime (Fig. 5.14).

The literature on channel patterns and their deposits is voluminous and complicated by an inconsistent use of terminology (see Church and Jones 1982, for a discussion). As such, an in-depth discussion is beyond our purposes here. Instead, we will focus on developing an understanding of the most important formative processes, and the basic characteristics of the resulting fluvial deposits.

5.5.1.   Channel Patterns

5.5.1.1.   Straight and meandering channels

In addition to possessing relatively linear banks, most straight channels exhibit accumulations of sediment, called *alternate bars* that are positioned successively downriver on opposing sides of the channel (Fig. 5.15a). Opposite the alternate bars are relatively deep areas called *pools* which are separated by shallower and

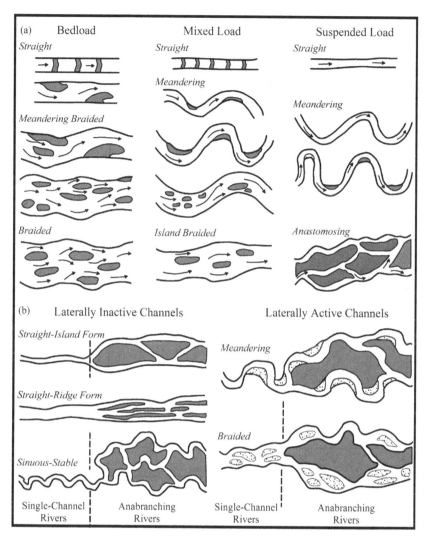

Figure 5.14. (A) Classification of channel pattern by Schumm (1981, 1985); (B) classification of anabranching channels adapted from Nanson and Knighton (1996) (From Huggett 2003)

wider reaches called *riffles* (Fig. 5.15c). The sequence of alternate bars, pools, and riffles is organized in such a way that the deepest part of the channel, referred to as the *thalweg*, migrates back and forth across the channel floor. A straight channel, then, possesses neither a uniform streambed nor a straight thalweg. Rather, the configuration mimics the sequence of features found in meandering rivers (Fig. 5.15b). The distinction between the two patterns (meandering and straight) is normally defined by a sinuosity of 1.5, where sinuosity is the ratio of stream length to valley length. The designated value of 1.5 is arbitrary and has no specific

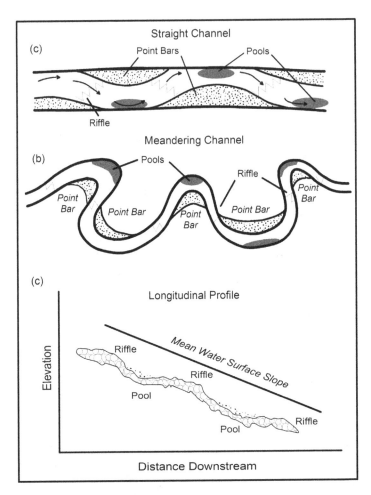

Figure 5.15. Planform morphology of straight (A) and meandering (B) channels. Longitudinal profile along thalweg (C)

mechanical significance. Channels with a sinuosity of 1.4 will likely exhibit similar flow patterns and sediment transport mechanics as channels possessing a sinuosity of 1.6 provided that all the other factors controlling channel form are the same.

Most rivers do not possess straight channels for long distances. In contrast, the meandering pattern is the most commonly observed planform. An important component of flow in meandering rivers is a secondary pattern of circulation that is oriented oblique to the downstream flow direction. Presumably, as water flows around a meander, centrifugal force leads to a slight elevation in the water surface on the outside of the bend (Fig. 5.16). The increase in elevation produces a pressure gradient that gives the flow circular motion. This corkscrew type motion, referred to as *helical flow*, has traditionally been thought of as a single rotating cell that

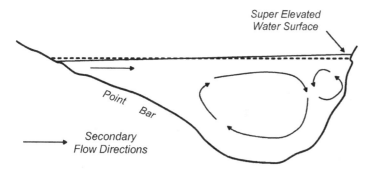

Figure 5.16. Secondary flow directions at a meander bend (From Markham and Thorne 1992)

reaches its maximum velocity within pools, immediately downstream of the axis of the meander bend. More recent studies have shown, however, that secondary circulation at meanders is more complex and may consist of several interacting, cells (Markham and Thorne 1992). Near the outer bank, the water interacts with bank materials causing flow to move upward along the channel margins before moving toward the center of the channel at the water surface (Fig. 5.16). The mid-section of the channel is characterized by the classical form of helical flow and may contain as much as 90% of the discharge (Markham and Thorne 1992). The flow direction of this cell is opposite to that occurring along the outer margins. As a result, the two cells meet to form a zone of flow convergence that reaches its maximum velocity along the base of the banks and at the channel bed, thereby promoting scour in these areas. Along the inside of the meander bend, there is commonly a component of flow that moves over the top of the point bar toward the opposing bank (Dietrich and Smith 1983; Dietrich 1987).

The direction of rotation of the secondary flow cells is reversed between successive meanders because the features within the channel are opposite to one another (Fig. 5.17). Originally, it was proposed that the reversal occurred between a pool and the next downstream riffle, where flow conditions promote channel bed deposition. However, the nature of the secondary currents existing between the pools appears to be more complex than originally thought. In fact, different flow patterns have been observed in different rivers for reaches located between two successive pools (Hey and Thorne 1975; Thompson 1986; Dietrich 1987; Markham and Thorne 1992). The observed differences in secondary circulation between rivers are apparently related to variations in channel planform, width/depth ratio, and flow stage (Dietrich 1987).

The pattern of pools, riffles and bars in meandering rivers is a manifestation of how flow, sediment transport and bedforms are interrelated. Successive riffles (or pools), for example, are usually spaced at distances of five to seven times the channel width. The spacing of the pool and riffle sequence is largely independent of the material forming the channel perimeter and, therefore, is thought to be predominantly related to larger scale flow patterns in meandering rivers. The sedimentology

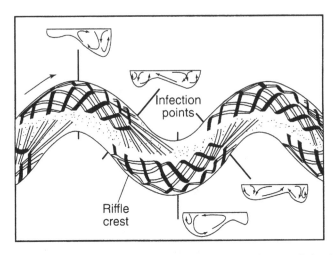

Infection
points

Riffle
crest

Figure 5.17. Model of secondary flow cells and flow lines in a meandering river (From Thompson 1986)

of the features also depends on the hydrologic conditions within the channel. At low flow, riffles tend to possess coarse grained sediments and are characterized by rapid, shallow flows and steep water surfaces; in marked contrast, the surface materials in pools are relatively fine grained, a trait consistent with the deeper, gentler flow conditions.

A significant question that has received considerable attention is how the pool and riffle sequence is maintained. Based on the flow patterns observed between periods of flood, one would expect the entrainment of the sediment at the coarse grained riffles to be transported downstream to the pools until the pools are destroyed by aggradation. Field measurements have shown, however, that the flow velocities during a runoff event increase at different rates between the pools and riffles (Fig. 5.18). These differences lead to a reversal in the flow rates so that the velocity within the pool is greater than that over the riffle during periods of high water. As a result, large particles eroded from the riffles can be transported through the pools during floods. During the recessional phase, the larger particles are deposited first on the riffle and the finer particles, eroded from the riffles, are redeposited in the pools (Fig. 5.18).

More recent studies have shown that the changes in velocity over pools and riffles with rising discharge may not be universally applicable to all rivers and that other mechanisms may account for the observed sedimentological and morphological patterns (Carling 1991; Carling and Wood 1994; Thompson et al. 1996, 1998). Nonetheless, there is little question that the competence of the flows within the pools must exceed that of the riffles during floods in order to maintain the pool and riffle sequence. In addition, it appears certain that the pool and riffle sequence is maintained by relatively large discharge events.

The scour and fill processes associated with the formation of pools and riffles can lead to the accumulation of channel lag deposits. These deposits are composed

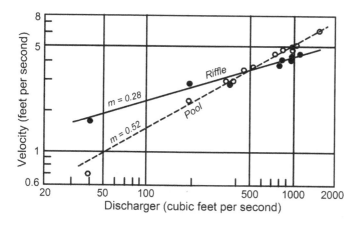

Figure 5.18. Change in velocity over a pool and riffle with increasing discharge, East Fork River, Wyoming (From Andrews 1979)

of coarse sediments that can only be moved during periods of flood. Typically the materials consist of coarse gravels, waterlogged plant debris, and consolidated aggregates and clasts of mud and clay from the eroded cut bank (Boggs 2001). Lag deposit are generally quite thin and laterally discontinuous. In fact, in rivers devoid of significant coarse sediment, they may be completely lacking. As we will see later in our discussion, channel lag deposits can contain accumulations of heavy minerals enriched in trace metals, particularly within rivers contaminated by mining debris.

During periods of low flow, channel lag gravels are likely to be buried by finer sediments which merge into the base of a downstream point bar. Point bars are the most significant features in meandering rivers. In cross section, they are characterized by a nearly horizontal surface on the inside of the meander bend that is positioned at the elevation of the adjacent floodplain. It then slopes gradually towards the thalweg until merging with the channel bed (Fig. 5.19). The point bar surface is the site of sediment deposition which occurs during the migration of the point bar across and downvalley (Fig. 5.20a). In the ideal case, decreasing velocities and flow depths over the point bar produce an upslope decrease in grain size that is accompanied by changes in bedforms (see Chapter 6). The actual character of the point bar, however, will depend on the size and composition of the sediment in transport. In rivers carrying a mixture of coarse and fine material, the upward progression is from basal gravels to sand to silt. The fine-grained silt cap at the top of the point bar is produced during high flow events by overbank deposition (Reading 1978). In rivers carrying fine-grained sediments, the upward transition often ranges from a fine sand layer near the base of the point bar to silty or clayey sediments near the top (Reineck and Singh 1980).

The ideal point bar rarely occurs in nature except perhaps along small meandering streams. A common complication to the ideal sequence is the flow of water across the point bar during floods which produces a pronounced channel called a *chute*. Chute channels tend to be accentuated on the upstream end of the point bar and

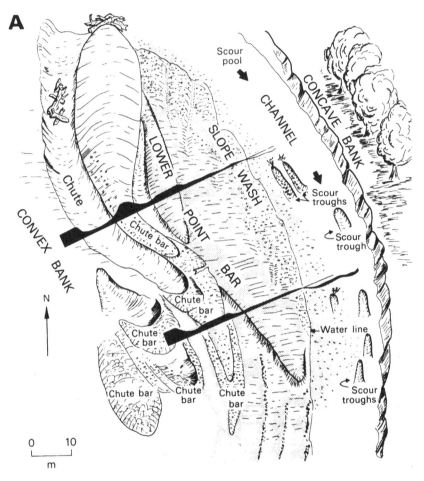

Figure 5.19. Topographic and sedimentology features of a coarse-grained point-bar (Original figure after McGowan and Garner 1970; reproduced from Reading 1978)

decrease in depth downstream, ultimately terminating in a bar composed of fine-grained sediment at the end of the chute (Fig. 5.19). In other cases, the chute may continue to erode, progressively capturing a larger portion of the flow. Eventually, flow diverted into the chute may completely dissect the point bar, resulting in the formation of a bar that subdivides the reach into two channels.

5.5.1.2. Braided rivers

The braided pattern is characterized by the subdivision of a single channel into a network of branches by islands of sediment (Fig. 5.20b) generically referred to as *braid* or *medial bars* (Fig. 5.21). In comparison to meandering rivers, braided channels tend to be highly dynamic; both the position and total number of channels

(A)

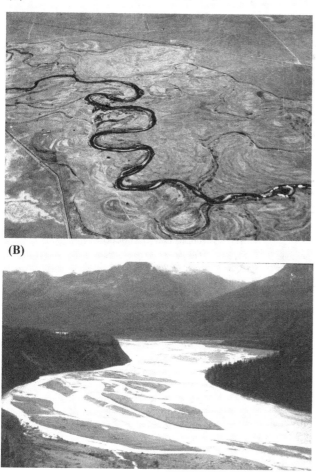

(B)

Figure 5.20. (A) Laramie River, Wyoming (photo by J.R. Balsley); (B) Matanuska River, Alaska (Photo by D. Germanoski)

within a braided reach can change significantly over a period of days. In addition, bars tend to be migratory and transient, inhibiting the development of a stable pool and riffle sequence (Church and Jones 1982). The dynamic nature of the braided system suggests that the residence time of contaminants stored within the channel bed will likely be shorter than within either point or alternate bars unless the river is aggrading (although data adequately supporting this contention is lacking).

Bar morphology and sedimentology in braided rivers is highly variable and depends in large part on the size and size distribution of sediment within the channel. Thus, braided rivers are commonly classified as either being coarse or fine-grained systems, while braid bars can be grouped into three basic types: longitudinal

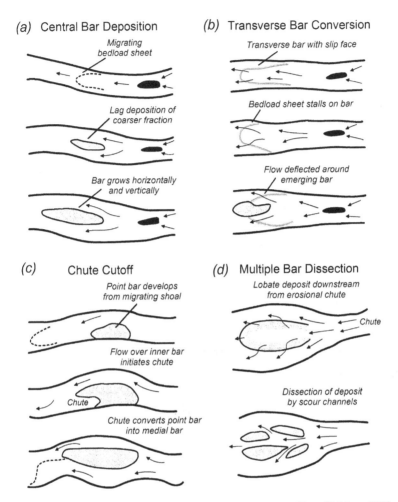

Figure 5.21. Mechanisms of braid bar development (From Knighton 1998)

bars, transverse bars, and lateral bars (Miall 1996; Boggs 2001). As the name implies, longitudinal bars are oriented roughly parallel to the general flow direction. Transverse bars possess avalanche faces oriented at an angle to stream flow (Boggs 2001). They are similar to subaqueous linguoid bars which exhibit a lobate or rhombic configuration and which resemble large dunes that migrate along the channel. Both linguoid and transverse bars are predominantly found in sandy rivers (although linguoid bars are perhaps more common in rivers with high bedload transport rates), and they are often associated with the capture of other migrating bedforms (Fig. 5.21) (Germanoski 2000; Boggs 2001). Lateral bars are attached to the margins of the channel and can be much larger than the other bar forms. Their formation is generally associated with the inability of the stream flow to transport sediment through low energy zones located along the channel banks. Growth of lateral bars primarily occurs by lateral accretion similar to that associated with point bars (Miall 1992). Thus, they can exhibit a fining upward sequence, particularly within sandy rivers.

Detailed studies have shown that where braided and meandering reaches occur along the same river, the divided segments typically possess steeper gradients, shallower flows, and greater total channel width, although the width of the individual channels may be less than that of the undivided reach (Fahnestock 1963; Smith 1970, 1974). The classical model of braid bar formation, presented by Leopold and Wolman (1957), is referred to as the Central Bar Theory because the process begins with the accumulation of sediment near the center of the channel during high flow conditions (Fig. 5.21). This local accumulation of sediment is likely to become an incipient bar since the re-entrainment of the particles requires a higher velocity than does either their transport or deposition (Fig. 5.2). As sediment moves through the reach, some particles will be deposited at the downstream end of the incipient bar where water depth dramatically increases and flow velocities decline. Continued bar growth decreases the total cross sectional area of the channel until it is no longer capable of containing the entire discharge. Flow is then diverted around the bars causing erosion of the bank materials. The channel bed may also be deepened by scour. The combined effect of bed and bank erosion is the lowering of water level, a process that allows the bar to emerge as an island separating two channel branches.

Braid bars are also formed in many instances by the formation and subsequent dissection of bedforms (Fig. 5.21). Examples of these formative mechanisms can be found in Ashmore (1991), Ferguson (1993), and Germanoski (1990, 2000). Other mechanisms of bar formation have little to do with bedforms or in-channel deposition of sediment, but involve the cutoff of point and alternate bars (Fig. 5.21c; Ashmore 1991), and the reoccupation of abandoned channels (Eynon and Walker 1974; Germanoski 2000). Thus, the braided configuration can be formed by multiple processes which may function simultaneously along any given braided reach, although one specific mechanism is likely to be dominant. In all instances, the formation of a braided river is promoted by erodible banks, abundant bed load which can be temporally stored in bars, and rapidly fluctuating discharge (Fahnestock 1963). Perhaps the most important of these is the occurrence of erodible banks; if erosion is prohibited by cohesive bank materials or vegetation it is unlikely that braid bars will form. Fluctuating discharge, on the other hand, is not essential as laboratory studies have produced the braided pattern under constant discharge conditions. Nonetheless, there is no question that variations in discharge enhance cycles of erosion and deposition of sediment which is an integral part of the braiding process.

5.5.1.3. *Anabranching channels*

Nanson and Knighton (1996) defined a anabranching river as an interconnected network of channels separated by relatively stable alluvial islands that subdivide the flow at discharges up to bankfull (Fig. 5.14). Perhaps the most common form is the anastomosing channel which is characterized by low gradients, low width/depth ratios, and fine-grained bed and bank sediment (Smith and Smith 1976; Smith 1986). Not all anabranching rivers, however, possess low-gradients or fine-grained particles (Miller 1991; Knighton and Nanson 1993). Nanson and Knighton (1996) defined six different types of anabranching channels ranging from low-gradient, fine-grained systems to steep, gravel bed rivers.

Anabranching rivers are rather uncommon and, until recently, the processes involved in their formation have received little attention. In fact, the mechanics of how the channel network is developed is still not fully understood, although it is now known that the network can be created by two fundamentally different processes. In one instance, channels are formed by avulsion in which channel bed aggradation leads to overbank flooding. The overbank flows cut a new channel into the existing floodplain deposits or scour out and reoccupy an abandoned channel. The other process involves the deposition of a ridge of sediment within the channel which subsequently becomes stabilized by vegetation and diverts the flow into two directions. Both of these formative mechanisms are promoted by the occurrence of: (1) stable, cohesive bank sediments that limit channel widening, (2) a hydrologic regime characterized by frequent, large floods, and (3) one or more mechanisms (e.g., ice or log jams, or channel sedimentation) which produces localized flooding (Nanson and Knighton 1996).

Although anabranching channels were once considered to be a type of braided river, it is important to recognize that the individual channels may acquire a straight, meandering, or braided configuration (Fig. 5.14). Thus, the sedimentology of anabranching channel deposits can broadly be described by the depositional models put forth for the other channel patterns. Deposits preserved within the floodplain sequence can, however, be distinctly different. We will examine these differences in the following chapter.

5.5.2.   Trace Metal Partitioning Mechanisms

5.5.2.1.   Grain size and compositionally dependent variations

Earlier it was argued that distinct morphological units within a river can possess significant differences in geochemistry. These differences are probably controlled by the types of features present and their sedimentological characteristics. For our purposes, it is important to understand what factors actually produce the observed variations in concentration. Perhaps one of the most important factors is the grain size distribution of sediment that comprises the morphological units of any given channel. The chemically reactive nature of silt- and clay-sized particles suggests that those containing an abundance of smaller particles should exhibit higher trace metal concentrations. In fact, grain size dependent variations in trace metal concentrations between morphological units have been recognized for a number of rivers. Ladd et al. (1998), for example, examined the variability in metal concentrations within and between seven different morphological units within a 500 m reach of Soda Butte

Creek, Montana. Five units were defined according to the nomenclature put forth by Bisson et al. (1982) for gravel-bed streams. They include lateral scour pools, eddy drop zones, glides, low gradient riffles (< 1% slope), and high gradient riffles (1–4% slope). Two types of bars were also defined according to Church and Jones (1982); they were attached bars and detached island bars. For all 12 of the metals examined, significantly different metal concentrations in the < 2 mm sediments were observed between the units. Eddy drop zones and attached (lateral) bars, which possessed the largest percentages of fine sediment, exhibited the highest concentrations of metals on average. In contrast, units with the largest quantity of coarse grained sediment, including glides, low gradient riffles, and high gradient riffles, tended to possess the lowest metal concentrations. Similar results have been derived for braided rivers, such as the Gruben located in the extreme arid environment of Namibia (Taylor and Kesterton 2002).

5.5.2.2. Density dependent variations

Not all of the differences in metal concentrations between the morphological units can be attributed to grain size, indicating that other factors must also be important. One of these factors is undoubtedly the association of metals with relatively dense particles which allows for their accumulation in high, rather than low energy environments. An example has been presented for the Carson River, Nevada by Miller and Lechler (1998). The Carson River was severely contaminated by historic mining operations which utilized Hg to extract Au and Ag from the Comstock Lode, one of the richest Au and Ag producing ore bodies in history. They found that point bar deposits exhibited higher Hg concentrations than the adjacent pools and riffles at any given site. The observed differences in concentration were attributed to the accumulation of sand-sized Hg–Au and Hg–Ag amalgam grains in point bars in the form of modern-day placer deposits.

Slingerland and Smith (1986) define a placer as "a deposit of residual or detrital mineral grains in which a valuable mineral has been concentrated by a mechanical agent," in our case, running water within the river. The concentrated minerals generally possess a density much greater than quartz, such as gold, diamonds, cassiterite, or platinum group elements, and tend to be relatively durable so that they are not broken down during repeated long-term reworking by the river. While sulfide minerals are denser than quartz, they tend to breakup and decompose in oxygenated waters and, thus, rarely form placers in natural (unaltered) environments (Guilbert and Park 1986). However, in rivers contaminated by mining wastes, the shear quantity of metal bearing sulfides released to a river may allow for their concentration in the form of a contaminant placer. A *contaminant placer* is defined here as a concentration of metal enriched particles by the hydraulic action of the river. Where they occur, trace metal concentrations will be locally elevated in comparison to other areas.

The concentration of heavy mineral grains from a heterogeneous mixture of sediment was originally thought to be created primarily by the more rapid settling of hydraulically heavier particles from the fluid. However, in a review of the processes involved in placer formation by Slingerland and Smith (1986), it becomes apparent that particle sorting through settling is probably not as important to placer formation as originally believed. Rather it involves three processes described earlier, including selective entrainment, differential transport, and selective deposition. Interactions between these sorting processes can form placers over a wide range of spatial scales ranging from zones of abrupt valley widening to point bars to migrating dune foresets (Table 5.5) (Slingerland and Smith 1986). We are primarily concerned here with placers concentrated at the immediate scale that are associated with distinct morphological features of the channel. At this scale, placers can be categorized as those associated with bedrock channels and those formed in alluvium. Bedrock placers tend to develop within crevasses that can trap denser particles (e.g., joints, fractures, and faults), or in pools behind obstructions to flow (e.g., outcrops of resistant strata or dikes) (Fig. 5.22a). Common sites of placer formation in alluvial channels, at the intermediate scale, include the head of mid-channel bars or islands, the basal zones of point bars, and tributary junctions (Fig. 5.22b). Depending on the environment, point bars can be particularly important because the concentration of heavy minerals during channel migration can lead to the development of pay streaks, or in our case hotspots, within the associated floodplain (Fig. 5.23).

Smith and Beukes (1983) argued that placer formation associated with bars and tributary junctions can be attributed to stable convergent flow patterns that

Table 5.5. Observed sites of water-laid placers organized by spatial scale

Site/Scale	Site/Scale
Large Scale $(10^4\ m)$	***Intermediate scale*** $(10^2\ m)$
Bands parallel to depositional strike	Concave sides of channel bends
Head of alluvial fans	Convex banks of channel bends
Points of exit of highland rivers onto alluvial plan	Heads of mid-channel bars
	Point bars with suction eddies
Regional unconformities	Scour holes, especially at tributary confluences
Strand-line deposits	Inner bedrock channels and false bedrock
Incised channelways	Bedrock riffles
Pediment mantles	Constricted channels between banks and bankward-migrating bars
Small Scale $(10^0\ m)$	
Scoured bases of trough cross-strata sets	
Winnowed tops of gravel bars	
Thin ripple-form accumulations	
Dune Crests	
Dune Forsets	
Plane parallel laminae	
Leeward side of obstacles	

Note: See Slingerland and Smith (1986) for references concerning accumulation sites. After Slingerland and Smith (1986)

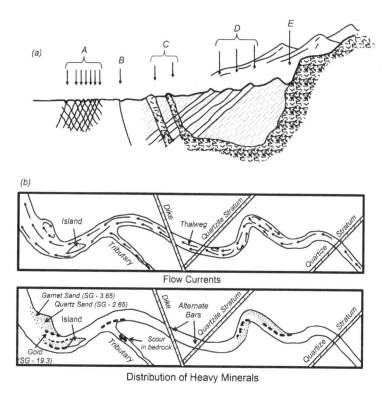

Figure 5.22. (A) Favorable sites for placer formation along a bedrock channel; (B) possible sites of heavy particle accumulation along alluvial channels (Modified from Guilbert and Park 1986)

allow for the efficient removal of hydraulically lighter materials from the channel bed sediment. For example, they examined the accumulation of magnetite within three sluiceways formed between a stable bank and a migrating sand bar in two rivers in South Africa. The zones were characterized by bar-top and sluiceway currents that are oriented across and downstream to the major flow direction, respectively. They found that at two sites magnetite concentrations were enriched over other parts of the river by approximately a factor of five. At both sites, bar growth and migration toward the bank had ceased because sediment swept into the sluiceway was immediately removed by the currents, leaving only the heavy mineral fraction. In contrast, heavy mineral concentrations were lacking at the third site where the bar was migrating toward the bank. In this case, the bar transported sediment was deposited in marginal foresets prior to being sorted by the sluiceway currents. The study by Smith and Beukes (1983) not only illustrates the need for the repeated scouring action associated with convergent flow, but demonstrates that even when heavy minerals are abundant, contaminant placers may not always develop.

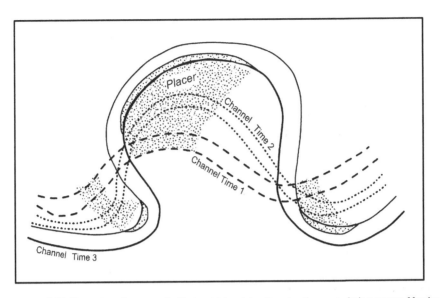

Figure 5.23. Formation of pay streaks (hotspots) in a laterally migrating meandering stream. Numbers represent different periods in time (Modified from Bateman 1950)

5.5.2.3.   *Variations dependent on time and frequency of deposition*

Both the timing of deposition, and the frequency of inundation, can lead to differences in metal concentrations between morphological units. Graf et al. (1991), for example, found that within Queens Creek, a dryland river in Arizona, the active channel deposits inundated a few times per year possessed higher Cu, Zn, V, Mg, Mn, and Ti concentrations than those inundated about once per decade (Fig. 5.24). Similar observations have been made along the Rio Pilcomayo in southern Bolivia. In this case, trace metal concentrations observed in high-water channel deposits which are inundated during the wet season were low in comparison to low-water channel deposits that are inundated throughout the year, in spite of the fact that the low-water deposits are coarser grained (Fig. 5.25) (Hudson-Edwards et al. 2001).

The influence of inundation on metal concentrations is primarily related to two factors. First, the more frequent inundation of the lower, in-channel, morphological units provides for a greater opportunity to accumulate contaminated sediment, particularly fine-grained particles which are most likely to be transported during small to moderate floods (Graf et al. 1991). Second, deposition on topographically higher morphological units occurs only during major floods when significant quantities of sediment are delivered to the river from the entire drainage basin. The addition of relatively clean sediment from uncontaminated tributaries acts as a dilutant which decreases the metal concentrations within deposits formed during high flow conditions (Graf et al. 1991; Hudson-Edwards et al. 2001). During low flow, tributaries are unlikely to deliver water or sediment to the channel in significant

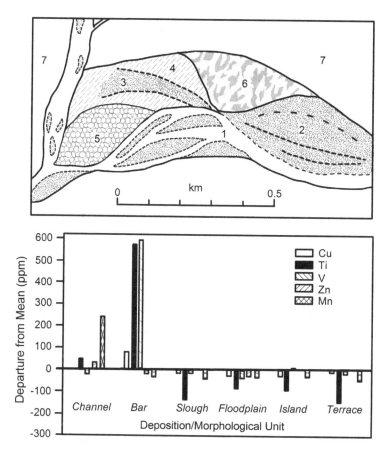

Figure 5.24. (A) Depositional environments along Queens Creek, Arizona, including: (1) active channel, (2) active bar, (3) slough, (4) foodplain, (5) island, (6) terrace, and (7) hillslopes; (B) Metal concentrations in selected depositional environments near Hewitt Canyon (From Graf et al. 1991)

quantities. Sediment-borne trace metals from pollutant sources, however, may enter the river year round with the discharge of waste waters. In the case of the Rio Pilcomayo, mentioned above, contaminated effluent from upstream ore processing facilities served as the primary source of water to the channel during the dry season.

5.5.2.4. *Geochemically dependent variations*

Differences in trace metal concentration between morphological units have also been attributed to localized variations in the deposits' physiochemical conditions, such as their ability to accumulate reactive coatings. A common example is the formation of Fe and Mn oxides and hydroxides on the surface of large immobile clasts. Once formed, the coatings may scavenge and accumulate metals from the water column (Chao and Theobald 1976; Tessier et al. 1982; Ladd et al. 1998). Moreover, scavenging by the coatings may be more significant within the channel bed deposits

Figure 5.25. Rio Pilcomayo, southern Bolvia near Uyuni. Photo taken in July during the dry season

than on the adjacent, topographically higher bars which are finer-grained and less frequently inundated. In other cases, differences in the inundation and exposure of morphological units can lead to fluctuations in geochemical processes that differ between morphological units through time. These geochemical variations can lead to distinct differences in metal speciation and their potential for remobilization. The Rio Pilcomayo of southern Bolivia serves as an excellent example. Sulfide minerals deposited in high-water channel deposits (Fig. 5.25) are oxidized during the dry season creating localized zones of acid drainage enriched in trace metals (Miller et al. 2002). Presumably, the acid waters are flushed from the high-water deposits during the onset of the wet season. In contrast, sulfide oxidation within the adjacent low-water channel deposits which are continuously inundated is thought to be minimal, allowing for their downstream transport with only minor alteration. Thus, the metals remain locked within the sulfide grains.

5.5.3.   Implications to Sampling

Collecting and analyzing multiple samples from a short-section of stream channel would reveal that metal concentrations within those samples vary from site to site. Birch et al. (2001) has referred to these local variations as *small scale* or *field variance*. The magnitude of field variance can be determined by the analysis of replicate samples from a given site; thus, it includes variations associated with the analytical methods. While analytical variation (error) is likely to be on the order of 5%, field variance is commonly on the order of 10 to 25% relative standard deviation for elements such as Cu, Pb, and Zn within fluvial systems (Birch et al. 2001).

   As we previously noted, the problem with field variance is that it can significantly hinder many types of analyses, such as the assessment of large-scale, spatial trends in contaminant levels used to identify pollutant sources or downstream dispersal rates. In other words, where field variation is substantial, the ability to decipher differences in contaminant levels between sample sites is considerably reduced, especially where concentrations approach background values (Birch et al. 2001). Thus, it is necessary to devise a sampling methodology that reduces field variance in order to document spatial trends over a broader area. While the need for such a program is widely understood, actually creating one is much more difficult. In fact, it has been argued that more money, effort, and time have been wasted because of poor sampling design than for any other reason (Keith et al. 1983; Horowitz 1991). The result has been the publication of numerous books on the topic (e.g., Watterson and Theobald 1979; Sanders et al. 1983; Keith 1988). Nonetheless, there is still no widely accepted strategy to the design of sediment sampling programs. Traditionally, many program managers have advocated a statistical approach in which samples are collected at a set distance (e.g., every 10 m across the channel), or on the basis of some form of gridded sampling design. The advantage of these methods is that they readily lend themselves to statistical manipulations (Provost 1984; Gilbert 1987; Mudroch and MacKnight 1991; Birch et al. 2001). However, our previous discussions show that trace metals are not likely to be randomly distributed within the channel bed sediment, but vary as a function of the sedimentology and associated hydraulic regime of the morphological units. In order to thoroughly characterize the concentrations that exist, each of the morphological units should be sampled. Performing this task using a random sampling approach would necessarily require a large number of samples to avoid missing some of the morphological units. The potential inefficiencies of the random sampling approach prompted Ladd et al. (1998) to suggest that investigators should initially conduct reconnaissance level surveys to determine the degree of variability that exists between morphological units before completing full-scale studies of metal distribution and their environmental impacts. If differences in concentration exist, they argue that the sampling program should stratify the collected samples (data) by morphological unit type.

Depending on the morphological complexity of the channel bed, it may be financially unrealistic to sample and analyze sediments from each of the morphological units at every sampling site. Thus, for some types of analysis, particularly those attempting to document spatial variations in metal concentrations, the sampling of only one morphological unit is advocated. Many exploration geologists use this approach when attempting to identify the location of potential ore bodies on the basis of spatial variations in the chemistry of the channel bed sediments (e.g., Day and Fletcher 1989). It should be recognized, however, that such an approach will not provide a full range of concentrations that exist (Ladd et al. 1998). Nonetheless, where this is necessary an understanding of the relationships between metal concentration and morphological unit may still lead to a more efficient sampling program than would be otherwise possible. Ladd et al. (1998) point out, for instance, that the full range of concentrations could be determined by sampling only two morphological units, those that possess the maximum and minimum concentrations within the river.

In addition to the variations in concentration that occur between morphological units, variance also exists within any given unit. This variance tends to increases with the sedimentological and morphological complexity of the deposits (Birch et al. 2001). In order to reduce the variance inherent within the morphological units, composite sampling is commonly utilized in which multiple samples from a given morphological feature are collected, combined, and mixed prior to analysis. The primary advantage of composite sampling is that it reduces small scale sampling variance while decreasing analytical costs. It is once again important to recognize, however, that the generated results represent an average concentration of the contaminants within the morphological unit, rather than the full range of concentrations that exist.

5.6.   PHYSICAL AND MATHEMATICAL MANIPULATIONS

In the preceding section it was shown how small scale variations in sediment-borne trace metal concentrations could be minimized by composite sampling of specific morphological units. In some cases, however, large scale variations in sediment size or composition cannot be removed by stratifying the data according to depositional environment. For example, a commonly observed trend along many rivers is for grain size of the channel bed material to decrease downstream. While sampling specific morphological units may adequately remove the local (reach scale) variations in grain size, and thus, metal concentrations, the larger scale, downstream variations will remain. These types of variations in sediment size and/or composition may reduce our ability to decipher spatial (or temporal) patterns in sediment geochemistry. It is therefore common to apply some form of stratification to the collected samples or data to compensate for variations in grain size or composition. Two approaches are frequently utilized. First, a specific grain size or sediment type can be removed and analyzed independently. The separation and analysis of a specific grain size fraction is widely advocated for environmental studies. The assumption inherent in this procedure is that trace metals are contained entirely within the analyzed fraction. While the actual size of the chemically active sediment will vary from river to river, many investigators have argued that the grain size of the analyzed sediment should be standardized to allow for comparison of different investigations conducted in diverse environments. In the U.S., most analyses, including those conducted by state and federal regulatory agencies, are based on the collection and analysis of the $< 63\,\mu m$ sediment fraction. The advantages of using the $< 63\,\mu m$ fraction over other size ranges are that: (1) this size fraction can be extracted from the bulk sample relatively quickly via sieving, a process that does not alter trace metal chemistry, (2) the $< 63\,\mu m$ particle size is most frequently carried in suspension in riverine systems, and may therefore be most readily distributed over the environment, and (3) it occurs in large enough quantities that it can generally be collected from both the water column and alluvial deposits (Miller and Lechler 1998). A significant disadvantage, however, is that the analyses do not provide for an understanding of the actual concentrations in the bulk sample which may be required for many risk assessments.

The second approach is to mathematically manipulate geochemical data obtained from the bulk sample using information collected from the analysis of a separate subsample of the analyzed material. The most common form of mathematical manipulation involves normalization of bulk concentrations. The first step in normalization is the calculation of a dilution factor describing the amount of a substance within the sample which is assumed to be devoid of trace metals; a normalized concentration can then be calculated by multiplying the dilution factor by the chemical concentration determined for the bulk sample. The following equations, for example, could be used to normalize data with respect to the proportion of sediment $< 63 \, \mu m$ in size:

(10)   Dilution Factor $= 100/(100 - \%$ of sample $> 63 \, \mu m$ in size)

(11) $= 100/(\%$ of sample $< 63 \, \mu m$ in size);

(12)   Normalized Concentration $=$ (dilution factor)

$\times$ (conc. metal in bulk sample)

In this case, sand-sized sediment is the diluting substance.

Although mentioned above, it is worth reiterating that calculated normalized data are used for clarifying spatial or temporal trends; they do not reflect the actual chemical concentrations within the alluvial sediments (Horowitz 1991). For example, Table 5.6 presents metal concentrations for the $< 63 \, \mu m$ sediment fraction determined by analyzing the fine materials separated from the bulk sample, and by normalizing the data according to the percentage of sediment $< 63 \, \mu m$ in size. Differences in the normalized and measured concentrations can be significant, especially when the samples contain $< 50\%$ silt and clay (Table 5.6). The observed differences may indicate that: (1) analytical errors are associated with the geochemical or grain size analyses, (2) not all of the sediment-borne trace metals are associated with the analyzed grain size fraction, and/or (3) chemical variations in trace metal concentrations are dependent on factors other than simply grain size (Horowitz 1991).

Normalization by grain size is by far the most frequently used form of data manipulation. Nonetheless, bulk concentrations have also been normalized according to the percentage of quartz, carbonate, or organic carbon within the sample (Table 5.7). For each substance, the assumption is that particles of a specific composition are free of substantial quantities of trace metals. Note, however, that organic carbon can act both as a dilutant and a concentrator of trace metals, depending on its form and the physiochemical conditions of the site.

A slightly different approach is to normalize bulk trace metal concentrations by the concentration of a conservative element such as Al, Ti, or Li. The underlying assumption is that the conservative elements are released from crustal rocks at a uniform rate when considered over a long enough timeframe. Thus, areas of trace metal concentration or dilution can be identified when normalized by the conservative element. In contrast to the methods used for grain size, quartz, etc., normalization in this case is performed by simply dividing the concentration of the

Table 5.6. Comparison of calculated and measured trace metal concentrations based on the $< 63\,\mu m$ sediment fraction; concentrations in mg/kg

Location	Percent $< 63\,\mu m$	Data	Cu	Zn	Pb	Ni
Georges Bank	1	Measured	104	55	22	20
(M8-5-4)		Calculated	100	500	500	100
Columbia Slough	18	Measured	47	225	72	31
		Calculated	145	811	239	145
Nemadji River	44	Measured	31	62	10	31
		Calculated	50	102	20	48
Patuxent River at:	54	Measured	35	116	39	32
Point Patience		Calculated	43	180	46	41
Yaharra River	66	Measured	22	45	28	22
		Calculated	20	41	33	17
Hog Point	76	Measured	25	131	26	31
		Calculated	26	147	29	36
George Bank	89	Measured	10	63	22	22
(M13A)		Calculated	15	67	22	25
Lake Bruin	95	Measured	24	104	23	29
		Calculated	26	108	23	31

Modified from Horowitz (1991)
Measured – actual analytical determination of $< 63\,\mu m$ sediment fraction; Calculated – based on product of bulk chemical concentration and a dilution factor; DF = 100/100-% of sample $> 63\,\mu m$ in size

Table 5.7. Selected dilution factors (DF) and normalization equations used to manipulate bulk geochemical data

Equation	Comment
Grain Size DF = 100/(100-% Grain Size > Size of Interest) NC = (DF)(Bulk Chemical Concentration)	Size is typically $> 63\,\mu m$ in US.
Carbonate DF = 100/(100-% Carbonate) NC = (DF)(Bulk Chemical Concentration)	Normalization to carbonate-free basis; used primarily in marine or karst environments
Organic Carbon Content DF = 100/(100-% Organic Carbon) NC = (DF)(Bulk Chemical Concentration)	Normalized to organic carbon-free basis; loss-on-ignition commonly used as estimate of % o. carbon; organic carbon does not serve as a dilutant in many systems, but can accumulate trace metals
Conservative Element NR = Concentration of trace element / Concentration conservative element	Produces a ratio rather than a value in concentration units; commonly used conservative elements include Al, Li, and Ti.

Adapted from Horowitz (1991)
NC – Normalized concentration of trace metal
NR – Normalized ratio

trace metal by the concentration of the conservative element. The resulting values are not in units of concentration, but represent a ratio. The use of a ratio can make comparisons with data from other regions difficult (Horowitz 1991). As a result it is probably fair to say that normalization to a conservative element is not frequently used in environmental studies.

It is extremely important to recognize that normalization procedures should only be used when there is some quantitative rationale for doing so (e.g., a strong correlation between trace metal concentrations and percent silt and clay in the sample). Even in this case, care must be taken in that misleading results can be produced. For instance, normalization by grain size is commonly inappropriate for many mining impacted rivers because the trace metals are distributed across a broad range of particle sizes (Moore et al. 1989). Unfortunately, many regulatory agencies require sampling protocols to be based on only the fine-grained sediment fraction, or some form of mathematical normalization, without ever determining if it is appropriate for the site of interest. The results can lead to a misunderstanding of both the quantity and the distribution of trace metals in the system.

5.7.   TEMPORAL VARIATIONS IN CONCENTRATION

Viganò et al. (2003) collected bed sediments along the Po River in Italy during two sampling campaigns: one during the summer, 1996, and the other during the winter of 1997. Comparison of the results revealed that trace metal concentrations at a given sampling site were similar and, therefore, largely independent of the season during which the samples were collected (Fig. 5.26). Their conclusions emphasize one of the commonly cited advantages of conducting geochemical surveys of channel bed sediment – that the data are largely devoid of the temporal variations in concentration observed for the dissolved or suspended load. Bed sediment, in other words, probably serves as an indicator of both river health and spatial variations in contaminant levels without requiring multiple sampling campaigns conducted on an event or seasonal basis. Although this assumption is founded on a surprisingly limited number of investigations, it is generally supported by the existing data, even though seasonal variations in trace metal concentrations in channel bed sediments have been identified along some rivers or river reaches. For example, Gaiero et al. (1997) found that in uncontaminated (pristine) areas of the Suquia River system of Argentina, differences in total non-residual trace metal concentrations between the spring and autumn were minimal. However, in contaminated areas affected by untreated sewage, trace metal concentrations were higher in the springtime than in the fall. The temporal variations were attributed to the type and abundance of organic matter and its effects via Eh on Mn and Fe hydroxide dissolution (see Chapter 2). The influence of organic matter cycling on seasonal variations in channel bed concentrations have also been suggested for other rivers (see, for example, Facetti et al. 1998). Nonetheless, it is important to recognize that while seasonal variations in concentrations at a specific site may occur for some constituents, the

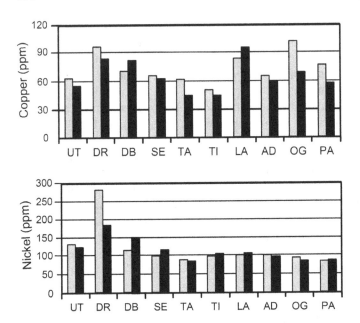

Figure 5.26. Concentrations of Cu and Ni in the < 63 μm fraction of channel bed sediments from the Po River, Italy. Samples were collected in the summer (grey bars) and winter (black bars). Acronyms along x-axis represent successive downstream sampling sites. Note minimal variations in concentration between seasons (Modified from Viganò et al. 2003)

spatial (downstream) trends typically remain consistent during the year, provided that the character of the contaminant sources remain relatively unchanged.

A number of studies have examined the impact of high magnitude, low frequency floods on channel bed concentrations. For the Carson River, Nevada, Miller et al. (1999) found that neither the concentrations of Hg at a site, nor the downstream trends in Hg concentration in the channel bed were altered during a 100-year event in 1997 (Fig. 5.8b). The observed stability of the Hg concentrations was apparently controlled by the overall geomorphic structure of the system, defined by its valley morphology, the location of tributaries that delivered "clean" sediment to the channel, and the distribution of Hg within the valley fill.

The alteration and rapid recovery of channel bed concentrations following a 100-year event was observed by Ciszewski (2001) along Biala Przemsza River of southern Poland. Here trace metal concentrations decreased by three fold over a 40 km reach of the river in response to the flood, but during the following period (August 1997–March 1998) concentrations recovered to values observed in 1993 prior to the event. A second, more moderate flood approximating the bankfull discharge, resulted in inconsistent downstream changes in channel bed concentrations; downstream reaches exhibited decreases in concentration, whereas upstream reaches lying adjacent to the influx of waste waters from a mine exhibited increases in concentration. The Biala Prezemsza study demonstrates that concentrations of trace metals in the channel bed can be altered by flood events, but may rapidly return to a mean condition.

While flood induced variations in trace metal concentrations in the channel bed appear to be limited, recognize that the depth of scour during a major flood can be significantly greater than at other times. The eroded materials can then be redistributed over the valley floor, thereby changing the spatial distribution of the sediment-bound contaminants observed prior to the event. The extent of metal redistribution can be particularly important in areas where significant quantities of trace metals have been stored beneath the channel as a result of aggradation. In this case, erosion of older, highly contaminated sediments can actually increase the concentrations observed at the surface of the adjacent floodplain and re-contaminate previously remediated areas (NRC 2005).

5.8. SUMMARY

The use of channel bed materials to determine ecosystem health, contaminant sources, and sediment-borne transport dynamics greatly simplifies site assessments as bed materials are easier to collect, can be obtained at most times throughout the year, and exhibit less variations in concentration through time than does suspended sediment. The ability of a river to entrain, transport, and deposit sediment and sediment-borne trace metals depends on the energy given to the water by velocity, depth, and slope and on the amount of energy consumed by the resistance to flow dictated by such elements as channel configuration, particle size, and sediment concentration. Contaminant sources and dispersal processes combine to produce longitudinal trends in trace metal concentrations that are river specific. Nonetheless, concentrations tend to decay quasi-systematically downstream of point sources. Observed decay patterns are dependent on the local importance of distinct, but interrelated dispersal mechanisms including: (1) hydraulic sorting, (2) sediment storage and exchange with the floodplain, (3) dilution associated with the mixing of contaminated and uncontaminated sediment, (4) biological uptake, and (5) geochemical remobilization or abstractions. Within many rivers, sediment transport and deposition is not uniformly distributed downstream. Rather, one or the other process is locally dominant creating alternating zones of transport and sedimentation that affect the general longitudinal trends in trace metal concentration. Sediment-borne trace metals may also move through the system as a series of sediment slugs, complicating geographical patterns predicted by regression or mixing models.

The in-channel storage of trace metals within stable channels is generally small, accounting for less than about 10% of the total load exported from the basin. This is in marked contrast to the storage of sediment within floodplains, where more than 40–50% of the total annual load may be deposited. The residence time of the contaminants also differs substantially, ranging from less than 5 years to decades, centuries, or millennia for channels and floodplains, respectively. At the reach scale, sediment-borne trace metals are selectively deposited within distinct morphological units (e.g., bars, riffles, pools, and chutes) as the result of hydraulic sorting by size and density, the timing and frequency of inundation, and various geochemical processes. These morphologic features are both created by and influence the mean flow configuration through the reach, and, therefore, have specific associations with a given channel pattern (e.g., point bars, pools, and riffles with meandering streams, braid bars with braided channels, etc.). Channel patterns can be broadly classified as straight, meandering, braided, and anabranching forms. However, the boundaries between the different patterns are indistinct and may change both along a given river and as a function of discharge at a specific site.

While protocols aimed at sampling specific morphological units may adequately remove local (field scale) variance in trace metal concentrations, larger scale (downstream) variations may remain as a result of differences in particle size or composition. The net result is that our ability to decipher spatial (or temporal) patterns in sediment geochemistry is greatly reduced. It is therefore common to apply some form of stratification to the collected samples or data to compensate for variations in grain size and/or composition. Two approaches are frequently used; a specific grain size or sediment type can be removed and analyzed independently, or bulk sample concentrations can be mathematically manipulated using data from a separate sample split to remove the effects of size and/or composition. The approach used necessarily depends on the sediment-trace metal relations within the channel bed sediment.

5.9.   SUGGESTED READINGS

Birch GF, Taylor SE, Matthai C (2001) Small-scale spatial and temporal variance in the concentration of heavy metals in aquatic sediments: a review and some new concepts. Environmental Pollution 113:357–372.

Graf WL (1990) Fluvial dynamics of Thorium-230 in the Church Rock event, Puerco River, New Mexico. Annals of the Association of American Geographers 80:327–342.

Helgen SO, Moore JN (1996) Natural background determination and impact quantification in trace metal-contaminated river sediments. Environmental Science and Technology 30:129–135.

Knighton D (1998) Fluvial forms and processes: a new perspective. London, Arnold.

Lewin J, Macklin MG (1987) Metal mining and floodplain sedimentation in Britain. In: Gardiner V (ed) International geomorphology, Wiley, Chichester, pp 1009–1027.

Slingerland R, Smith ND (1986) Occurrence and formation of water-laid placers. Annual Review of Earth and Planet Science 14:113–147.

Walling DE, Owens PN, Carter J, Leeks GJL, Lewis S, Meharg AA, Wright J (2003) Storage of sediment-borne nutrients and contaminants in river channel and floodplain systems. Applied Geochemistry 18:195–220.

CHAPTER 6

FLOODPLAINS

6.1.   INTRODUCTION

In the previous chapter we examined the mechanics of sediment and sediment-borne trace metal transport within river channels. Rivers, however, are more than natural sluiceways through which sediment and contaminants are dispersed from point and non-point sources; rather, they mold the landscape into specific topographic forms. This sculpturing process is accomplished through the erosional capability of the sediment-laden waters and through the deposition of debris once the transporting ability of the water falls below that required to keep it in motion. As a result of this duality in process, some of the topographic forms produced by rivers are entirely erosional, and the features consist of little, if any, unconsolidated debris. In other cases, the surface topography is the product of deposition which buries the underlying geological materials. Most features, however, are produced by a combination of erosional and depositional processes.

In this chapter we are concerned with the erosional and depositional processes that lead to the formation of floodplains. Floodplains are, perhaps, the most intensively studied of any landform to be found along contaminated rivers. Their assessment is particularly well-suited to the geomorphological-geochemical approach because sediment-borne trace metals follow the same transport pathways as any other sedimentary particle, and therefore their depositional patterns can be directly related to floodplain processes (Macklin 1996; Macklin et al. 2006). It is a mistake, however, to assume that the investigation of floodplains is performed only to determine the location, depth, and magnitude of contamination as required for site remediation. While this is clearly of importance, data obtained from floodplain deposits can also yield insights into (1) natural elemental concentrations inherent in the studied river, (2) changes in contaminant flux to the river over periods greatly exceeding historical records, (3) natural rates of system recovery following the removal or containment of the contaminant source, (4) contaminant sources and the changes in those sources through time, and (5) the potential for sediment-borne trace metal remobilization along the river valley.

We will begin our discussion by examining the processes of floodplain formation and the types of sedimentary bodies associated with floodplains formed by the various channel patterns. We will then turn our attention to deciphering depositional patterns of sediment-borne trace metals on and within floodplains and the use of floodplains to determine the timing, loading rates, and source of contaminants to the river. The analysis is limited, however, to stable rivers in which sediment-borne trace metals are transported as part of the river's natural load. Lewin and Macklin (1987) referred to this style of transport as *passive dispersal*. In the following chapter, we will examine the alternative to passive dispersal, known as *active transformation*, in which the channel and floodplain are rearranged in response to external or internal perturbations, such as might be caused by changes in climate or land-use.

6.2.   FLOODPLAINS: DEFINITION

Our first image of a floodplain is likely to be that of a flat, low-lying surface located immediately adjacent to a river channel. It should come as no surprise, then, that a flood-plain could be simply defined on the basis of topographic position and morphology as the flat-lying terrain residing along and connected to a river. At other times, however, we may be more concerned with the potential for flood waters to damage housing developments, industries, or other types of infrastructure. From this perspective, floodplains can be defined on the basis of the frequency with which the surface is inundated during rainfall-runoff events. This has led to the field of flood hydrology in which sophisticated models are routinely applied to identify zones of the landscape which are inundated on average once within the specified time period (e.g., every 10-, 50-, or 100-years). Neither of these two definitions are concerned with the composition or nature of the underlying geological materials. For our purposes it is important to link the definition of a floodplain to the mechanisms that are respon-sible for its formation. This is generally referred to as the genetic definition of a floodplain. It has been most accurately expressed by Nanson and Croke (1992) as a "largely horizontally-bedded alluvial landform adjacent to a channel, separated from the channel by banks, and built of sediment transported by the present flow regime." According to this definition floodplains are created by a combination of deposi-tional and erosional processes that must somehow be related to the present activity of the river. They are, then, a product of the modern river environment. It is also true, however, that they serve as important storage areas for both water and sediment generated within the watershed. Thus, floodplains are not only produced by the river, but are a functional part of the riverine environment that regulates the downstream transport of constituents including contaminated particles to the basin mouth.

Nanson and Croke (1992) went on to present a detailed classification of flood-plains based on the erosional resistance of the alluvial materials that form the channel perimeter and the stream's ability to erode and transport sediment. The latter was based on the available specific stream power (i.e., stream power per unit width, Chapter 5) because it has been shown to be a predictor of bed and bank erosion, sediment transport dynamics, and depositional processes. The classification scheme includes three primary categories: high-energy non-cohesive floodplains, medium-energy non-cohesive floodplains, and low-energy cohesive floodplains. These three

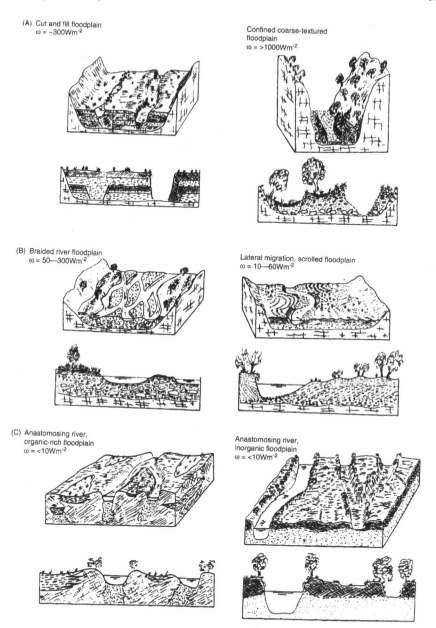

Figure 6.1. Examples of floodplain types defined by Nanson and Croke (1992). (A) high-energy, non-cohesive floodplains, (B) medium-energy, non-cohesive floodplains, and (C) low-energy cohesive floodplains (Modified from Nanson and Croke 1992)

classes are further subdivided on the basis of nine parameters into 13 orders and suborders. Figure 6.1 presents examples of the delineated floodplain types for various ranges of specific stream power. The classification system clearly shows that there is a wide range of floodplain types which fall along a continuum between high- and low-energy environments and which reflect the amount and texture of the alluvial materials (Nanson 1986; Nanson and Croke 1992).

Although floodplains may be genetically defined and classified according to the contemporary processes which are operating along the river system, it is generally recognized that floodplains are time-transgressive features. This means that they contain elements or deposits that were developed at times when the flow regime was different than it is today. The formation and character of these features, when combined with those formed under current hydrologic conditions, provides a record of the past history of erosional and depositonal processes within the basin (Brown 1997). As we will see later, our ability to read this history can provide important insights into the changes in contaminant loading rates, storage processes, and dispersal mechanics through time.

6.3.   THE FORMATIVE PROCESSES

The variability in floodplain morphology and sedimentology described in the literature, and which is inherent in the classification scheme presented by Nanson and Croke (1992), is the product of the combined interactions of a suite of distinct depositional processes. The most important of these processes include lateral accretion, vertical accretion, and braided-channel accretion. The nature and significance of lateral and vertical accretion has been most thoroughly investigated for meandering streams, particularly those in humid temperature environments. In this case, lateral accretion is produced by a combination of erosional and depositional processes as the channel migrates across the valley floor (Fig. 6.2). Although erosion may occur anywhere within the channel, it is concentrated along the outer bank, just downstream of the axis of curvature of the meander bend. During floods, materials eroded from the cut bank are combined with the rest of the channel's load and deposited, in part, on downstream point or alternate bars. Detailed measurements from a wide range of meandering rivers have shown that for stable rivers

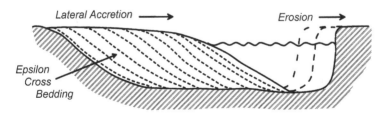

Figure 6.2. Formation of lateral accretion deposits during channel migration. Dotted lines represent epsilon cross-bedding

(i.e., those in a state of equilibrium), the volume of material eroded is equal to that deposited. This allows migration to occur without a change in cross-sectional shape or channel dimensions. The primary product, then, of lateral migration is a thin-sheet of sediment deposited on point bars which is left in the wake of the progressively shifting channel (Fig. 6.2). The thickness of this laterally accreted sheet depends largely on the depth to which the river can scour during floods. Although the scour depth can be highly variable, it is commonly estimated for many perennial river systems to be 1.75 to 2 times the water depth (Leopold et al. 1964).

Vertical accretion is produced by the periodic, overbank deposition of sediment during flood events. In most, but not all, cases, the deposited sediment is composed of relatively fine-grained particles suspended in the flood waters (Fig. 6.3). Significant attention has been given in past years to determining and predicting the rates of overbank deposition that can occur. We will examine the primary factors controlling overbank deposition later in the chapter. Here, however, it is important to recognize that if there is no change in the elevation of the channel bed, then the rate of vertical accretion by overbank flows should change through time. Initially, depositional rates should be quite rapid because the frequency of overbank flooding is relatively high. With continued growth, the height of the floodplain above the channel will increase. This reduces the frequency of inundation and therefore the rate of vertical accretion (Fig. 6.4).

The development of floodplains along meandering streams as described above raises the question as to the relative importance of lateral and vertical accretion in floodplain development. There is no universal answer to this question; both

Figure 6.3. Vertical accretion deposits along the Carson River, Nevada. Dark layer near center of photo is a buried soils

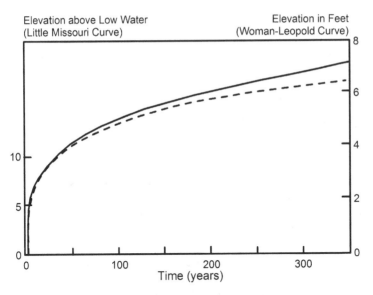

Figure 6.4. Increase in elevation of floodplain surface with time. Lower dashed curve from data collected for the little Missouri River; upper solid curve derived theoretically for Brandywine Creek, Pennsylvania by Wolman and Leopold (1957) (From Everitt 1968)

have been found to be the predominant process in certain environments, depending primarily on the rates of lateral channel migration (Ritter et al. 2002). Nonetheless, for many meandering streams, floodplains are dominated by lateral accretion deposits because vertically accreted sediments are reworked during channel migration (Wolman and Leopold 1957). Take, for example, a river possessing a 1 km wide floodplain which is migrating at a rate of 2 m/yr. The channel can move completely across the valley floor in 500 years. Although a meter or two of sediment could be deposited by vertical deposition during this time, most of the overbank deposits would be reworked by bank erosion and replaced by laterally accreted sediment as the channel moved across the valley.

In reality, the above example is an oversimplification of floodplain development in that rivers general do not migrate across the entire width of the valley floor. Rather, they tend to occupy a relatively small belt, which may help explain the cross-sectional asymmetry commonly observed along some valley reaches (Brown 1997). In these situations, those portions of the floodplain which are not reworked during lateral migration are likely to be dominated by overbank deposits (Kesel et al. 1974), whereas floodplains within the preferred belt(s) will consist predominately of lateral accretion deposits. Along other rivers, particularly in low-energy environments, the channel may experience little, if any, channel migration over long periods of the time. A part of the Delaware River, for instance, has been stable for at least 6,000 years, allowing the unimpeded growth of the floodplain by overbank deposition (Ritter et al. 1973).

Some investigators have argued that the association of vertically accreted deposits with overbank deposition and lateral accretion deposits with in-channel deposits is misleading. This follows because not all overbank flood deposits are laterally synchronous, and within channel features may contain a significant component of vertical deposition (Fig. 6.5) (Taylor and Woodyer 1978; Brown 1997). Moreover, it is important to remember that not all lateral and vertical accretion is associated with meandering streams in humid temperature regimes. Where the geomorphic setting is different, lateral and vertical accretion combine to produce highly variable floodplain types. A case in point is the Powder River in southeastern Montana. Here floodplain development following a flood event in 1978 did not occur by the lateral accretion of sediment on point bars (Moody et al. 1999). Rather, the floodplain was formed by vertically accreting sediment on sand and gravel surfaces located within the channel that were created during channel widening. The floodplain deposits,

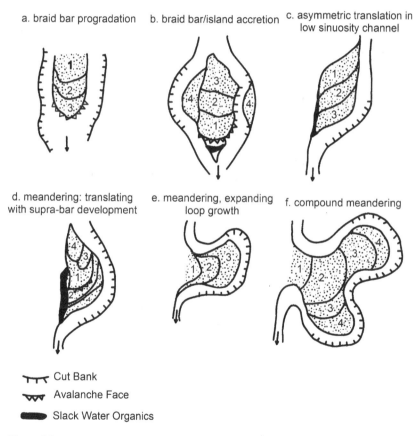

Figure 6.5. Types of accretion observed within braided and meandering channels. Relative unit age is shown numerically, one being the oldest deposit. Unit ages for meandering streams (diagrams e and f) may be much longer than those shown for bar growth (a and c) (From Lewin et al. 2005)

then, formed narrow stratigraphic units that paralleled the active channel and were inset into or overlapped older floodplain materials. Moody et al. (1999) also found that the rate of vertical accretion decreased with a decline in the frequency of inundation as the elevation of the floodplain above the channel bed increased.

Floodplains along many arid systems, or rivers with transitional-braided configurations, are also much more variable than the meandering channels described for humid, temperature regimes. For example, flood discharges associated with arroyos (deep, flat-floored trenches) of the American southwest are highly variable because of an uneven distribution of rainfall through time and over the basin. During rare, high-magnitude runoff events, the flood waters erode the non-cohesive sediments of the trench walls. Thus, the excess water is accommodated by an expansion in the width of the trench, rather than spilling onto the valley floor. The created floodplains, if they exist, are confined to the trench and tend to be rather discontinuous along the channel (Fig. 6.6). Moreover, the large fluctuations in discharge may not allow an equilibrium state between process and floodplain form to develop (Graf 1988), producing floodplains with transient characteristics.

Braided-channel accretion has been less thoroughly studied than accretion formed by meandering rivers. In general, floodplains along braided rivers are the product of channel and braid bar abandonment, followed by deposition on the stabilized surface. Nanson and Croke (1992) argue that channel/bar abandonment and stabilization can by driven by three processes: (1) the stabilization of braid-bars and channel deposits by the movement of the river to another section of the valley floor, (2) the formation of abandoned braid-bars by local aggradation and subsequent channel incision, and (3) the development of extensive, elevated bars during large floods, a process which produces a stable surface above the reach of more moderate flood events. Once abandoned and stabilized, the channels and islands can gradually

Figure 6.6. Floodplain preserved within the arroyo trench of the Rio Puerco near Gallop, New Mexcio

coalesce into a continuous floodplain surface. In most instances, the floodplain deposits created by braided-channel accretion are thinner and more irregular than those associated with other systems.

6.4. FLOODPLAIN SEDIMENTS

Fluvial sedimentologists have expended enormous effort during the past several decades in describing the range of sediments that are found in floodplains and the processes that form them. Their studies have resulted in a complex and somewhat confusing vocabulary that is, nonetheless, useful to characterize the variety of sedimentary deposits that exist. Of importance to our brief overview are *architectural elements* which essentially consist of distinct arrangements of sediment packages called facies (for greater detail, see Miall 1996, 2000). For our purposes, a *facies* can be defined as a unit of sediment that reflects a distinct environment of deposition and can be distinguished on the basis of distinct lithologic characteristics, such as composition, grain size, bedding, and sedimentary structures (Miall 2000). Facies may range in size from a single layer of sediment that is only millimeters thick to the succession of beds hundreds of meters thick.

The typical sequences in which facies are found in outcrop (e.g., along a stream bank) have been documented for a wide range of riverine settings to create models, called *facies models*, that are used to interpret the environment in which the deposition occurred. In other words, the nature of the sediment is used to determine the environmental setting in which it was deposited.

Historically, facies models have relied heavily on the vertical sequence of facies as described at one or more locations. It became apparent, however, that facies models could only be improved by using a three-dimensional view of the deposits. This three-dimensional thinking is incorporated in the concept of *fluvial (or alluvial) architecture*, a term that describes the geometry and the internal arrangement of channel and overbank deposits in a fluvial sequence (Allen 1965; Miall 1996). In other words, the alluvial architecture of a floodplain defines how the architectural elements, and the facies that they contain, fit together into a three-dimensional mosaic.

Most textbooks on sedimentology contain cross sections or block diagrams depicting the alluvial architecture of specific stream patterns. Examples for meandering, braided, and anastomosing channels are presented in Fig. 6.7. It could also be argued that the diagrams presented in Fig. 6.1 by Nanson and Croke (1992) represent alluvial architectural models by relating fluvial processes to deposit characteristics. Recognize, however, that while these models provide important insights into the types of deposits that are likely to be associated with floodplains of a given river system, the relationship between the pattern of sedimentation and channel planform is much more complicated than suggested by these block diagrams (Nanson and Croke 1992; Brown 1997). In fact, Nanson and Croke (1992) argue that the range of fluvial processes involved in floodplain

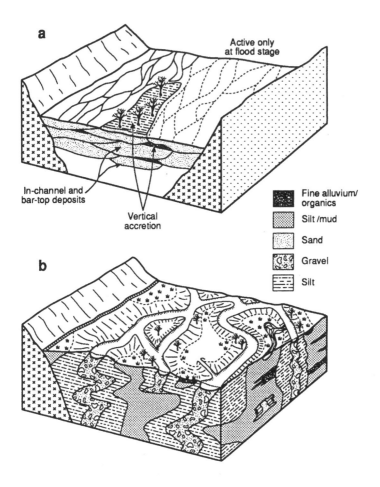

a

Active only
at flood stage

In-channel and
bar-top deposits

Vertical
accretion

Fine alluvium/
organics

Silt /mud

Sand

Gravel

Silt

b

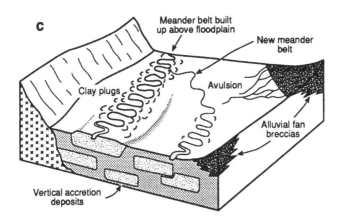

c

Meander belt built
up above floodplain

New meander
belt

Clay plugs

Avulsion

Vertical accretion
deposits

Alluvial fan
breccias

formation is so large, and the types of floodplains produced so diverse, that a new model is essentially required for almost every river system.

6.4.1.   Floodplain Deposits

In the previous chapter it was demonstrated that sediment-borne trace metals are often partitioned between distinct morphological (depositional) environments within the channel (e.g., point bars, pools, chutes, etc.). It follows, then, that variations in trace metal concentrations should also be observed between the sediments that comprise these deposits when they have become part of the floodplain sequence. This relationship between trace metal concentration and architectural elements is a fundamental component of the geomorphological-geochemical approach because it suggests that spatial variations in sediment-borne trace metal concentrations can be determined and/or predicted on the basis of floodplain stratigraphy and sedimentology. Moreover, it indicates that much can be gained in terms of reducing analytical costs by examining the types of architectural elements that exist in modern floodplains before attempting to document geographical patterns in trace metal concentrations within the floodplain deposits. Unfortunately, architectural elements have been classified by different investigators in different ways and there is currently no one universally accepted classification scheme for fluvial systems. Fielding (1993) even argued that those studying fluvial deposits may need to construct their own classification system to more accurately characterize the riverine environment being investigated.

Rather than examine the components of one classification system in detail, we will examine the most important architectural elements (or sediment packages) that are inherent in nearly every classification system. Those interested in specific methods of categorizing fluvial deposits should turned to more detailed discussions found in Miall (1978, 1985, 1996, 2000), Cowan (1991), and Brierley (1991), among others.

Most classification schemes broadly categorize floodplain architectural elements (deposits) into channel types and overbank types. Within meandering rivers, both types of deposits are often intertwined to create (in the ideal case) a fining-upward sequence produced by the combination of lateral migration and overbank flooding. From the base to the top of the floodplain are found coarse-grained channel lag deposits, a progression of large-scale, cross-stratified to finer grained, cross-laminated lateral accretion deposits, and, at the top, various fine grained deposits formed by overbank vertical accretion (Brown 1997).

Channel lag deposits commonly rest on an erosion surface and consist of coarse debris from which the fine particles have been winnowed. The debris is often

---

Figure 6.7. Idealized floodplain deposits associated with (A) a low-sinuosity sandy-braided channel, (B) an intermediate sinuosity anastomosing floodplain, and (C) a sinuous avulsion-dominated floodplain (From Brown 1997)

composed of partially decomposed organic materials as well as aggregates of sediment derived from slump blocks that have accumulated in the deepest parts of the channel. The lag deposits are laterally discontinuous and grade upward into sediments associated with lateral accretion.

Lateral accretion deposits, as noted in our discussion on floodplain formation, can form significant portions of the floodplain in some environments. The actual sedimentology of the deposits can vary significantly between rivers, depending primarily on the size of the sediment in transport and the variability of the discharge regime (Miall 1996). Generally, however, they are composed of sand or gravel, and are more coarse-grained than the overlying vertically accreted overbank materials. Perhaps the most diagnostic feature of lateral accretion is the occurrence of large-scale, gently dipping beds that correspond to successive increments of lateral deposition. The angle of dip on the beds, traditionally referred to as *epsilon cross-bedding*, can reach about 25° in fine-grained point bars (Fig. 6.2) (Miall 1996). In gravel-bed rivers characterized by higher width-depth ratios the dip of epsilon cross-bedding can be significantly less, and the bounding surface between the beds is often difficult to identify.

Epsilon type cross-bedding has also been observed to dip roughly parallel to flow, indicating bar growth in the downstream direction (Fig. 6.5). They generally indicate bar growth in the downvalley direction, such as might be associated with downstream migration of meander bends. In fact, it is not uncommon for point bar deposits to contain both lateral and downstream accretion deposits (Fig. 6.5d).

The predominant deposits at the top of our idealized floodplain sequence are formed by vertical accretion. As noted earlier, vertical accretion involves the deposition of suspended particles over a relatively flat-surface during a discrete overbank flood event. Thus, vertical accretion deposits often occur as sheet-like layers of sediment consisting of fine sands, silts, and clay. The layering, however, may be obscured by bioturbation, creating an indistinct, mass of fine-grained sediment. Where layering is visible, the sedimentology of the deposits can be quite variable, owing to differences in flood magnitude and small, but important, changes in depositional processes through time (Miall 1996).

Vertical accretion is not only produced by deposition over a flat surface, but is also associated with several morphological features found outside of the river channel, particularly along meandering streams. A commonly observed feature is a low ridge, referred to as a *natural levee*, that parallels the river (Fig. 6.8). Levees are usually highest immediately adjacent to the channel and slope gradually away from the channel toward the hillslopes. They are often much wider than one might think; natural levees along the lower Mississippi River are up to 3 km wide and as much as 9 m high (Farrell 1987). Along smaller rivers, levees are typically less than 100 m wide (Smith et al. 1989; Miall 1996). The observed morphology of levees is due to the retardation of flow velocity as the water leaves the channel, a process that causes the largest particles in suspension of settle out adjacent to the channel margins (Middelkoop and Asselmann 1994). The depositional pattern is

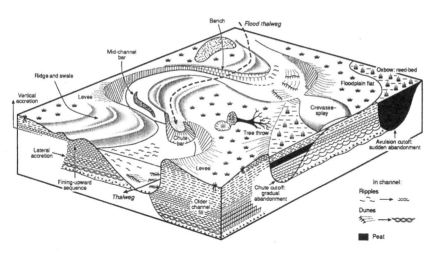

Figure 6.8. Deposits associated with specific landforms along a meandering river
(From Brown 1997)

most pronounced where the contrast between the flow velocity in the channel and
the adjacent floodplain are greatest.

Sedimentologically, levee deposits typically consist of a wedge-shaped accumulation of particles that are slightly larger than other overbank sediments and which
are deposited more rapidly (Fig. 6.8). Where they have not been extensively
disturbed by bioturbation, the sediment is often characterized by layers of ripple-
laminated, silty sands which may be covered by thin layers of fine silt and clay.
Each layer usually corresponds to a separate flood event, and the cross-laminations
are oriented away from the channel (Miall 1996).

Occasionally, natural levees may be broken by channels called *crevasse-channels*.
These channels flow across the floodplain obliquely to the main flow direction and
lead into *crevasse-splays* (Fig. 6.9), which are lobate or delta-like deposits formed
along the margins of the river (Fig. 6.8). Both crevasse-channels and crevasse-
splays are composed of sediment larger than the other overbank deposits, including
natural levees. In fact, particles within the coarsest units may be similar in size to
the sediment found in the main channel. The coarse-grained nature of these deposits
results from either (1) sediment carried in the upper portions of the river and transported directly onto the floodplain, or (2) the building of temporary ramps along the
channel margins which allow bedload to roll or saltate (bounce) onto the floodplain
during high flow conditions. Crevasse-channel deposits usually form ribbon shaped
bodies of sediment that can be a few tens to hundreds of meters wide (Miall 1996).
In contrast, crevasse-splays form lobate-shaped bodies of sediment that interfinger
at their margins with other finer-grained overbank deposits. The lobate pattern of
deposition is produced by flow expansion and a loss of stream power as waters
leave the confines of the crevasse-channels. Sedimentologically, splay deposits are

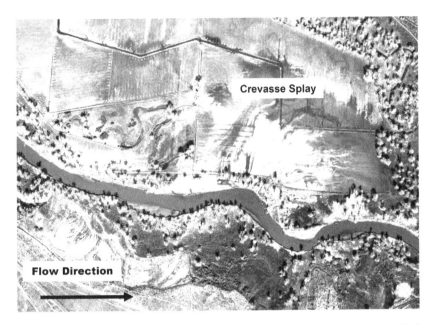

Figure 6.9. Crevasse splay formed along the Carson River, Nevada during the 1997 flood. Dark areas beyond splay are composed of fine-grained overbank deposits which buried the former floodplain surface

generally composed of trough cross-bedded and/or ripple-laminated fine to medium sand that decreases in size with distance from the channel. Although the deposits are generally considered to represent a form of overbank vertical accretion in the geomorphological literature, they may internally exhibit low-angle accretion surfaces indicative of growth by lateral progradation (Miall 1996). Moreover, it is not uncommon for a splay to be dissected by its associated crevasse-channel(s), creating various morphological forms that may evolve from one form to another (Fig. 6.10) (Smith et al. 1989).

Abandoned channel fills are common along a number of stream patterns. In the case of meandering and anastomosing rivers, channel fills are often associated with the deposition of sediment in meander bends that have been cut off from the rest of the channel. It is common for this process to involve the gradual filling of the mouth of the cutoff channel (i.e., the upstream end) by fine-grained sediment, leaving the downstream portion of the channel to become progressively less disturbed by high velocity flows during floods. The downstream end of the channel may also become filled as floodwaters back up into the channel during high stage. The net result is the formation of a pond that has been given various names, such as oxbow lake in the U.S. or billabong in Australia (Fig. 6.11). These lakes are subsequently filled with silts and clays deposited from suspension until the surface is level with the surface of the surrounding floodplain. The sediments which fill abandoned channels generally change in character through time (Fig. 6.11) and

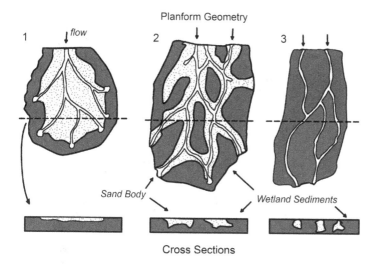

Figure 6.10. Types of crevasse splays present in the Cumberland Marshes anastomosing river system, Saskatcheqan, Canada. Splays may evolve from type one to type three (From N.D. Smith et al. 1989)

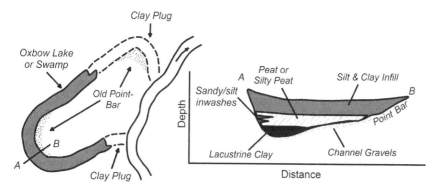

Figure 6.11. A meander cutoff and the nature of the subsequent fill (Modified from Erskine et al. 1982; Brown 1997)

can initially be rich in organic particles as a result of plant growth associated with the lacustrine environment. The net result of these processes is the formation of a clay-plug within the floodplain sequence.

Other types of channel fill are also present, such as those associated with an episode of channel aggradation or channel avulsion. Perhaps the most diagnostic feature of these channel fills is a concave-up erosional base that depicts the morphology of the former channel (Fig. 6.12). The cutbank slopes are often greater than 45° and may even be vertical. The nature of channel fill deposits can vary significantly depending on the grain size distribution of the river, but are often

Figure 6.12. Paleochannel fill, Indian Creek valley, Nevada

composed of a poorly sorted admixture of silt, sand, and gravel. Those associated with aggradation usually exhibit a fining-upward sequence.

Channel fills are also common along high-energy braided rivers. These deposits can be produced by a variety of processes, including the braided-channel accretion referred to earlier in the section on floodplain formation. The primary deposits include bedded gravels associated with bar formation, cross-bedded sands and gravels produced during bedform migration, and laminated mud formed during low-flow conditions (Brown 1997). Other characteristics of these deposits include abundant erosion surfaces produced by extensive and rapid scour, and fill associated with highly fluctuating discharge regimes.

In addition to the architectural elements produced by fluvial processes, some are associated with mass wasting processes operating on adjacent hillslopes. The primary deposit types include colluvial sediment resulting from the downslope movement of particles under the influence of gravity. These deposits tend to consist of coarse, angular particles and are primarily located along the valley margins. Debris flow deposits may also be present, particularly along high gradient streams in mountainous terrain, such as the southern Appalachians (Kochel 1987). The primary distinguishing characteristics of debris flows is that they are usually coarse-grained, matrix supported (i.e., the large clasts do not touch one another), and inversely graded (exhibiting a coarsening upward sequence) (Costa and Jarrett 1981; Costa 1984). They may also differ geometrically from alluvial deposits, forming narrow sediment packages characterized by abrupt, but non-erosional boundaries at their base.

6.5. TRACE METAL STORAGE AND DISTRIBUTION

Mass balance calculations have demonstrated that between 10 to 60% of the sediment delivered to a drainage network fails to reach the basin outlet (Walling et al. 2003). There is no question that the bulk of these sediments, and the contaminants that they contain, are deposited and stored within floodplains for periods ranging from decades to millennia (Table 6.1). The magnitude of contaminant storage depends on multiple factors including the proximity to the source of pollutants, the contaminant content of the sediment, the sediment-contaminant relationships (e.g., grain size to which contaminants are attached), the spatial extent of the floodplain, the mechanisms and rates of floodplain formation, and the hydrologic regime (Macklin and Lewin 1989; Marron 1989, 1992; Owens et al. 2001; Walling et al. 2003). As a result, downstream patterns of sediment-borne trace metal storage in floodplains is often highly variable. In many cases, however, it tends to increase with greater valley width and decrease with greater cross-sectional stream power (Fig. 6.13).

The change in storage along the Belle Fourche River of South Dakota (Marron 1992) provides an excellent example of the types of variations that can be expected, particularly in mountainous terrains. The Belle Fourche is contaminated by As enriched mine tailings that were released into Whitewood Creek, a tributary of the Belle Fourche. Marron (1992) estimated that approximately 13% of the 110×10^6 Mg of tailings discharge by the mine was deposited and stored within the Whitewood Creek floodplain. Immediately below the discharge area, storage of As-enriched sediment was limited because the channel was incised into the

Table 6.1. Typical percentages of total load stored in floodplain deposits

Locality	Storage (% of total suspend sediment load)	Reference
River Culm, UK	28 %	Lambert and Walling (1987)
River Waal, Netherlands	19 %	Middelkoop and Asselmann (1994)
River Severn, UK	23 %	Walling and Quine (1993)
Coon Creek, Wisconsin	50 %[1]	Trimble (1976)
Lower Mississippi	24 % of sediment input to river from 1880 to 1911	Kesel et al. (1992)
Amazon	10 %	Mertes (1990)
Lower Amazon	12.5 %	Meade (1994)
Belle Fourche, South Dakota	29 % of mine tailings	Marron (1992)
Ouse Basin	39 %	Walling et al. (1998)
Wharfe Basin	49 % (1995–1996)	
River Tweed	~ 40%	Owens et al. (1999b)

[1] % of eroded hillslope sediment

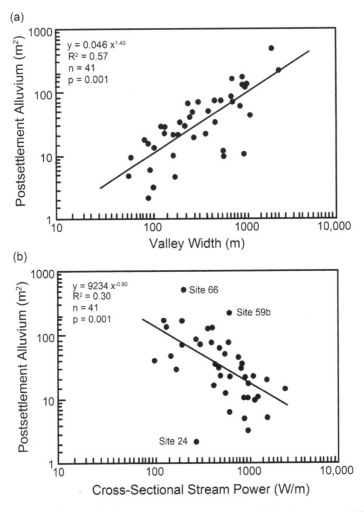

Figure 6.13. Relationship between cross-sectional area of alluvial storage and (A) valley width and (B) stream power for the Blue River, Wisconsin. Sediment-borne trace metal storage is likely to mirror the observed trends in alluvial sediment storage (From Leece 1997)

local bedrock and possessed a steep-gradient with only limited valley widths. This configuration made the reach an effective conduit for particle transport. Further downstream, floodplain width increases dramatically and sediment-associated As was deposited and stored in vertically accreted overbank deposits that locally extend for more than 100 m from the channel. Near its confluence with the Belle Fouche River, the floodplain of Whitewood Creek is affected by meander abandonment (cutoff) and channel incision. The abandoned channels are filled with As-contaminated sediments, most of which consist of nearly pure mine tailings.

In contrast to the downstream reaches of Whitewood Creek, the Belle Fourche River has experienced limited lateral migration and meander cutoffs were rare. Thus, the majority ($\sim$ 60%) of the contaminated sediment was stored in vertically accreted deposits, rather than in paleochannel fills. Moreover, about 40% of the total area consisting of contaminated, laterally accreted, point bar deposits was associated with just two, rapidly migrating reaches that constituted only about 20% of the total length of the investigated floodplain.

Although the two parameters are commonly correlated, it is important to recognize that contaminant storage is not analogous to contaminant concentration; the former represents the total amount of sediment-borne trace metals that occur along a given river reach, the latter the amount of trace metals per unit weight of sediment. Both are important parameters in site assessment. Concentrations, for example, will largely dictate potential ecological and human health affects and will be used to set action thresholds above which remediation may occur. Storage influences the areal extent of contamination and the volume of material that potentially requires remediation. Thus, it is not only important to document the quantity and geographical distribution of sediment-borne trace metals stored along the channel, but the spatial variations in concentration that exist within the floodplain deposits.

All of the factors that influence sediment-borne trace metal concentrations within channel bed sediment, described in the previous chapter, also influence concentrations within the floodplain. Thus, sediment-borne trace metal concentrations typically decrease downstream of point sources within floodplain deposits for the same reasons that they do within the channel sediments (i.e., as a result of hydraulic sorting, dilution, contaminant storage, biological uptake, and remobilization by biogeochemical processes). Within a given river reach, the magnitude of the variations in concentration closely reflects the processes of floodplain formation. Along rivers or river reaches characterized by rapid rates of channel migration and meander cutoff, the alluvial architecture of the floodplain is likely to be complex, resulting in significant spatial variations in trace metal concentrations. In contrast, less dynamic rivers characterized by floodplains consisting of vertically accreted overbank deposits tend to exhibit much less variability. Miller et al. (1998), for example, found that the architecture of the valley fill within the Carson River valley of Nevada was enormously complex along unstable channel reaches, consisting predominately of filled paleochannels inset into and overlapping vertically accreted deposits (Fig. 6.14). The filled paleochannels largely resulted from rapid channel migration, meander abandonment and subsequent channel infilling during flood events. This process is illustrated in Fig. 6.15. Between 1953 and 1965, three meanders were cutoff presumably during a flood event in 1963. During and following cutoff, the ends of the abandoned paleochannels were filled with sediment, progressively blocking the flow of water into the cutoff sections of the channels as illustrated in Fig. 6.15b. By 1991, each of the abandoned meanders had filled to the surface of the floodplain, forming a filled paleochannel that had become part of the valley fill. As described in Chapter 5, the Carson River was contaminated by the

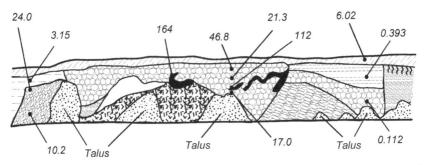

Hg Concentration in ppm

Figure 6.14. The alluvial architecture of the valley fill of the Carson River is locally complex. In this particular section, the complex nature of the stratigraphy led to Hg concentrations that vary by three orders of magnitude. Section is not to scale, but is approximately 75 m in length
(From Miller et al. 1998)

introduction of large quantities of Hg enriched mill tailings between approximately 1860 and 1890. The complexly structured floodplain (valley fill) downstream of the Comstock mills exhibited Hg concentrations that varied by over three-orders of magnitude (Fig. 6.14), greatly exceeding that found within the channel bed deposits. In contrast, deposits formed by overbank vertical accretion processes along relatively stable reaches possessed significantly less variation in Hg concentrations. Nonetheless, Hg variations in the vertically accreted deposits were still greater than those observed within the channel bed sediment.

A significant question arises from the data collected along the Carson River and other riverine environments as to why floodplain materials tend to exhibit much higher variations in trace metal concentration than the adjacent channel bed deposits. After all, the floodplain is created by the same river and with materials transported in the river channel. The answer rests on several characteristics of the floodplain that differ significantly from that of the channel bed. These include (1) a wider range of sediment sizes and compositions that vary over relatively short distances, (2) longer-sediment residence times and a wider range of deposit ages, (3) less sediment reworking and homogenization between flood events, and (4) more extensive post-depositional mobility.

6.5.1.   Grain Size Variations

We know from earlier discussions that trace metal concentrations are closely linked to particle size; thus, much of the variation in contaminant concentration observed within the floodplain can be explained by differences in the quantity of fine-grained sediment contained within the deposits. In most cases, floodplain sediment is characterized by larger percentages of silt and clay, and higher trace metal concentrations, than the axial channel. Moreover, the range of grain size variations observed within a floodplain tends to be significantly greater than within the channel. The greater

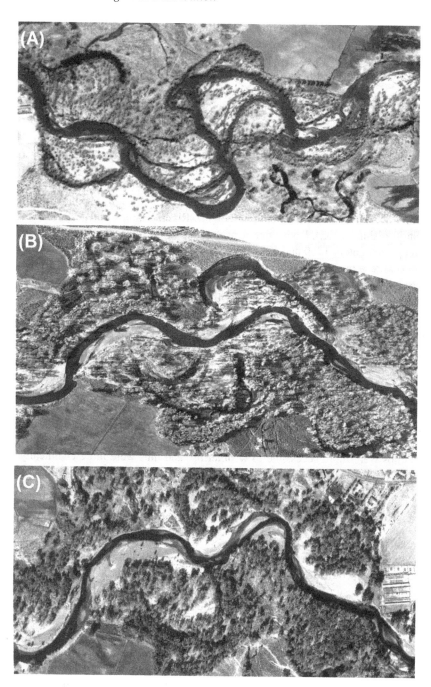

Figure 6.15. Abandonment and filling of meanders along the Carson River, Nevada. Photographs taken in (A) 1953, (B) 1965, and (C) 1991. Cutoff presumably occurred during a flood in 1963 (After Miller et al. 1998)

variability results from the deposition of sediment by multiple processes, ranging from lateral accretion on point bars, to vertical accretion during overbank flooding, to lacustrine deposition in oxbow lakes. Grain size is also influenced by rapidly changing hydrologic conditions on the floodplain surface caused by topographic irregularities as well as changes in vegetation and other roughness elements. The net result is that changes in unit sedimentology can be extremely abrupt, occurring over distances of a few centimeters. These sedimentological changes can, in turn produce abrupt alterations in trace metal concentrations.

While the style of floodplain deposition typically depends on system dynamics, the grain size of the contaminated particles introduced to a river system can also affect depositional processes and the resulting alluvial architecture of the floodplain. Bradley and Cox (1987), for example, examined the spatial distribution of trace metals within floodplains of the Rivers Hamps and Manifold in North Staffordshire (U.K.). Both catchments contained Cu mines that were worked between approximately 1760 and 1790. However, tailings piles within the Manifold catchment were reprocessed, reducing the size of the sediment introduced to the river system. The reprocessing also depleted the sand-sized sediment fraction of trace metals. The fine sediment released within the Manifold catchment accumulated on downstream floodplains primarily through vertical accretion. However, the coarse sediment within the Hamps basin were deposited primarily as gravel splays on the floodplain surface. Thus, the nature of the contaminated debris and the transporting processes dictated the distribution of the trace metals within the floodplain. Moreover, Bradley and Cox found that pedogenic alteration of the sediment, which controlled the chemical remobilization of the trace metals, was grain size dependent. Only the surface of the fine-grained vertical sequences described for the Manifold were affected by pedogenic (weathering) processes. In contrast, fine particles serving as the matrix of gravel splays within the floodplain of the River Hamps were more accessible to weathering, and thus were more chemically mobile.

6.5.2.   Residence Time and Deposit Age

In contrast to sand-sized or smaller *channel bed* sediment which may reside in a given location for a period of a few years or less (Walling et al. 2003), the residence time for sediment within a *floodplain* is usually on the order of centuries to millenia. The net result is that floodplains contain sediment of widely varying age, representing deposits formed almost instantaneously (e.g., during an individual flood event) to periods exceeding several decades (e.g., the filling of an abandoned channel or the progressive deposition on laterally migrating point bars). Floodplains, then, are composed of sediments that record variations in contaminant transport rates over long periods of time, and the trace metal content of the sediment reflects changes in the contaminant load that occur over the period of deposition. A nice illustration of the importance of deposit age on sediment-associated trace metal concentration was provided by Rang and Schouten (1989). They were able to constrain the

Table 6.2. Variations in metal concentrations (mg/kg) as a function of deposit age

Location	Age (years)	Zinc (mg/kg)	Cadmium (mg/kg)	Lead (mg/kg)	Copper (mg/kg)
Former Channel	350	154	1.0	84	27
Former Channel	120	8000	24.7	1406	67
Former Channel in gravel pit	10	1660	27.0	530	160
Modern Channel	1	1600	21.8	390	90

From Rang and Schouten (1989)

depositional age of selected channel fills preserved within the floodplain of the River Meuse and found that trace metal concentrations varied with the timing of deposition (Table 6.2). More specifically, they were able to deduce that the peak in metal pollution most likely occurred in the middle of the 19th century when Pb and Zn mining was prevalent in the basin.

Variations in transport rates and contaminant concentrations also occur over much shorter time periods than described by Rang and Schouten (1989). Figure 6.16 provides a case in point from the Carson River valley of Nevada. The light colored sediment is Hg enriched mill tailings which were deposited during one or more episodes of overbank flooding. The tailings are surrounded both above and below by less contaminated materials indicating significant temporal variations in the nature

Figure 6.16. Floodplain along a tributary to the Carson River, Nevada. Light colored sediments in mid-section of floodplain are composed of nearly pure mill tailings

of the sediment load transported along the Carson River during construction of the floodplain. These types of variations are often associated with differences in influx rates from the contaminant source as well as from uncontaminated upland areas which influence the degree to which the contaminated materials are diluted. In many cases, inter-event variations in dilution reflect spatial variations in precipitation intensity and magnitude, particularly in arid climatic regimes. The tailings rich deposits shown in Fig. 6.16, for example, could be the product of localized rainfall concentrated over tributary basins containing mining and milling wastes, whereas the less contaminated deposits may have been produced by either larger scale events, or local storms positioned over relatively uncontaminated tributary basins. Other factors, among many, that may influence the contaminant load through time include (1) fluctuations in contaminant releases from point sources as a result of changes in production, production technologies, or environmental regulations, (2) changes in the nature of the released contaminants as a result of changing industrial processes or the use of waste treatment techniques, and (3) changes in channel dynamics that increase or decrease sediment transport rates.

6.5.3. Sediment Mixing and Homogenization

In Chapter 5, it was suggested that temporal variations in trace metal concentrations within the channel bed are often limited between flood events. This conclusion may seem to run contrary to the observation that the geochemistry of the suspended sediment load during specific flood events can vary significantly as a result of differences in discharge, precipitation patterns, contaminant influx rates, etc. At least part of the answer to this apparent contradiction is that with the exception of armored rivers, the channel bed is characterized by repeated episodes of scour and fill associated with rising and falling flow conditions. The effect of this process is to mix sediment-borne trace metals transported during multiple events, thereby homogenizing concentrations within the channel bed sediment. In the case of floodplains, however, the deposited sediments are less likely to be reworked by a subsequent flood before they are semi-permanently incorporated into the sequence of floodplain deposits. As a result, variations in trace metal content of the sediment is not only dependent on the timing and age of the deposits, but on the degree to which the sediment is reworked and homogenized following their initial deposition.

6.5.4. Post-Depositional Processes

Up to this point we have primarily been concerned with variations in sediment-borne trace metal concentrations which are associated with depositional processes and patterns. It is possible, however, for spatial variations in trace metal content resulting from deposition to be altered by post-depositional processes. One of the most thoroughly documented examples was provided by Taylor (1996) for a 6.5 km reach of the River Severn in the U.K. The studied reach of the River Severn was characterized by a dynamic floodplain in which meanders were cut off and subsequently filled. Taylor argued that historically deposited floodplain units should possess higher trace metal contents than older units as a result of metalliferous mining activities in the 19th century. In order to determine if such vertical trends existed, he measured the metal content as a function of depth in eight separate cores extracted from different depositional environments, but which appeared from the surface topography to be similar. The obtained geochemical data revealed that metal concentrations did not follow a predictable pattern as defined by deposit depth or age, nor did metal concentrations show any systematic variation with unit grain size or organic matter content. This led him to suggest that the observed vertical variations in concentrations were produced by post-depositional processes including metal leaching, the downward translocation of contaminated fine-grained particles, and the redristribution of metals by the lateral migration of metal enriched groundwaters (Table 6.3).

Taylor's (1996) work holds several implications that are worth highlighting. First, it is unlikely that vertical trends in trace metals can be adequately determined from a single sampling or coring site. Rather sampling strategies should be based on the extraction of multiple cores (Macklin et al. 1994) from the various depositional environments. The use of multiple cores will provide a more accurate understanding of the spatial variations in sediment-borne contaminants, particularly within dynamic floodplains. Second, the magnitude through which post-depositional processes have modified the depositional patterns in trace metal concentrations need

Table 6.3. Summary of controls on trace metal patterns within floodplain deposits of the River Severn

Depositional environment	Grain-size	Trace metal pattern invertical profile			
		Pb	*Zn*	*Ni*	*Cu*
Braided Outwash	Gravel	↔	↑ A	↑ A	↓ L, T
Paleochannel fill	Gravel overlain by silt	↑ A	↓ L, T, Q	↓ L, T, Q	↔
Paleochannel fill	Gravel overlain by silt	↔	↔	↔	↔
Paleochannel fill	Sandy-silts	↑ M	↑ M	↔	↑ M
Inchannel & overbank sed.	Sandy-silts	↑ M	↑ M	↑ M	↑ M
Paleochannel Fill & Overbank sed.	Gravel overlain by silt	↓ M, Q	↔	↔	↔
Paleochannel fill	Silts	↑ M	↔	↔	↑ M
Paleochannel fill	Gravel overlain by silt	↓ M, Q	↔	↑ M	↑ M

From Taylor (1996) *Trends in trace metal concentration*: ↑ – upward increases; ↓ down profile increases; ↔ – no recognizable trend. *Potential causes of trace metal distribution in profile*: A – atmospheric; L – leaching; T – translocation; Q – lateral migration; M – mining to be taken into account. In the most severe cases, linkages between floodplain deposits and trace metal concentrations may be weak or non-existent. Where post-depositional processes are significant, an alternative approach is to subdivide and map the floodplain units on the basis of age, such as whether they were deposited before and after the onset of pollution (Davies and Lewin 1974; Macklin 1996). In doing so, the most severely contaminated sediment is delineated, and areas in need of more detailed investigation can be identified (Macklin 1996). Moreover, it reduces sampling and analytical costs because it eliminates the need for an all-inclusive sampling program that targets all of the alluvial deposits that exist along the valley floor.

6.6.   OVERBANK SEDIMENTS

The distribution of sediment-borne contaminants in vertically accreted overbank deposits has received considerable attention in recent years, so much so that they deserve a more detailed examination than was provided for the other types of floodplain deposits. The interest in these fine-grained sediments is driven in part by their potential to be used as a medium for regional geochemical surveys, providing information on both natural variations in elemental concentrations as well as those associated with human activities (Ottesen et al. 1989; Bølviken et al. 1996). Perhaps more importantly, the semi-continuous accumulation of overbank sediment often produces deposits in which the age of the sediment varies semi-systematically with depth below the floodplain surface. Thus, the analysis of core samples from multiple depths provides a historical record of the influx of trace metals to the river, and enables an assessment of the changes in water and sediment quality through time.

Vertically accreted deposits are most prevalent along low-gradient channels such as those found within the coastal plain of the eastern U.S. or the low-lands of the U.K. In these areas, floodplains generally exhibit poorly developed point bars, few if any abandoned cutoff channels, and minimal rates of lateral movement. Howard (1996) notes that many of these rivers are also characterized by (1) minimal bedload, (2) a low, wide floodplain relative to the width of the channel, (3) erosionally resistant banks produced by cohesive, heavily vegetated sediment, (4) low valley gradients, and possibly (5) a low frequency of major floods. Fine-grained overbank deposits can, however, be found locally along almost any river valley where lateral channel migration has not reworked the floodplain deposits (Ritter et al. 1973; Nanson and Young 1981). It follows, then, that the potential to reconstruct pollution histories exists along a great number of river systems. In order to effectively construct these histories, it is important to understand the processes involved in overbank deposition and the patterns of deposition that typically occur. Moreover, we need to recognize the typical variations in trace metal concentrations that occur across the floodplain as a result of the depositional processes associated with overbank flows. We will tackle both of these subjects below, before examining vertical patterns in metal concentrations.

6.6.1. Depositional Rates and Patterns

Determining the rates and patterns of deposition on a floodplain is no easy task. Consider, for example, the difficulties of gaining access to a flat surface that may be kilometers wide and inundated by a thin layer of water that is incapable of floating a boat. Moreover, the infrequent and random occurrence of overbank flooding makes direct sampling of floodwaters difficult, particularly where the recurrence of floodplain inundation is measured in years. Nevertheless, the rates and patterns of floodplain deposition have been reliably determined by (1) using data from sediment traps, (2) quantifying the thickness and areal extent of flood deposits immediately after the event, and (3) documenting the depth to which dated horizons within the floodplain have been buried.

Traditionally, rates of vertical floodplain accretion have been expressed in terms of the thickness of sediment deposited during a specific time interval (e.g., mm/yr). In the case of contaminant studies, thickness (or depth) may not be the most appropriate form in which to express depositional rates because we are often concerned with the total amount of contaminated debris at the site (perhaps as part of a sediment budget) and changes in the total influx of the contaminant to the river through time. Both of these parameters are more effectively expressed in terms of the volume or weight of sediment accumulated per unit area per unit time (e.g., $kg/km^2/yr$). Unfortunately, rates of sedimentation expressed in terms of volume are more difficult to obtain because most of the techniques used to date specific layers within a floodplain provide age estimates for a given point (e.g., from a core or site of exposure). Thus, while average sedimentation rates measured in terms of sediment thickness can be easily calculated, volume or weight estimates require extrapolation of the age data over the area and an understanding of the three-dimensional geometry of the deposits above the dated horizon. An exception is the use of fallout radionuclides which allow depositional patterns to be determined over broad areas of the floodplain by measuring nuclide inventories in cores obtained from multiple locations (see Chapter 3 and Walling et al. 1996, for details).

Regardless of whether sedimentation rates are measured in terms of sediment thickness or volume, any interpretation of the data must consider the potential affects of changing channel conditions on overbank processes. Rumsby (2000), for example, demonstrated that vertical accretion rates estimated at Broomhaugh Island within the Tyne River basin (U.K.) using the two different methods (thickness versus volume) led to vastly different conclusions. In this particular case, the divergent results were associated with channel bed incision during the 17th century that produced a 76% increase in channel capacity. The enlargement of the channel, combined with elevated bank heights, resulted in fewer overbank flows capable of depositing sediment over the floodplain. As would be expected, the volume of vertically accreted floodplain deposits subsequently declined. However, reduction in the areal extent of the floodplain surfaces available for deposition resulted in an increase in the annual thickness of the vertically accreted sediment where the floodplain existed. The point is that without having examined the alterations in channel configuration, and the change in the volume of sediment deposited on the floodplain through time, it would have appeared from the thickness measurements alone that the amount of sediment stored within the floodplain had increased substantially during the 17th century.

Assuming that the influences of channel change can be filtered out, most investigations have shown that rates of vertical accretion when averaged across the width of the floodplain are dependent on how much sediment is yielded from a watershed, the grain size and concentration of sediment suspended within the flood waters, and the magnitude and frequency of overbank flooding (Walling and He 1998; Ritter et al. 2002). These factors are controlled by the basin's geological and morphometric characteristics as well as the region's climatic regime (Lajczak 1995; Brown and Quine 1999). One would guess, for example, that rates of floodplain sedimentation would be higher in a basin underlain by easily erodible shales than in one composed of resistant sandstones if the other controlling factors are similar. The effects of human activities, such as changes in land-use, have also been shown to significantly influence rates of vertical accretion; in fact, increases in sedimentation rates of more than an order of magnitude have been documented for some rivers in response to human activities (Fig. 6.17; Table 6.4) (Orbock Miller et al. 1993).

While average depositional rates are commonly reported in studies of floodplain formation and evolution, it is clearly recognized that vertical accretion tends to

Figure 6.17. Barn located on floodplain along Drury Creek, southern Illinois. Floor has been buried by vertically accreted sediment produced by clear-cutting in the headwaters of the basin near the turn of the 20th century

Table 6.4. Summary of overbank floodplain sedimentation rates

Locality	Period (YBP)	Sedimentation rates (cm/year)	Reference
Lower Ohio	7000-6000	0.027	Alexander and Prior (1971)
	3000-2000	0.060	
Washita River, Oklahoma	1760-1000	0.080	Gross et al. (1972)
Ti Valley, Oklahoma	2350-1850	0.020	Ferring and Peter (1982)
	720-320	0.200	
Jackfork Creek, Oklahoma	3555-1091	0.039	Vehik (1982)
Fourche Maine Ck, Oklahoma	4500-1940	0.050	Galm and Flynn (1978)
Carnegie, Oklahoma	3200-2600	0.640	Hall and Lintz (1984)
Delaware, Oklahoma	2750-1900	0.650	Ferring (1986)
Galena Watershed, Wisconsin	3610-~ 150	0.040	Knox (1987)
Horseshoe Lake, Illinois	6969-350	0.12	McGovern (1991)
	350-Present	3.05-5.08	
Delaware River, PA	3468-2718	0.115	Ritter et al. (1973)
	5590-5198	0.077	
Coon Creek, Wisconsin	1853-1938	2.35	Trimble (1976)
Belle Fourche, South Dakota	~ 1876-1978	2.0	Marron (1987)
Tonalli River, NSW, Australia	50-Present	0.26	Harrison et al. (2003)
	~ 240-50	0.07	
Upper Axe Valley, UK	1864-1982	0.24	Macklin (1985)
	1858-1863	1.60	
	1711-1857	~ 0.46	
	1670-1710	~ 0.88	
Yorkshire Ouse Basin[1]	~ 1750-P	0.85	Hudson-Edwards et al. (1999b)
Catterick Reach	1525-~ 1750	0.11-0.31	
	~ 1750-P	0.45	
Myton-on-Swale	1350-~ 1750	0.49-0.70	
	3275-1350	0.10-0.12	
	~ 1750-P	0.24	
Myton-On-Swale	1120-~ 1750	0.51-0.63	
	1165-P	0.19-0.25	
Beal	610-1165	0.13-0.16	
	2145bc - 610	0.047-0.05	

Updated from Ferring (1986) and Orbock Miller et al. (1993)
[1] Age in calendar years

decrease with distance from the channel (Kesel et al. 1974; Pizzuto 1987; Howard 1996; Walling et al. 1996). Walling and He (1998), for example, demonstrate that sedimentation rates based on ^{137}Cs inventories were several times higher immediately adjacent to the channel than at greater distances from the river along selected

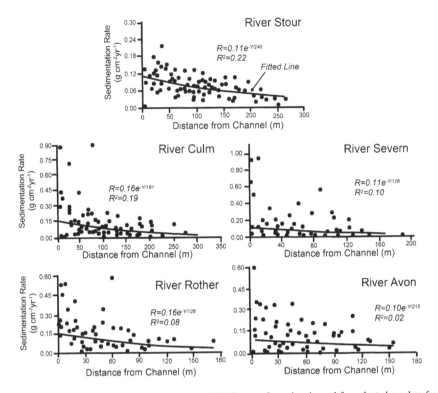

Figure 6.18. Change in sedimentation rate with distance from the channel for selected reaches from five different rivers in the U.K. (From Walling and He 1998)

reaches of five different river systems in the U.K. (Fig. 6.18). Moreover, they suggested that the decline in sedimentation can be described by:

$$(1) \qquad R = R_0 e^{\frac{-Y}{Y_0}}$$

where R is the sedimentation rate, Y is distance from the channel, R_0 and Y_0 are constants and e is the base of the natural logarithm. The general decline in depositional rates with distance from the river is consistent with the transport of sediment by particle diffusion. As described by Allen (1985), the diffusion process is driven by differences in the flow conditions between the channel and the floodplain once overbank flooding has occurred. More specifically, flow within the channel is relatively fast and deep in comparison to that over the floodplain. The abrupt change in flow produces turbulent eddies along the margins of the channel which result in the transfer of momentum from the channel to the floodplain as flow velocities change (Fig. 6.19). In addition, the concentration of suspended particles is higher within the channel than over the floodplain because of the channel's deeper, faster flowing waters. The differences in suspended sediment concentration sets up a gradient that leads to the diffusion of sediment from zones of high

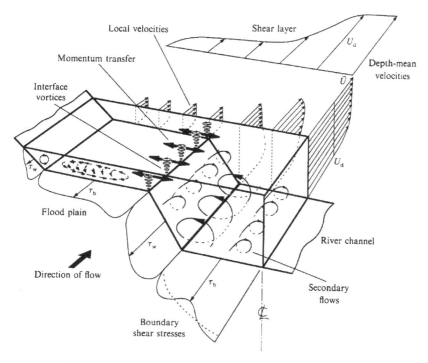

Figure 6.19. Hydraulics of overbank flow (From Shiono and Knight 1991)

to low concentration, or away from the channel and onto the floodplain surface. Sediment transferred away from the channel by diffusion may overload the relatively incompetent flows over the floodplain, causing some sediment to be deposited (Pizzuto 1987). The deposition of these suspended particles, particularly coarse particles, close to the channel further reduces the sediment concentration, leading to an outwards decrease in suspended sediment concentration across the floodplain. The channel, then, serves as a linear source of suspended particles, which through the action of turbulent eddies are diffused from the channel and spread across the adjoining flat-lying surface in such a way that depositional rates are greatest along the channel margins (Allen 1985).

The transfer of sediment from the channel to the floodplain by particle diffusion is closely related to the channel's planimetric shape. In general, diffusion associated with eddy turbulence is most pronounced along straight channel reaches where the flow is predominantly oriented downvalley. Where the flow is oriented oblique to the channel, sediment may be transported onto the floodplain by convection and/or macroturbulent flow. The latter can often produce coarse-grained overbank deposits such as the sand and gravel feature shown in Fig. 6.20 (Ritter 1988). Convection, however, is much more common and, when associated with clockwise and anticlockwise helicoidal flow cells, can lead to considerable overbank deposition

Figure 6.20. Overbank sand and gravel deposited (light colored materials) on the floodplain of the
Gasconade River near Mt. Sterling, Missouri in December 1982. Lobe of sediment is 1–2 m thick and
100 m wide (Photo courtesy of D.F. Ritter)

on the inside of meander bends (Marriott 1996). In fact, vertical accretion can
be dominated by convective flow. Along the Waipaoa River of New Zealand, for
example, Gomez et al. (1999) argued that advective flow, rather than diffusion,
was the dominant process responsible for transporting sediment onto the floodplain.
As a result, sediment accumulation on the floodplain varied downstream as a
function of floodplain width because it influenced the transport capacity of the
overbank flows.

The data provided in Fig. 6.18 demonstrates that while a general decrease in
sedimentation rates with distance from the channel exists, there is considerable
variation in the data. The observed variation indicates that other factors are also
important in controlling local rates of deposition. One of the more significant of
these is the microtopographic form of the floodplain surface and its influence on
water depth. Topographic variations usually lead to higher rates of vertical accretion
within localized depressions. It is, perhaps, obvious that the enhanced rates of
deposition within a depression could result from the long-term ponding of water in
the feature after a flood, allowing all of the sediment in the overlying water column
to settle out. What may be less obvious is that the amount of sediment deposited
over a given area is proportional to the total mass of sediment within the overlying
water column. Thus, the deeper the water, the higher the total mass of available
sediment and the higher the rates of sediment accumulation, provided that the other
controlling factors are constant (Walling and He 1998).

Microtopographic variations, combined with changes in channel orientation, can
also affect depositional rates by influencing flow patterns across the floodplain

surface. For example, in some instances depositional rates in depressions are low as a result of the movement of relatively deep waters capable of transporting sand sized or larger particles over the surface (Walling et al. 1996). The complex influence of microtopography on depositional patterns is well illustrated by the work of Steiger et al. (2001). They found that overbank deposition was limited in low-lying areas because relatively deep flows were capable of transporting the fine-grained particles out of the area. Deposition was also limited within zones residing at relatively high elevations. Here, however, the low sedimentation rates were related to shorter periods of flood inundation. The peak deposition rates, then, were associated with parts of the floodplain that were characterized by moderate inundation times and transport capabilities.

A final factor which is known to significantly influence floodplain deposition is the type, density, and age of riparian vegetation. In fact, one of the most important functions of the riparian zone is thought to be its ability to keep sediment and sediment-borne contaminants from either entering the channel from the adjacent hillslope or from moving downstream (Castelle et al. 1994; Naiman and Décamps 1997; Steiger and Gurnell 2002). Surprisingly few field studies, however, have been conducted on the effects of vegetation on overbank deposition, in part because of the difficulties of assessing the complex influence of vegetation on overbank flow hydraulics and depositional processes. Those that have been conducted primarily document differences in depositional rates and styles between distinct vegetational zones during individual flood events (Miller et al. 1999; Steiger and Gurnell 2002) (Fig. 6.21). Clearly, then, this is an area in need of additional study.

6.6.2. Geographical Patterns in Contaminant Concentrations

Grain size of overbank deposits within a given river reach is of course closely linked to that of the suspended sediment from which they are derived. In most cases, however, there is considerable variability in the grain size distribution of vertically accreted deposits across the floodplain, particularly where the suspended particles encompass a wide range of sizes. Perhaps the most commonly observed trend is for grain size to decline with distance from the channel (Fig. 6.22), a trend which is often attributed to the more rapid settling of coarser particles from the overbank flows. When combined with higher sedimentation rates in these same areas, the coarser sediment results in the natural levees described earlier. Actual studies of the variations in grain size across the floodplain have found, however, that there is considerably more variability in the trend than may have been expected. Part of this variation can be attributed to the deposition of clay-sized particles from standing water within depressions on the floodplain surface following a flood event. It is also attributed to the simple fact that fine-grained particles are often transported as aggregates (Droppo and Ongley 1994; Walling et al. 1996). In other words, the effective size of the suspended sediment (which describes the dimensions of the particles actually in motion) is considerably larger than the same material measured by using standard analyses (which rely on the use of chemical reagents

(A)

(B)

Figure 6.21. Overbank sediments deposited during the 1997 flood along the Caron River, Nevada in (A) zones of low-lying vegetation and (B) coarsely vegetated areas. The nature of the vegetation had a significant affect on both depositional styles and deposit thickness

to disperse the particles prior to size determination) (Fig. 6.23). The significance of these findings is that substantially larger quantities of fine grained particles can be deposited immediately adjacent to the channel than would normally be expected.

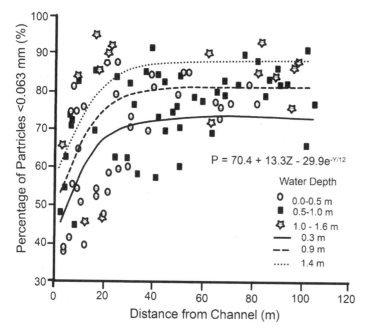

Figure 6.22. Change in grain size distribution as a function of water depth and distance from the channel of the Dorest Stour, U.K. (From Walling et al. 1996)

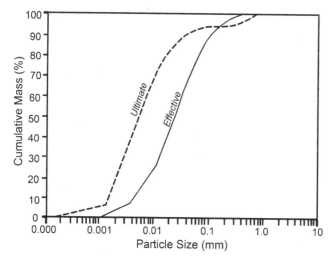

Figure 6.23. Comparison of the effective and ultimate grain-size distribution for suspended sediment from the River Culm, U.K. The effective particle size represents the actual size of the particles in motion; the ultimate is the size of particles following disaggregation by chemical reagents (From Walling et al. 1996)

The geographical pattern in trace metal concentrations within vertically accreted deposits is closely linked to grain size and the way in which the metals are partitioned between the various grain size fractions. Put more simply, metal concentrations tend to mimic spatial patterns in the grain size distribution of the overbank deposits, the highest concentrations being associated with those sediments in which the metals are concentrated. Miller et al. (1999), for example, found that Hg concentrations in overbank deposits formed during the 1997 flood along the Carson River of Nevada roughly varied as a function of the percentage of sediment <63 μm in size within the deposits. The highest concentrations of silt and clay were found immediately adjacent to the channel and at significant distances from the river. They noted, however, that not all of the Hg was attached to fine-grained particles, but was associated with larger and denser Hg–Au and Hg–Ag amalgam grains. Thus, metal concentrations were a function of both grain size and particle density. In contrast, metal concentrations were found to be elevated immediately adjacent to the River South Tyne because the majority of the sediment-borne metals were associated with the sand-sized sediment fraction of the suspended load rather than the finer materials (Fig. 6.24). In yet other situations, such as the River Derwent, spatial variations in metal concentrations are minimal, in part because of a more uniform distribution of the sediment sizes with which the metals are associated (Fig. 6.25) (Bradley and Cox 1990; Macklin 1996).

Another factor influencing metal concentrations in vertically accreted overbank deposits is the frequency with which the floodplain surface is inundated by flood waters. Table 6.5, for example, demonstrates that along the River Muese there is a strong correlation between Zn, Cd, Pb, and Cu concentrations and the recurrence interval over which the deposits are inundated (Rang and Schouten 1989). The influence of inundation on metal concentrations is primarily related to two factors. First, the more frequent inundation of the floodplain surface allows for a greater opportunity to accumulate contaminated sediment, particularly fine-grained particles which are most likely to be transported in suspension during small to moderate floods. Second, deposition on topographically higher morphological units occurs only during major rainfall-runoff events. As demonstrated in Chapter 4, trace metal concentrations within the suspended load are likely to change as a function of discharge because of alterations in sediment source. Thus, topographically higher surfaces may exhibit either higher or lower concentrations than more frequently inundated surfaces, depending on the nature of the changes in trace metal concentrations with discharge. An example comes from the decrease in concentration proposed by Marron (1987) for portions of the floodplain along the Belle Fourche River of South Dakota. Here As concentrations were higher in vertically accreted deposits immediately adjacent to the river than in more distant, and topographically higher, overbank deposits. The spatial differences in concentration resulted from the dilution of the suspended load as the result of erosion and incorporation of relatively clean alluvial deposits into the river during major flood events. Because the higher surfaces could only be inundated during these large floods, the concentrations of the overbank deposits reflected the lower concentrations that occurred at high flow.

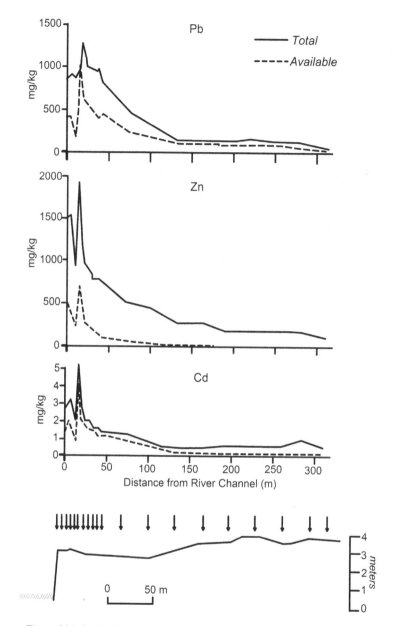

Figure 6.24. Lead, Zinc, and Cadmium concentrations in surficial floodplain sediments of the River South Tyne, England (Modified from Macklin 1996)

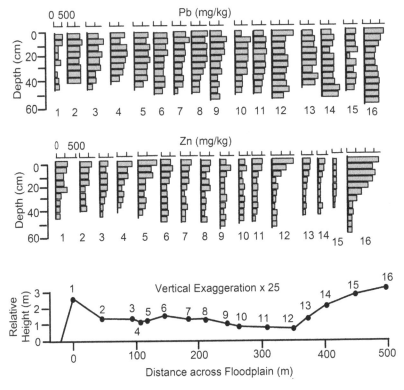

Figure 6.25. Lead and Zinc concentrations in surficial and subsurface floodplain sediments from the River Derwent, Derbyshire, England (Adapted from Bradley and Cox 1990)

6.6.3.   Documenting Pollution Histories

It is an unfortunate fact that monitoring data for most contaminated rivers is either absent or of poor quality prior to a site assessment. Where sound monitoring data are available, the records are usually of short-duration, encompassing a period of a few

Table 6.5. Influence of flood recurrence interval on metal concentrations along the Meuse River

Recurrence interval	Zinc[1] (mg/kg)	Cadmium[1] (mg/kg)	Lead[1] (mg/kg)	Copper[1] (mg/kg)
2 year	888 (681)	8.2 (4.8)	264 (140)	75 (35)
5 year	760 (601)	5.4 (3.9)	270 (130)	54 (47)
10 year	332 (350)	2.2 (2.1)	118 (116)	28 (67)
50 year	220 (203)	1.6 (1.3)	76 (76)	22 (13)
250 year	314 (139)	1.4 (0.5)	71 (58)	20 (12)

From Rang and Schouten (1989)
[1] mean (standard devation)

years to perhaps a few decades. The difficulty of extracting long-term trends from short-term data has received considerable attention in the environmental literature (e.g., Lomborg 2001). It is now clearly recognized that data extrapolation either forward or backward in time from a given period can lead to erroneous results. For example, examine the changes in river length in England and Wales which are in fair to good condition in terms of their chemical quality between 1990 and 2001 (Fig. 6.26). If we were to utilize data only between 1994 and 1996, it would appear that the chemical quality of the riverine waters had declined. In contrast, data collected between 1994 and 1999 indicates that there has been little change in water quality (either positively or negatively) in the monitored rivers. However, during the entire monitoring period water quality has, on average, improved. The point is that temporal trends in water and sediment quality are closely linked to the duration and interval over which the samples are collected. The longer and more closely spaced the data, the better the results are likely to be in terms of identifying temporal trends in water or sediment quality.

One means of increasing our understanding of changes in sediment quality through time is to construct histories of contaminant influx to the river using the chemical content of the overbank deposits as a proxy for the chemical quality of the suspended sediments. This not only provides an understanding of the changes in chemical health of the river where monitoring records are absent, but allows short-term monitoring records to be placed into an historical context of the changes in sediment geochemistry over periods of decades to millennia. Moreover, the persistent nature of metals in riverine environments often leads to the situation in which the contaminating source is no longer present. The development of pollution histories for multiple sites along the river allows for the assessment of contaminant sources and their relative importance as a contributor of pollutants for discrete

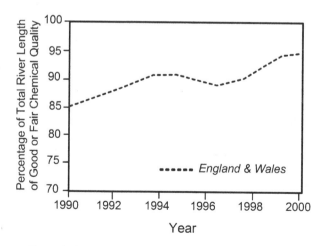

Figure 6.26. Changes in the chemical quality of rivers in England and Wales from 1990 to 2001 (Modified from U.K. General Quality Assessment; Graphs from UKEA 2003)

time intervals, even if the contaminating sources have long since vanished. Such analyses are of particular significance in areas where the polluter pays principle is in affect. Finally, the data may provide highly valuable insights into whether water and sediment quality is recovering on its own, without any human intervention, once the contaminant source has been mitigated.

While the procedure for documenting pollution histories has been widely used by the academic community, it is less-frequently applied in site assessments. This primarily stems from the fact that many site managers do not understand how such data can be used, and because the process can be time consuming and expensive. In general, the development of pollution histories from overbank deposits involves three steps: (1) a stratigraphic analysis in which the relative age and mode of deposition are determined, (2) the absolute age dating of the deposits or specific intervals within the deposits, and (3) the geochemical analysis of the sediment for metal content. These data are then combined to produce a chronology of the changes in sediment geochemistry for one or more trace metal.

The tasks outlined above are well illustrated in the analysis of metal contamination of the upper Vistula basin in southwestern Poland. In this case, Macklin and Klimek (1992) analyzed three to four alluvial units at multiple sites to document the history of pollution for the basin. The sampled materials included sediment from the channel bed, overbank sediments associated with levees and inset fills confined by a dike constructed near the turn of the 19th century, and overbank deposits that occurred outside the dike-protected area which were deposited prior 1800. All of the identified deposits were dated using radiocarbon and other techniques and analyzed for Pb, Zn, and Cd. The age and geochemical data were then correlated to provide a general history of contamination within the basin that could be linked to anthropogenic activities. The results showed that sediment quality progressively declined through time within the upper Vistula basin. The decline was particularly pronounced after about 1850. Initially, contamination was limited to sites located immediately downstream of mining districts within the Przemsza River, a tributary to the Vistula. However, during the past 100 years contamination from mining and industrial activities became much more widespread, affecting the entire basin.

The work by Macklin and Klimek (1992) illustrates that the spatial distribution of Pb, Zn, and Cd in the basin, and the identification of contaminant sources, was made possible by developing pollution histories for multiple sites along the Vistula and its tributaries. Moreover, the older (pre-industrial) sediment could be used to assess the natural background concentrations of the metals within the overbank deposits, thereby allowing the extent of anthropogenic metal enrichment in the upper Vistula to be determined. In fact, Macklin and Klimek (1992) argued that the pre-industrial overbank deposits provided a sound medium for background determinations because the sediment was derived from the same catchment and were of similar grain size and composition.

The sampling increments used in analysis of pollution histories varies widely as a function of the study's objectives, the rates of sedimentation, and the available financial resources, among other things. Hudson-Edwards et al. (1998) note that previous investigations fall into two general categories, those conducted at a coarse-scale in which broad temporal trends in pre-, peak, and post-metal contamination are determined (e.g., Davies and Lewin 1974; Rang and Schouten 1989; Sear and Carver 1996), and those performed at a fine-scale whereby the sampling interval is measured in centimeters, and the contaminated units are subdivided into discrete periods (e.g., Klimek and Zawilinska 1985; Taylor and Lewin 1996; Harrison et al. 2003). In both instances, variations in metal enrichment are commonly derived, at least in part, from sediment cores which can be taken back to the laboratory, described in detail, and then sampled using clean-lab procedures to reduce the possibility of cross contamination.

The use of core data is based on the fact that sediments are deposited over the floodplain surface in near horizontal layers, and that age of the sediment increases with depth. Stratigraphers refer to these assumptions as the principles of original horizontality and superposition, respectively. If taken without question, the principles suggest that the choice of sampling locations is infinite, particularly when one is standing on a nearly flat, floodplain surface. Unfortunately, deposition across the floodplain surface is rarely uniform or continuous as we learned earlier. Thus, the deposition of overbank sediment at one site during an event may or may not occur just a few meters away. This fact prompted Macklin and Klimek (1992) to argue that variations in sediment-borne trace metal concentrations need to be based on sediment age (rather than depth below the floodplain surface) and that multiple cores are needed for an accurate assessment of a given site.

Two important assumptions inherent in the development of pollution histories are that (1) the observed changes in sediment-borne trace metal concentrations are a reflection of the metal content within the suspended sediment at the time of deposition, and (2) the primary depositional signatures correlate with the influx of trace metals to the river. Both assumption can be violated at a site and therefore require validation through other means. In the case of the first assumption, for example, Bradley and Cox (1987) found along the River Hamps that Cd had been leached over distances of several tens of centimeters. Similarly, Hudson-Edwards et al. (1998) demonstrated using detailed speciation and mineralogical data from the River Tyne that trace metals often migrate over depths of a few centimeters within the floodplain. The magnitude of migration in both areas was metal specific; the more tightly bound elements such as Pb exhibited the least mobility. Moreover, it is now clear that the degree of mobility can be enhanced by environmental change, such as channel incision which leads to a drop in water table elevations and the weathering of the exiting minerals (e.g., sulfides). It follows, then, that the results from fine-resolution data are more likely to be affected by post-depositional processes than those associated with coarser-resolution data sets.

Violation of the second assumption (that the primary depositional signatures reflect contaminant loading rates) can be caused by an exceedingly large range of processes that cannot be exhaustively listed here. Nonetheless, changes in channel or floodplain morphology and their effects on the frequency of overbank inundation often influence temporal trends in trace element content. To illustrate this point, let us look at Fig. 6.4. The illustration demonstrates that the rate of deposition on the floodplain decreases through time because progressively larger floods are required to inundate the surface. In essence, then, the pollution history preserved at the site is being censored in that it is recording the metal content of increasingly larger floods. Thus, if trace metal content of suspended sediment varies as a function of flood magnitude, as is commonly the case, semi-systematic changes in sediment-associated metal concentrations may reflect changes in bank height rather than alterations in metal influx. Short-term variations in metal concentrations can also be associated with flood magnitude. Harrison et al. (2003), for example, showed that localized changes in Pb, Zn, As, and Cu concentrations within sediment cores along the Tonalli River of New South Wales, Australia were coincident with heavy rainfall-runoff events that followed periods of drought (Fig. 6.27). The two most significant changes in metal content (which occurred between 1935–1953 and 1993–2000) were associated with La Niña.

Another commonly encountered problem when interpreting geochemical data from overbank deposits is that grain size and organic matter content of the analyzed sediment vary throughout the materials. Although the observed variations are often random, progressive changes in both grain size and organic matter content are not uncommon. This is particularly true for organic matter content which tends to increase toward the floodplain surface as a result of the decomposition of plant materials. In the previously mentioned study of the Tonalli River, Harrison et al. (2003) found that the percentage of both fine sediment and organic matter increased

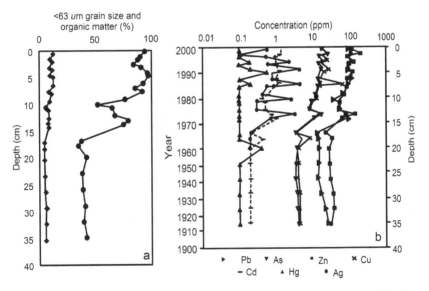

Figure 6.27. Core data from a floodplain along the Tonalli River, Australia showing (A) changes in organic matter and < 63 μm sediment content with depth below surface, and (B) metal concentrations versus sediment depth and age (Modified from Harrison 2003)

toward the top of the core (Fig. 6.27). The abrupt change in grain size and composition were attributed in this case to the affects of a dam on the flow hydraulics. In other areas, changes in grain size and composition have been attributed to alterations in channel form such as channel incision (see, for example, Macklin and Klimek 1992). Where variations in either sediment size or composition occur, their effects may be minimized by using one or more of the normalization methods described in the preceding chapter before analyzing the data.

6.7.   SEDIMENT AND CONTAMINANT SOURCE DETERMINATION

An important aspect of a site assessment is to identify the source of the observed contaminants. It is, of course, impractical to develop a remediation strategy if it is unclear where the contaminants originated, and if they are continuing to enter the system, perhaps on a sporadic basis. In other situations, source identification may be required to address legal issues associated with the polluter pays principle. Historically, source identification has relied heavily on spatial patterns in metal concentrations within selected media (e.g., water, channel and floodplain sediments, and biota). In Chapter 5, for example, it was shown how downstream changes in trace metal concentrations within the channel bed sediment could be used to identify potential points of metal influx to the river. This basic concept was expanded upon in the previous section of this chapter by demonstrating how the construction of pollution histories at multiple downstream sites could be used to identify contaminant sources, and the changes in sources through time. In many situations, however, spatial trends in concentration data alone do not provide a sound understanding of contaminant source. The confusion occurs because anthropogenic pollutants can be derived from multiple point and non-point sources that create complex and often overlapping geographical patterns in metal values. Moreover, at low contaminant loadings, the concentration of metals from anthropogenic sources may be masked by those found naturally in rocks, sediment, and soils. This is particularly true where concentrations in the natural geological materials are highly variable, such as in the case of basins possessing mineralized rocks.

The problems associated with the use of concentration data to identify contaminant sources have led to the development of various physical and geochemical tracers and tracer methods. In general, a *tracer* represents some unique characteristic of the source material (contaminated or otherwise) that allows it to be distinguished from other sediment in the basin. Tracers are now widely accepted as an important tool for identifying contaminant sources and for distinguishing the relative contributions of sediment and sediment-borne contaminants to river systems. The types of tracers used for environmental purposes encompass a wide and growing range of parameters, as illustrated in Table 6.6. The methods through which they are applied fall into two broad categories: (1) multivariate geochemical fingerprinting methods (primarily used to identify non-point sources of sediment and sediment-associated contaminants), and (2) isotopic methods (used to track specific substances from

Table 6.6. Summary of common tracers used to identify sediment sources

Tracer	Example of application
Physical Properties (e.g., grain size distribution)	Fenn and Gomez (1989), Stone and Saunderson (1992), Kurashige and Fusejima (1997)
Mineralogical Composition	Grimshaw and Lewin (1980)
Mineral-Magnetics (e.g., magnetic susceptibility, anhysteretic remanent magnetization, saturation isothermal remanent magnetization)	(Oldfield et al. (1979), Yu and Oldfield (1989, 1993), Caitcheon (1993), Walden et al. (1997), Dearing (2000), Slattery et al. (2000)
Radionuclides (e.g., ^{210}Pb, ^{137}Cs, ^{7}Be)	Walling and Woodward (1992), Wallbrink and Murray (1993), Olley et al. (1993)
Geochemical Composition (e.g., trace metals)	Forster and Walling (1994), Passmore and Macklin (1994), Collins et al. (1997a, 1997b), Bottrill et al. (2000), Owens et al. (2000)
Stable Isotopes (e.g., Pb, Rb, Sr, Sm, Nd)	Salomans (1975), Douglas et al. (1995, 2003a, 2003b), Borg and Banner (1996), Miller et al. (2002)
Rare Earth Elements (e.g., Nd, Pu)	Morton (1991)
Organic substances	Hasholt (1988), Owens et al. (2000), Douglas et al. (2003a)
Biogenic Properties	Peck (1973), Brown (1985)

both point and non-point sources). The importance of both of these techniques for site assessments is likely to grow because with ever more stringent controls on contaminant releases to river systems, a more forensic approach will be required to assess the complexities of the contaminant source-to-sink relationships (Macklin et al. in press).

6.7.1.   Non-point Source Multivariate Fingerprinting Methods

Although the earliest studies date back several decades, the use of geochemical fingerprinting techniques to indirectly determine the relative contributions of sediment from distinct non-point sources within a basin has grown enormously in popularity and sophistication in recent years. The basic premise of the approach is that the complex processes involved in the erosion, transport, and deposition of sediment ultimately result in alluvial deposits that represent mixtures of sediment derived from definable source areas within the watershed. It is then theoretically possible to characterize both the source areas and the alluvial sediment for a suite of parameters and quantitatively compare their parameter characteristics to unravel the proportion of material that was derived from each source type. The approach seems to be most beneficial where there is a general lack of water quality or upland erosion data that can be used to directly identify the predominant sediment sources. Moreover, it is important to recognize that while the above fingerprinting techniques have historically been used to identify the primary sources of sediment within a basin, there is no reason that they cannot be used to elucidate the primary sources of sediment-borne contaminants from either point or non-point sources. In fact, ongoing studies by the Commonwealth Industrial Research Organization (CSIRO) in Australia are currently using this type of geochemical fingerprinting to identify the predominant sources of both sediment and nutrients in coastal catchments to the Great Barrier Reef.

Figure 6.28 provides a schematic diagram of the components typically involved in the fingerprinting approach as related to the erosion of sediment from upland areas. A significant question that must be addressed is how to define the source areas within the basin. To date, source areas have primarily been delineated according to land cover category (Collins 1995; Slattery et al. 1995; Wallings and Woodward 1995; Russell et al. 2001; Miller et al. 2005) or by the geological units which underlie the watershed (Fig. 6.28) (Collins et al. 1997a; Walling et al. 1999; Douglas et al. 2003a; Miller et al. 2005). Other subdivisions have also been used, however, including differentiation by tributary sub-catchments, and the depth from which the sediment originates (Collins et al. 1997a; Bottrill et al. 2000).

Once sources have been defined, it is necessary to determine the property or properties which most effectively distinguish between sediment from each of the source types. Early studies commonly relied on a single parameter, but more recent investigations have clearly demonstrated that erroneous sediment-source area associations may occur when only a single fingerprinting property is used (Collins and Walling 2002). As a result, more recent investigations generally utilize multiple parameters to generate a *composite fingerprint*. The types of parameters

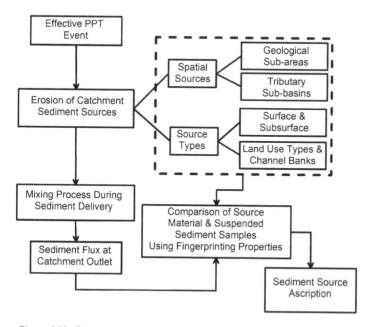

Figure 6.28. Components of the source ascription process (Modified from Collins and Walling 2002)

that may be used to form a composite fingerprint include all of those shown in Table 6.6. Thus, a composite fingerprint can consist of a single group of properties (e.g., different measures of mineral-magnetics), or can be derived from different groups of properties (e.g., mineral-magnetics, acid-soluble trace metals, and short-lived radionuclides). Unfortunately, exactly which of the properties or groups of properties will be most effective for discriminating between the sediment sources in a particular basin cannot be determined prior to the analysis. It is therefore common for a wide range of physical and chemical properties to be measured for each of the sediment sources and for various types of multivariate statistical techniques, such as discriminate analysis, to be subsequently applied to the data to define the properties to be included in the composite fingerprint (Collins et al. 1997a, b).

In most instances the relative contributions of sediment to the channel or alluvial deposit is estimated using a multivariate sediment mixing model. The precise nature of these sediment mixing models varies between investigators and has become extremely complex (see, for example, Palmer and Douglas, in press). In general, however, the relative contribution from each sediment source is estimated by using a computerized iterative process in which the relative proportion of sediment from each source type is adjusted until the differences between the measured parameter values in the source materials and the deposits is minimized for all of the utilized fingerprinting properties (Yu and Oldfield 1989; Collins et al. 1997a; Miller et al. 2005).

The fingerprinting/mixing model approach has most often been applied to suspended sediments (e.g., Collins et al. 1997a, 1998). However, it has also been used to determine the provenance of floodplain deposits (Collins et al. 1997a, c, d; Owens et al. 1999; Rowan et al. 1999; Bottrill et al. 2000), lake and reservoir materials (Yu and Oldfield 1993; Foster and Walling 1994; Kelley and Nater 2000; Miller et al. 2005), and estuarine sediments (Yu and Oldfield 1989). One of the advantages of applying the technique to floodplain deposits is that sampling and dating of discrete intervals within the deposits as a function of depth allows the changes in sediment and contaminant sources through time to be determined (Fig. 6.29).

6.7.2.   Isotopic Tracing Methods

The benefits of using isotopic tracers to decipher the source and dispersal pathways of sediment-borne trace metals in floodplain and other alluvial materials has only recently been recognized. Nonetheless, it is clear that the approach has the potential to dramatically alter the means through which contaminated rivers are characterized. *Isotopes* are atoms of an element that have a different number of neutrons, and therefore, different masses. An isotope is designated by its mass number which is placed as a superscript to the left of the element's two-letter symbol (e.g., ^{206}Pb). The mass number is equal to the number of protons and neutrons in the atom. Isotopes may be broadly classified as being stable or radioactive (unstable). *Radioactive isotopes* are those that spontaneously disintegrate into other isotopes. The rate of

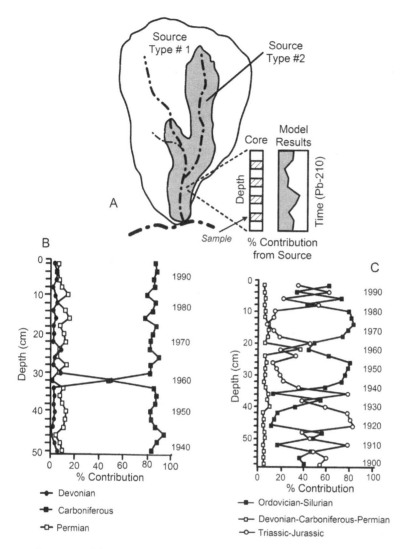

Figure 6.29. (A) Schematic diagram of source ascription for floodplain deposits made using a fingerprinting/sediment mixing model approach. Sources are determined for each subsample from the dated core. Lower diagrams represent contributions of sediment in floodplain deposits from bedrock units underlying the Exe (B) and Severn (C) basins. Variations in the contributions are thought to reflect changes in intrinsic and extrinsic controls on sediment generation such as extreme hydrologic events and changes in land use (B and C from Collins et al. 1997)

decay is not affected by changes in environmental conditions and is time-invariant. In contrast to radioactive isotopes, *stable isotopes* do not decay to form other isotopes over geological time scales. It is possible, however, that they were produced by radioactive decay, in which case they are referred to as being *radiogenenic*.

The most extensive application of isotopic tracers has been in hydrology where they have been used to determine sources of water and solutes, characterize water and contaminant flow paths, and assess the biogeochemical cycling of nutrients and other substances (Kendall and McDonnell 1998; Kendall and Doctor 2004). The application of isotopes to the dispersal of sediment-borne trace metals has been slower to develop, but is now expanding. The growth in isotopic analyses is due in part to: (1) advances in geochemical instrumentation including Thermal Ionization Mass Spectrometry (TIMS) and high resolution Inductively-Coupled Plasma Mass Spectrometry (ICP-MS), and (2) the realization that isotopic tracers can lead to a more detailed understanding of contaminant sources and cycling than could be obtained by examining concentration data alone.

With regards to trace metals, the isotopes of Pb have been most extensively utilized for environmental studies. They have been shown to be particularly effective for determining both the potential sources of Pb contamination, and their relative contributions to the particulate and dissolved load of river systems (Petelet et al. 1995, 1996; Allégre et al. 1996; Luck and Othman 1998; Steding et al. 2000), alluvial floodplain and terrace deposits (Hudson-Edwards et al. 1999; Miller et al. 2002; Villarroel et al. 2006), lacustrine sediments (Shirahata et al. 1980; Petit et al. 1984; Moor et al. 1996), soils (Gulson et al. 1981; Steinmann and Stille 1997; Hansmann and Köppel 2000), and peat bogs (Shotyk et al. 1998; Weiss et al. 1999) as well as within air and aerosols (Chow and Johnstone 1965; Ault et al. 1970; Chow and Earl 1972), ice (Rossman et al. 1993), and tree rings (Marcantonio et al. 1998; Wartmough et al. 1999). A number of isotopes of the transition metals and semi-metals also hold great promise, but to date have been underutilized. These include Cr, Mo, Cu, Zn, Se and Ge (Kendall and Bullen 2003).

Most of the Pb isotopic tracing studies have focused on the use of four isotopes. They include ^{204}Pb, which is stable and has no known radiogenic parent, and ^{206}Pb, ^{207}Pb, and ^{208}Pb which represent the daughter products of ^{238}U, ^{235}U, and ^{232}Th, respectively (Table 6.7). The use of Pb as a tracer of anthropogenic contamination rests on the fact that it is obtained from ore deposits and subsequently

Table 6.7. Selected properties of Pb isotopes used in environmental studies

Lead (Pb)
Abundance
^{204}Pb – 1.48 %
^{206}Pb – 23.60 %
^{207}Pb – 22.60 %
^{208}Pb – 52.30 %
Half-life and parent isotope
^{204}Pb (stable)
^{238}U to ^{206}Pb; $^{1}/_{2}$ life of 4.5×10^{9} years
^{235}U to ^{207}Pb; $^{1}/_{2}$ life of 7.1×10^{8} years
^{232}Th to ^{208}Pb; $^{1}/_{2}$ life of 1.4×10^{10} years

incorporated into manmade items without its original isotopic composition being affected (Ault et al. 1970; Graney et al. 1995). Because Pb ores typically exhibit high Pb/U and Pb/Th ratios in comparison to most other rock types (Hansmann and Köppel 2000), the Pb within ore bodies tends to be relatively unradiogenic and possess an isotopic ratio that can be distinguished from other geological materials. It is this unique isotopic signature that allows the isotopes to serve as tracers of anthropogenic and natural Pb in the near surface environment.

A significant benefit of using Pb isotopes as a tracer of sediment-borne contaminants is that the material's isotopic composition can be used to estimate the contribution of Pb in the sample from a specific source. The upper reaches of the Rio Pilcomayo of southern Bolivia, for example, has been extensively contaminated by trace metals including Pb as a result of more than 450 years of mining activity at Cerro Rico. The Pb isotopic composition of the pre-mining (uncontaminated) alluvial floodplain and terrace deposits were found to plot along a linear trend when $^{206}Pb/^{207}Pb$ ratios were plotted against $^{206}Pb/^{208}Pb$ ratios (Fig. 6.30). These types of linear trends can be interpreted as mixing lines in which the samples that form the trend are composed of a mixture of Pb from two distinct sources. The isotopic composition of those sources possesses isotopic compositions that plot at the opposing ends of the line and are often referred to as isotopic end-members. In this case, the source of the lead appears to be Pb found in a subset of Mesozoic rocks that underlie the basin, and Pb occurring in a variety of geological media including Ordovician rocks (Fig. 6.30). In marked contrast to the pre-mining alluvial deposits, samples collected in 2002 along the modern channel bed plot along a separate

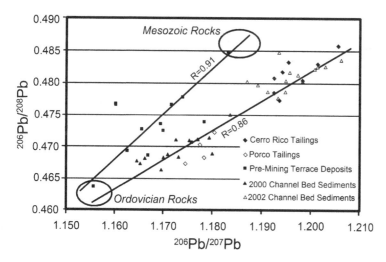

Figure 6.30. Comparison of Pb isotopic ratios in the modern channel bed sediments of the Rio Pilcomayo, Bolivia to uncontaminated, pre-mining terrace deposits. Change in isotopic ratios reflect the influx of waste materials of two distinct isotopic compositions from the mines at Cerro Rico

trend (Fig. 6.30). The difference in slope between the two trends indicates that there has been a change in the dominant source of Pb to the river since mining operations began. The new Pb source, found in the modern channel deposits, exhibits $^{206}Pb/^{207}Pb$ and $^{206}Pb/^{208}Pb$ ratios that are consistent with those found in mine and mill tailings from Cerro Rico. Thus, the isotopic data demonstrated that the change in Pb geochemistry was most likely due to the influx of mining and milling wastes to the river. Moreover, the isotopic data could be used to semi-quantitatively

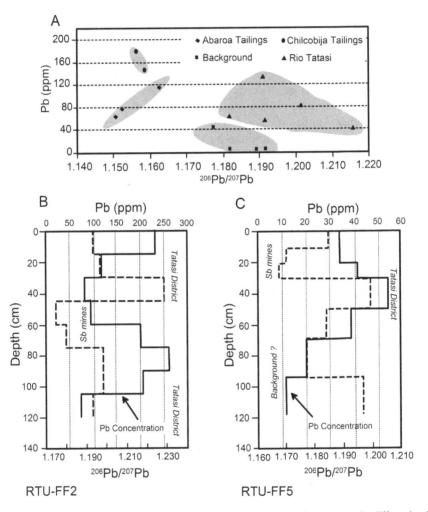

Figure 6.31. (A) Lead from three mining districts and the background sources can be differentiated on the basis of total Pb concentration and Pb isotopic ratios within the Rio Tupiza basin, Bolivia, (B) differences in Pb concentrations and isotopic abundances can be used to assess changes in Pb contributions from the Chilcobija and Abaróa Sb mines and the Tatasi polymetallic ore deposits within the floodplain (From Villarroel et al. 2006)

determine the proportion of the Pb found naturally within the underlying rocks and that were derived from two different types of ore (consisting of Porco and Cerro Rico type Pb fingerprints) from the mines. Similar methods have also been used to determine the sources of Pb in floodplain deposits from different mining operations along the Rio Tupiza of southern Bolivia (Fig. 6.31) (Villarroel et al. 2006), and the relative contributions of mining and milling wastes to agricultural soils developed in floodplain and terrace deposits along the Rio Pilcomayo and its tributaries.

In the case of the Rio Pilcomayo, the relative contributions of Pb to channel bed sediments were determined using a sediment mixing model similar to that described in the previous section. Other investigators, however, have used analytical methods when only two or three Pb sources are important. In this latter case, the relative contributions of lead can be determined using the expression:

$$
(2) \qquad Pb_{con}\% = \frac{[(\frac{^{206}Pb}{^{207}Pb_{bg}} - \frac{^{206}Pb}{^{207}Pb_{meas}}) x 100]}{(\frac{^{206}Pb}{^{207}Pb_{bg}} - \frac{^{206}Pb}{^{207}Pb_{con}})}
$$

where $Pb_{con}\%$ is the percent Pb from the contaminant source, $^{206}Pb/^{207}Pb_{bg}$ is the isotopic ratio of the background Pb, $^{206}Pb/^{207}Pb_{meas}$ is the ratio measured within the sediment sample, and $^{206}Pb/^{207}Pb_{con}$ is the ratio or fingerprint associated with the contaminant source (after Kersten et al. 1997). While the equation is presented using $^{206}Pb/^{207}Pb$, any other Pb isotopic ratio could be used in the above expression.

6.8. PHYSICAL REMOBILIZATION

Up to this point we have been examining the storage of sediment-borne trace metals within floodplains, and the means to determine where those trace metals origi- nated. From this perspective the floodplains serve as contaminant sinks, reducing the amount of trace metals which are exported to downstream reaches. It is also true, however, that contaminated particles can be remobilized from floodplains by physical processes. In fact, in many riverine environments floodplain deposits serve as an important, if not the predominant, source of sediment-borne contaminants to the aquatic environment (Macklin and Klimek 1992; Miller et al. 1998). The importance of floodplains as a contaminant source is particularly pronounced where the original influx of trace metals to the system has been reduced by environmental regulation, changes in industrial and waste stream technologies, or alterations in anthropogenic activities, and where bank erosion is significant. The latter follows from the fact that the primary method through which sediment-borne trace metals are remobilized is by the varied processes of bank erosion. Quantification, then, of the regulating effects of floodplains on contaminant transport not only requires an understanding of how much sediment will be deposited on the floodplain in a given period of time, but the rate at which it will be remobilized by erosion.

6.8.1.   Bank Erosion Processes

Bank erosion has attracted considerable attention from both the geomorphological and engineering community (Thorne 1982; ASCE 1998; Simon et al. 1999) in part because it can negatively impact agricultural land while adding considerably to the sediment load of the river. Moreover, through its controls on channel width, it exerts a significant influence on other channel processes (Ritter et al. 2002). The outcome of these detailed investigations over the past two decades is that bank erosion is rarely, if ever, the product of a single process, but is accomplished by multiple, interacting processes that are dependent on the local geomorphic setting. In general terms, however, bank erosion is conducted by either corrasion or mass wasting. *Corrasion* represents the grain by grain removal (entrainment) of particles from the bank surface by forces generated within the river flow. In this case, erosion is primarily dictated by the flow velocities in the near-bank environment as well as by the composition of the bank materials and the type, density, and rooting structure of riparian vegetation. *Mass wasting* is the movement of material, either in bulk or as an individual grain, under the influence of gravity. It ranges from the downslope movement of material at rates so slow that it is difficult to discern (commonly called *creep*) to the rapid failure of a sediment mass along a planar surface and its fall to the channel bed.

Bank erosion is often the product of both corrasion and mass wasting. For example, many floodplains are characterized by a fining-upward sequence as described earlier in the chapter; thus, non-cohesive sand and gravel deposits are overlain by more cohesive silts and clays. In this type of situation, corrasion often produces an overhanging mass of cohesive sediment along the banks because the underlying non-cohesive materials are eroded faster than the cohesive sediment (Thorne and Tovey 1981). These overhanging masses of bank materials (Fig. 6.32), called *cantilevers*, then fail under the influence of gravity and drop to the surface below (Fig. 6.33). A slightly different process involving both corrasion and mass wasting occurs where vertical fractures, called *tension cracks*, exist in the floodplain sediment. In this case, lateral erosion of the bank near the water surface by corrasion intersects a tension crack, thereby initiating the movement of material along a fracture plane (Fig. 6.33) (Ritter et al. 2002). This type of failure process has been given various names including slab failure (Hagerty 1980), earth fall (Twidale 1964), soil fall (Brunsden and Kesel 1973) and shallow slip (Thorne 1982).

The likelihood that mass wasting will occur is dependent on the balance between the shear force promoting the downslope movement of bank material and the shear strength of the sediment which resists that motion. The shear force is produced by gravity and indicates the ability of a sediment mass to move. Actually, it is equivalent to that portion of the gravitational force acting on a particle or mass of sediment which is oriented parallel to the slope (Fig. 6.34). The steeper the slope and the heavier the mass of material, the greater the shear force and the more likely that the material will move into the channel. The shear strength (S) of the bank material can be derived from three components contained within the Coulomb equation:

(3) $S = c + \sigma' \tan \varphi$

Figure 6.32. Corrasion of non-cohesive bank sediments, followed by failure of the overlying cohesive materials along Kingston Creek, Nevada

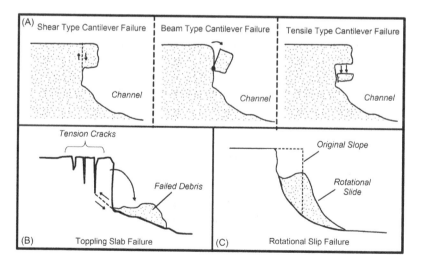

Figure 6.33. Types of bank failure commonly found along rivers with cohesive sediments (Adapted from Thorne 1982; Ritter et al. 2002)

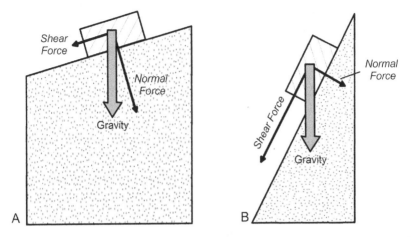

Figure 6.34. Changes in shear and normal force components as a function of slope. Arrow length correlates to force magnitude

where c is cohesion, σ' is the effective normal stress, and φ is the angle of internal friction, a parameter which describes the material's overall frictional characteristics. The effective normal force in Eq. 3 is dependent on two elements, the total normal force (σ) and the pore pressure (μ). The latter represents the pressure exerted by air and/or water within the void spaces in the bank materials. Its affects on mass wasting is expressed in relation to effective normal stress:

$$(4) \quad \sigma' = \sigma - \mu$$

The total normal stress is that component of the gravitation force acting perpendicular to the slope (Fig. 6.34). Its magnitude influences the capacity to hold the material together; thus, the greater the total normal force the greater the shear strength. An examination of Fig. 6.34 shows how an increase in slope tends to increase the shear force while reducing the effective normal stress and therefore shear strength of the material.

If all of the force acting perpendicular to the sloping surface was absorbed by grain to grain contacts within the bank materials, then the effective and normal stress would be equivalent. In most instances, however, the pore spaces between the grains are partially or completely filled with water. The degree to which they are filled influences the normal force which holds the mass together. Within the unsaturated zone (above the water table) the pores will contain both air and water, the latter attached to particles by capillarity. The attached water increases the weight of the soil and alters the pore pressure, μ, in such as way that it is negative ($\sigma' = \sigma - (-\mu)$). In this case, pore pressure actually increases the effective normal stress. Below the water table, pore pressure is positive, lowering the effective normal stress ($\sigma' = \sigma - \mu$).

In light of the effects of water on the effective normal stress it should be clear that the soil moisture condition of the bank materials serves an extremely important control on mass wasting processes. Where saturated banks are found in poorly drained, cohesive sediment, positive pore pressures can lead to bank failure. This tends to be particularly true in high steep banks after prolonged rainfall events or in the case of rapid drawdown in river levels as the result of dam closure or water diversions.

Although the effects of water are most pronounced under saturated conditions, changing moisture regimes may have a significant effect on the shear strength of the bank sediments even when the materials are unsaturated. During precipitation events, for example, infiltrating water tends to decrease *matrix suction*, defined as the difference between the air pressure and the water pressure in the unsaturated pore spaces. A decrease in matrix suction reduces the apparent cohesion of the bank materials and, therefore, its shear strength. In some cases, the effects of the infiltrating rainwaters on matrix suction and shear strength may be sufficient to induce bank failure (Simon and Curini 1998; Simon et al. 1999).

Another way in which water can promote bank erosion is through a process referred to as sapping or piping. This process tends to be most pronounced where the bank materials exhibited a layered structure consisting of cohesive sediment that overlies non-cohesive materials. Usually, the non-cohesive sediment is composed of highly permeable sands and gravels in marked contrast to the cohesive sediments which are usually composed of silts and clay. Because of the differences in permeability, the non-cohesive layers at the base of the banks tend to serve as avenues of pronounced seepage of groundwater. In some cases, the seepage can be significant enough to actually transport sediment away from the bank creating an overhanging mass of cohesive, upper bank material. Eventually, sediment removal by sapping will lead to bank instability and failure, commonly along tension cracks that form immediately after a mass wasting event (Ullrich et al. 1986; Hagerty and Hamel 1989; Odgaard et al. 1989).

Efforts to measure the rates of bank erosion have shown that mass wasting processes do not usually occur during peak flood conditions, but reach a maximum during the waning stages of the event. The occurrence of bank failure during flood recession has been attributed to two primary factors: (1) the movement of water into unsaturated bank sediments, thereby increasing both the weight of the materials and the pore pressure, and (2) the removal of pressure exerted horizontally by the river water on the banks as the flood flows recede. Many investigations have also found a distinct seasonality in erosion rates with the maximum rates occurring during the winter and spring months (Wolman 1959; Thorne and Lewin 1979; Simon et al. 1999). The enhanced rates are presumably related to increased moisture content of the bank materials during winter and spring and its influence on pore pressure.

It should be clear from the above discussion that bank erosion along many streams is intimately tied to mass wasting processes that have little to do with particle entrainment. It can be argued, however, that if the river failed to remove the debris from the base of the banks following a mass wasting event, the accumulated debris would stabilize the bank and inhibit further mass movements (Pizzuto 1984). The rates of bank erosion, then, can be minuscule or enormous depending both on the mechanisms of bank failure and the ability of fluvial entrainment to remove bank sediment and the failed debris. In general, banks consisting of fine-grained sediment erode by undercutting and subsequent block failure (Stanley et al. 1966), while coarse-grained non-cohesive bank sediment tends to erode by corrasion.

Almost any bank may erode rapidly, depending on the dominant process operating along the channel. Even cohesive, heavily vegetated banks can be quickly removed when erosion is initiated beneath the root zone.

6.9.   SUMMARY

Floodplains can be defined in varying ways, depending on the primary purpose for which they are being investigated. Herein we are primarily concerned with the genetic classification in which floodplains are created by a combination of depositional and erosional processes that reflect the river's current hydrologic and sedimentologic regime. From this perspective, floodplains serve as important temporary storage sites for both water and sediment generated within the watershed. Thus, they are not only produced by the river, but are a functional part of the riverine environment that regulates the downstream transport of constituents including contaminated particles to the basin mouth. The primary processes of floodplain development include lateral accretion, vertical accretion, and braided-channel accretion. The rates and magnitudes at which these processes operate create a continuum of floodplain types extending from those formed by fine-grained, low-energy meandering or anastomosing rivers to coarse-grained, high-energy braided rivers. Floodplains along many meandering streams, particularly in humid, temperate regions, are predominately produced by lateral and vertical accretion. Lateral accretion involves the progressive accumulation of sediment on point bars and other features as the river shifts its position across the valley floor. Vertical accretion is produced by the periodic, overbank deposition of sediment during flood events. The total amount of vertically accreted sediment is controlled by the rate of lateral channel migration which reworks the existing floodplain deposits. Because meandering rivers tend to move rather rapidly across their valley floors, the extent of vertically accreted sediment tends to be limited along many rivers. Floodplains located along braided rivers are typically associated with channel and braid bar abandonment and the subsequent deposition of sediment on the stabilized surface. In most instances, floodplain deposits associated with braided rivers are thinner and more irregular than those associated with other channel patterns.

The distribution and sedimentology of floodplain deposits is often described by a floodplain's alluvial architecture which defines how distinct packages of sediment, called the architectural elements, fit together in three-dimensional space. Most classification schemes broadly categorize floodplain deposits, and the architectural elements of which they are composed, into channel types and overbank types.

Representative examples include lateral accretion deposits, natural levees, crevasse splays and channels, and abandon channel fills.

The distribution of sediment-borne trace metal concentrations in floodplain deposits depend on the same factors that influence geographical patterns in concentration within the channel bed sediments. Thus, trace metal concentrations typically decrease downstream of point sources within floodplains as a result of hydraulic sorting, dilution, contaminant storage, biological uptake, and biogeochemical remobilization processes. Within any given river reach, geographical patterns in trace metal concentrations reflect the processes of floodplain formation and the alluvial architecture of the resulting deposits. This relationship between trace metal concentration and alluvial architectural is a fundamental component of the geomorphological-geochemical approach because it indicates that spatial variations in sediment-borne trace metal concentrations can be predicted on the basis of floodplain stratigraphy and sedimentology. Thus, significant savings in analytical costs can be obtained while gaining a more in-depth understanding of trace metal distributions by examining the alluvial deposits of a floodplain. Comparisons between floodplain and channel deposits have demonstrated that floodplains usually exhibit much higher trace metal concentrations, and a wider range of trace metal values, than the adjacent river bed deposits. These observations can be attributed to several floodplain characteristics including (1) a wide range of sediment sizes and compositions that vary over relatively short distances, (2) long-sediment residence times and a wide range of deposit ages, (3) limited sediment reworking and homogenization between flood events, and (4) extensive post-depositional remobilization of trace metals.

The distribution of sediment-borne contaminants in vertically accreted overbank deposits has recently received considerable attention. Interest in these fine-grained sediments reflects their potential to provide information on both natural variations in elemental concentrations as well as those associated with human activities. More importantly, the semi-continuous accumulation of overbank sediment often allows for the development of an historical record of trace metal influx to the river. The development of such pollution histories for multiple sites provides for the assessment of contaminant sources and their relative importance as a pollutant source for discrete time intervals. Changes in pollutant sources can also be determined using isotopic or multivariate geochemical fingerprinting methods. Such analyses are particularly useful for sites where the polluter pays principle exists.

Floodplain deposits often serve as an important, if not the predominant, source of sediment-borne contaminants to the river system. The primary method in which sediment-borne trace metals are remobilized is through the varied processes of bank erosion. Bank erosion is primarily conducted by corrasion, mass wasting, or a combination of the two processes. Corrasion represents the grain by grain removal (entrainment) of particles from the bank surface by forces generated within the river flow. In contrast, mass wasting is the movement of material under the influence of gravity. This latter process is significantly influenced by the moisture content of the floodplain or bank materials and tends to occur during the recessional phase of a flood event.

6.10. SUGGESTED READINGS

Anderson MG, Walling DE, Bates PD (1996) Floodplain Processes. Wiley, Chichester.

Collins AL, Walling DE (2002) Selecting fingerprint properties for discriminating potential suspended sediment sources in river basins. Journal of Hydrology 261:218–244.

Hudson-Edwards KA, Macklin MG, Curtis CD, Vaughn DJ (1998) Chemical remobilization of contaminated metals within floodplain sediments in an incising river system: implications for dating and chemostratigraphy. Earth Surface Processes and Landforms 23:671–684.

Harrison J, Heijnis H, Caprarelli G (2003) Historical pollution variability from abandoned mine sites, Greater Blue Mountains World Heritage Area, New South Wales, Australia. Environmental Geology 43:680–687.

Macklin MG, Klimek K (1992) Dispersal, storage and transformation of metal-contaminated alluvium in the upper Vistula basin, southwest Poland. Applied Geography 12:7–30.

Nanson GC, Croke J (1992) A genetic classification of floodplains. Geomorphology 4:459–486.

Rang MC, Schouten CJ (1989) Evidence for historical heavy metal pollution in floodplain soils: the Meuse. In: Petts GE (ed) Historical change of large alluvial rivers: Western Europe. Wiley, pp 127–142.

Ritter DF (1988) Floodplain erosion and deposition during the December 1982 floods in southeast Missouri. In: Baker V, Kochel RC, Patton PC (eds) Flood Geomorphology. Wiley, New York, pp 243–259.

Walling DE, He Q (1998) The spatial variability of overbank sedimentation on river floodplains. Geomorphology 24:209–223.

CHAPTER 7

RIVER METAMORPHOSIS

7.1.   INTRODUCTION

A basic principle of geomorphology is that landforms reflect the interactions between a set of driving and resisting forces to change. In Chapters 5 and 6, we were primarily concerned with the transport and fate of sediment-borne trace metals in rivers characterized by a state of balance where the driving forces were exactly offset, on average, by the resisting forces. In this case, sediment-borne trace metals are transported as part of the river's load in such a way that the shape of the channel and floodplain is not significantly altered. You may remember from the previous chapter that this style of transport was referred to by Lewin and Macklin (1987) as *passive dispersal*. In many instances, however, either natural or anthropogenic disturbances, such as climate change, tectonic activity, urbanization or mining, may alter the driving and resisting forces such that the system is stressed beyond the limits of stability. When this happens, the river system will be in a temporary state of disequilibrium. Lewin and Macklin (1987) referred to contaminant dispersal in these types of disequilibrium systems as *active transformation* because it is characterized by the transport of material through a channel that is being altered or metamorphosed into a form that is dictated by a new set of prevailing conditions.

In this chapter, we will examine the impact of river instability on the transport and storage of sediment-borne trace metals in alluvial deposits located along the valley floor. First, however, we must examine the means through which river channels adjust to natural and anthropogenic disturbances. This topical area is now a major thrust of fluvial geomorphology, and we cannot possibly cover all of its aspects in this pages that follow. Additional information, however, can be obtained from a number of textbooks, including Knighton (1998), Ritter et al. (2002), and Downs and Gregory (2004).

7.2.   THE BALANCED CONDITION

The idea that some form of balanced condition exists between landform morphology and the geomorphic processes that created them is not new (Ritter et al. 2002). Gilbert (1877), for example, clearly proposed that landforms reflect the interactions between the dominant geomorphic processes operating in the area and the local geology (Gilbert 1877). There are few concepts, however, that are more controversial than equilibrium. Some geomorphologists have even argued that equilibrium is so difficult to define (Thorn and Welford 1994), or so imprecise if it is defined, that it cannot serve as an effective paradigm (Phillips and Renwick 1992). Nonetheless, we believe that the equilibrium concept has significant merit if applied carefully in that it provides a sound basis for interpreting the relations between process and form, and the changes in those relationships through time.

An inherent component of the equilibrium concept as applied to rivers is the belief that fluvial processes function in such a way as to establish and maintain the most efficient conditions for transporting water and sediment. This idea has its roots in the concept of a *graded river* which was defined by Mackin (1948) as a river:

in which, over a period of years, slope is delicately adjusted to provide, with available discharge and with prevailing channel characteristics, just the velocity required for the transportation of the load supplied from the drainage basin. The graded stream is a system in equilibrium; its diagnostic characteristic is that any change in any of the controlling factors will cause a displacement of the equilibrium in a direction that will tend to absorb the effect of the change.

Although Mackin's definition overstates the role of slope, the concept of a graded river is highly valuable in understanding fluvial mechanics. Of most importance is its suggestion that a change in the hydrologic or sedimentologic regime will ultimately lead to an adjustment in channel form (e.g., width, depth, slope, etc.) in order to reestablish the most efficient transporting condition. It should be remembered, however, that discharge and load are not in themselves independent variables. They are controlled by such factors as climate, the underlying geology, tectonic activity, and vegetation (see Chapter 3). It follows, then, that a change in one or more of these external controlling factors will not only alter sediment load and discharge, but may result in changes in channel morphology. As an example, let us look at the potential effects of climate change on a river system. Temperature and precipitation are known to influence both mean annual runoff (Fig. 7.1) and the magnitude of erosion (Fig. 7.2). Thus, a change in climate may alter the amount of water and sediment that reaches the channel. If the resulting alterations of sediment load and discharge to a change in climate are significant enough, they will result in a change in channel shape by altering the rate and magnitude at which erosional and depositional processes function within the channel (Fig. 7.3). The resulting alterations in channel shape will bring the system back into equilibrium by changing the driving or resisting forces, or both until they are once again in a balanced state. This change in channel morphology as a result of the altered erosional and depositional processes is often referred to as *process-response*.

Process-response phenomena are not, of course, limited to the potential effects of climate change, but apply to a wide range of natural and anthropogenic perturbations (Table 7.1). An easily visualized example resulting from an anthropogenic disturbance is an increase in sediment load in the channel as a result of forest clearing. In this case it may not be possible for the river to carry all of its newly

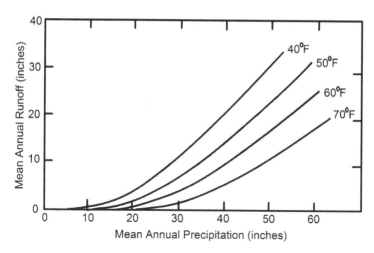

Figure 7.1. Change in mean annual runoff as a function of mean annual precipitation. Influence of temperature is related to enhanced evapotranspiration at higher temperatures
(After Langbein et al. 1949)

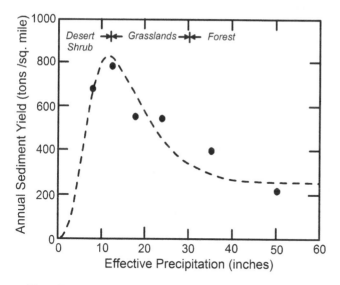

Figure 7.2. Average annual sediment yield as a function of effective precipitation and vegetation
(From Langbein and Schumm 1958)

acquired load. Some, then, will be deposited on the channel bed, causing aggradation and a steepening of gradient. As the channel gradient increases, it will reach a point at which it can transport all of the added load and aggradation will cease.

Equilibrium, as described above, implies that the external controls on sediment load and discharge must be constant for a balanced condition to exist. In reality, however, the

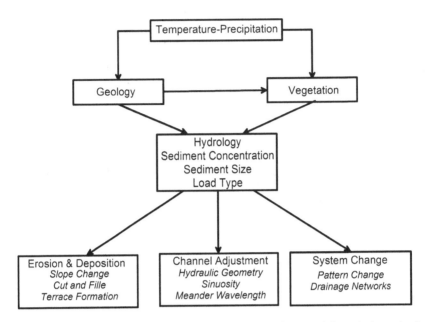

Figure 7.3. Flow diagram showing how climate influences river morphology. A change in climate alters the rivers hydrology and sediment load, size, and type, thereby requiring a response in channel morphology (From Ritter et al. 2002)

Table 7.1. Selected disturbances which may initiate adjustments in channel morphology

Changes in Climate	**Channel Changes**
Alterations in the frequency, intensity and duration of rainfall	Channel straightening and meander cut-off
Changes in storm characteristics	Sediment removal (gravel mining)
Temperature changes	Sediment additions
Vegetation change	Invasion by exotic vegetation
Changes to the Drainage Basin	Bank protection and stabilization
Deforestionation/afforestation	Dredging
Fire, burning	Embankments
Land-use change, such as agriculture, urbanization and building construction, and road construction	Diversions of flow
	Dam construction
	Weirs
Changes to Drainage Network	Return flows
Development of irrigation networks, drainage ditches, and storm drains	Bridge Crossings
	Culverts
Tectonic Activity	Restoration modifications
Uplift or fault development	

controlling factors do change, and the magnitudes of those changes vary through time. This led to the realization by Schumm and Lichty (1965) that equilibrium is closely linked to the timeframe over which the analysis is conducted (Table 7.2). They argued that different types of equilibrium can be defined for specific time intervals, which they referred to as cyclic, graded, and steady (Fig. 7.4). During steady time (days to months), they suggested that landforms are in a state of static-equilibrium in which their morphology does not change; they are therefore time-independent features. Landforms considered over graded-time (encompassing 10 to 1,000s of years) exhibit steady-state equilibrium. Landform equilibrium during this time interval is characterized by changes in morphology which revolve around some kind of average condition. Systems may be in a state of dynamic equilibrium during cyclic time (measured over periods of millions of years) (Schumm 1977). In this case, the average condition of the river system is continuously changing (Fig. 7.4).

A significant component of Schumm and Lichty's (1965) seminal work which is often overlooked is that the specific parameters used to describe fluvial geomorphic systems may act either as dependent or independent variables according to the time span being considered (Table 7.2). Channel morphology, for example, is a dependent variable, adjusted to the prevailing conditions of discharge and load, during the graded time interval. However, it is an independent variable during steady time because channel morphology exerts a direct control on the hydraulics of flow.

Table 7.2. Summary of cause and effect relations for different timeframes

Drainage basin variable	Steady time	Graded time	Cyclic time
Time	Not relevant	Not relevant	Independent
Initial Relief	Not relevant	Not relevant	Independent
Geology	Independent	Independent	Independent
Climate	Independent	Independent	Independent
Relief or volume of system above base level	Independent	Independent	Dependent
Vegetation (type and density)	Independent	Independent	Dependent
Hydrology (runoff and sediment yield per unit area within the system)	Independent	Independent	Dependent
Drainage network morphology	Independent	Dependent	Dependent
Hillslope Morphology	Independent	Dependent	Dependent
Hydrology (discharge of water and sediment from the system)	Dependent	Dependent	Dependent

After Schumm and Lichty (1965)

Steady time – 1 year or less; Graded time – hundreds of years; Cyclic time – millions of years

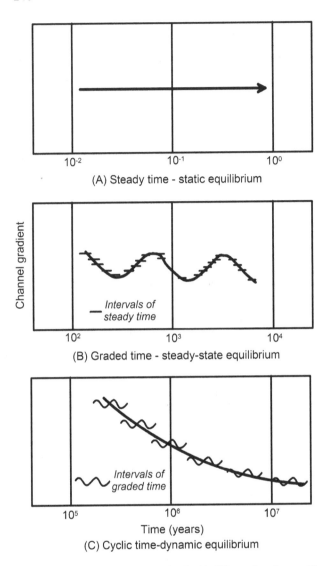

Figure 7.4. Equilibrium states associated with different time frames. Time intervals of steady-state and graded equilibrium are set within cyclic time scale (After Chorley and Kennedy 1971; Schumm 1977; figure reproduced from Ritter 1978)

Our primary interest with regards to the transport and fate of sediment-borne trace metals is the adjustments a river might make to counterbalance changes in discharge and load that occur over periods of years to decades. This timeframe falls within the graded interval as defined by Schumm and Lichty (1965). Thus, for our purposes, the equilibrium condition can be thought of as one in which both processes and form continuously fluctuate around a definable mean.

7.3. THRESHOLDS, COMPLEX RESPONSE, AND PROCESS LINKAGES

Any paradigm involving equilibrium implies that the contrasting state of disequilibrium also exists (Ritter et al. 2002). Disequilibrium is generally considered to be that period during which river processes and form readjust to a change in the sedimentologic and hydrologic regime. As alluded to earlier, changes in regime can be brought about by changes in climate and land-use. Disequilibrium can also result from the direct modification of the river channel, such as occurred during drainage improvement projects of the 1960s and 1970s in which low-gradient meandering channels were transformed into straight, higher gradient ditches. In this case, the driving and/or resisting forces operating within the channel were altered to such a degree that the limits to equilibrium, called *thresholds* by Schumm (1973), were exceeded. Thus, a threshold crossing event occurs when a river system moves from a state of balance to a temporary condition of disequilibrium. This disequilibrium phase is gradually corrected as the system develops a new state of balance adjusted to a different set of environmental conditions (Ritter et al. 1999).

Threshold crossing events are common enough to be considered as a fundamental characteristic of fluvial systems (Ritter et al. 2002). It should come as no surprise, then, that significant effort has been devoted to defining thresholds (i.e., the limits of equilibrium) on the basis of real parameters. One of the best examples pertains to the formation gullies (Fig. 7.5). In this case, slope and drainage area of both gullied and ungullied valley floors were determined for basins in northwestern Colorado. Except where drainage areas are small, the gullied and ungullied valleys possessed distinct slope-basin area relationships. Patton and Schumm (1975) argued that the division between the gullied and ungullied systems represents a threshold in that an increase in basin area, or its surrogate, discharge, within an ungullied system was likely to lead to gully initiation. Similarly, for a given basin area, an increase in slope on an ungullied valley floor as a result of localized deposition could produce gullying. The solid black line in Fig. 7.5, then, provides an estimate in real terms of the threshold condition for which the transition from a balanced to unbalanced system will occur.

If we assume that equilibrium is time dependent, the problem for our "graded-time" analysis becomes one of determining when a change in channel form (e.g., width, depth, and slope) represents a fluctuation around a mean condition of a river in equilibrium, and when it represents a significant change in channel form and process associated with a threshold crossing event. Determining this transition is not as easy you might think because floods can significantly alter the river's topographic form, but the channel will eventually return to its pre-event morphology. The time needed for this to occur has been referred to as the *recovery time* (see Pitlick 1993, for alternative definitions). Thus, if the river returns to its original state before another flood event of the same magnitude is repeated, a threshold has not been crossed even though channel morphology had been temporarily altered. The key, then, to identifying thresholds is whether the disruption caused by floods or some other disturbance is temporary or leads to a

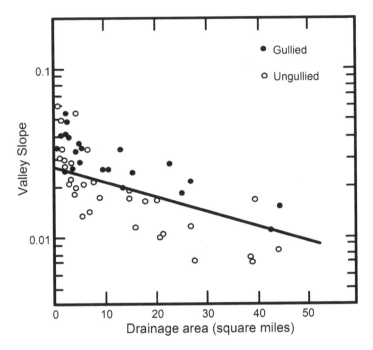

Figure 7.5. Threshold defined by gullied and ungullied valley floors in drainage basins of northwestern Colorado (From Patton and Schumm 1975)

new set of average parameter values that are significantly different from those that existed prior to the perturbation (Ritter 1988; Ritter et al. 1999).

When a threshold is crossed, it is typically assumed that the alteration in channel form will initially be rapid and decrease with time until a new equilibrium state is attained (Fig. 7.6). The changes in river morphology will not coincide with the onset of the disrupting event because the system requires time to react to the disturbance, a period referred to as the *reaction time*. It is now clear that the reaction time will vary between system components (Bull 1991), in part because the responses are linked through a sequence of cause and effect relations, which Ritter (1986) referred to as *process linkage*. For example, the first component of the system to respond to a change in climate may be upland vegetation (Fig. 7.6b). Only after the changes in vegetation have become significant will rates of hillslope erosion be altered, and adjustments along the river may lag even further behind, eventually responding to changes in hillslope processes (Fig. 7.6b). Moreover, alterations in channel form (e.g., width, depth, and roughness) that require only limited expenditures of energy, and relatively minors transfers in sediment may occur quite rapidly. In contrast, adjustments requiring large expenditures of energy (e.g., the reconfiguration of the longitudinal profile) will require long-periods of time (Fig. 7.7). Thus, the initial adjustments in channel form and process to a disrupting event may not be the final response.

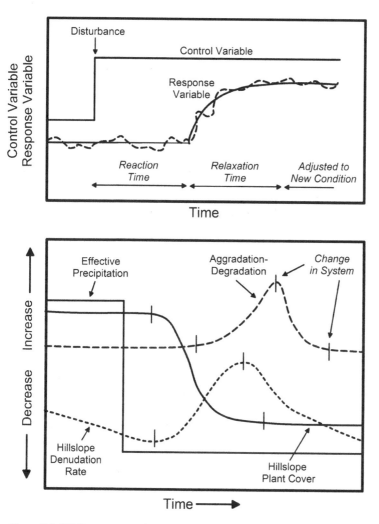

Figure 7.6. (A) Response to an instantaneous change in a control variable; solid line of the response curve represents the mean conditions about which the form variable fluctuations, (B) Hypothetical responses of different systems to a decrease in effective precipitation, moving the system from a semi-arid to an arid environment. Each system is linked and responded at different times and at different rates (Adapted from Bull 1991; figure from Knighton 1998)

Another significant observation of readjusting rivers is that threshold crossing events may produce a series of interrelated reactions. This suite of reactions has been referred to as *complex response*. Complex response was first recognized in the laboratory by Schumm and Parker (1973) while studying the development and evolution of drainage basins. During their experiments, a lowering of base level (the level to which a channel can erode) at the mouth of the artificial catchment

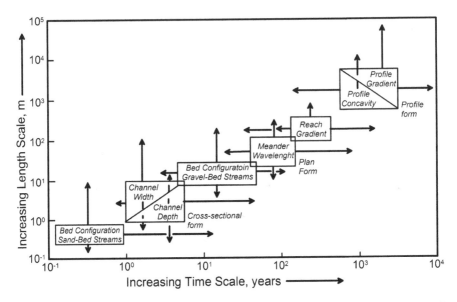

Figure 7.7. Timescales of adjustment for selected channel form parameters; developed for basins of intermediate size (From Knighton 1998)

caused an episode of channel incision and terrace formation along the trunk channel. Tributaries to the main stem of the drainage network were unaffected. However, with time channel incision migrated upstream, thereby affecting the base level of each of the tributaries. Channel incision then propagated through the tributaries, producing substantial quantities of sediment which were delivered to the main stem of the drainage network (Fig. 7.8). In fact, so much sediment was produced that the flow could not transport all of it to the basin mouth and channel aggradation occurred. Eventually, the tributaries adjusted to the new (lower) base level and sediment delivery to the trunk channel declined, producing yet another period of channel incision along the main stem of the drainage network as the trunk stream responded to the decreased sediment load from upstream areas.

The experimental studies demonstrated two important aspects of complex response which are of importance to contaminated rivers. First, at any given time the processes functioning along the channel may be out-of-phase; erosion and channel incision occurring in one part of the basin and deposition and channel bed aggradation in another. Thus, the transport and storage of sediment-borne trace metals may differ dramatically along rivers adjusting to some form of disturbance. Second, the initial response of the channel at a given location may not be the final response, indicating that over periods of years to decades, variations in sediment-associated trace metal transport and storage at a given site may change significantly through time following a pattern that is distinctly non-linear.

Perhaps the obvious question is whether complex response occurs at scales larger than those associated with laboratory studies. The answer is yes. However, it is more

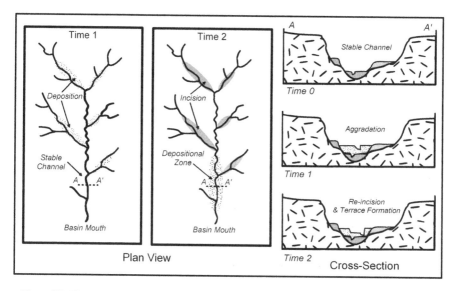

Figure 7.8. Illustration of complex response. Erosion and deposition may occur in different parts of the basin at the same time. In addition, multiple responses are possible for a given disturbance at any given location (After Schumm 1973, 1977)

difficult to directly observe complex response in natural systems because of the long time spans required for each component of the system to readjust. Nonetheless, a number of studies, based either on long-term observations or interpretation of the stratigraphic record, have demonstrated that complex response is a common phenomenon within drainage systems. Moreover, it is clear that the disequilibrium state does not need to be initiated by a lowering of base level, but can be produced by changes within upland areas as well.

One of the most thoroughly documented examples of complex response was provided by Gilbert (1917), although he did not refer it as such. Gilbert was charged with examining the effects of hydraulic gold mining during the late 1800s along the west flank of the Sierra Nevada. The primary concern was that hydraulic mining operations were generating abnormally high sediment loads within the adjoining rivers thereby causing an increased incidence of downstream, overbank flooding. Gilbert found that the rivers were unable to transport all of the debris from the mining operations and responded by aggrading their channel beds as the coarse-sediment was deposited. With continued aggradation, however, channel slope increased until the river acquired the capacity needed to transport the sediment farther downstream. The amount of aggradation that occurred ceased at different times along the channel, depending on the distance from the mines and the source and amount of load. Spatial differences in geomorphic processes clearly existed at the time; upstream areas were undergoing aggradation, whereas the effects had not yet reached downstream sections of the channel.

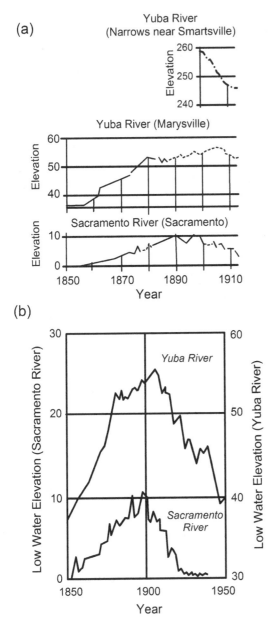

Figure 7.9. Changes in elevation of the low-flow channel bed along the Sacramento and Yuba Rivers, California. (A) Data from Gilbert 1917; (B) Data from Gilbert as updated by Graves and Eliab (1977). Episodes of channel bed aggradation and degradation form a semi-symmetrical pattern through time indicating that the introduced sediment moves downstream as a sediment wave

In 1884, mining operations were terminated as a result of legal action intended to remedy the environmental impacts of the mines. The decreased sediment loads associated with the reduction of mining debris to the headwaters of the drainage systems resulted in an excess of energy which was used to entrench the formerly aggraded channel bed. Sediment generated during the entrenchment process was transported downstream where it contributed to channel aggradation. As the system readjusted to yet another change in sediment load, a set of responses consisting of aggradation and subsequent incision moved downstream through the drainage network in wave-like pattern (Fig. 7.9). The height of the wave form decreased downstream growing longer and flatter away from the mines as sediment was lost to storage along the channel margins.

The type of river response demonstrated for the Sierra Nevada has been observed in a variety of geological and hydrologic settings, although minor differences in the depositional processes exist (James 1989; Knighton 1991; Miller et al. 1998). A significant lesson to be gained from these studies is that the river systems continued to respond through a progression of aggradation and degradation for decades after the termination of mining. Along the Yuba River at Marysville, California, for example, the river bed reached its highest level in 1905, 11 years after mining had ceased (Fig. 7.9). Thus, the movement of contaminated particles may require years to decades to move through the drainage network after the source of the contaminants has been removed. In fact, James (1989) argued that the wave-form described by Gilbert did not acquire a symmetrical form with respect to time along the Bear River, a tributary to the Sacramento. Rather it is skewed (Fig. 7.10) because sediments stored in the floodplain, which were not considered by Gilbert, are continually reworked and added to the river's load, and will be for decades or centuries to come. James (1989) also suggested that the wave form would exhibit a

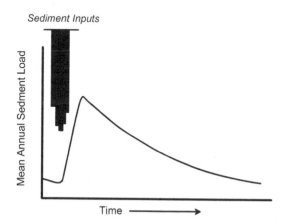

Figure 7.10. Sediment wave depicting affects of sediment storage following an episodic influx of sediment. The skewed nature of curve is the result of sediment remobilization of sediment stored in and along the channel (From James 1989)

"saw-toothed" pattern that was superimposed on the overall trend as a result of the episodic nature of sediment transport commonly associated with adjusting channels.

7.4.   ADJUSTMENTS IN CHANNEL GRADIENT, SHAPE, AND PATTERN

Ideally, we would like to be able to predict how a river system will respond to a given perturbation, and how this might affect future trace metal loads. Predicting the exact nature of a response has proven to be a difficult process. Part of the problem rests with the fact that the adjustment to a change in the controlling factors may be accomplished by altering any one of a wide range of parameters, including channel form (width, depth, hydraulic roughness, or width/depth ratio), gradient, planimetric characteristics (sinuosity, meander geometry), and pattern (straight, meandering, braided, anabranching). In some cases, only one variable may be involved in the adjustment process; others may involve multiple parameters and, as we learned earlier, the initial parameter to respond may not be the final parameter. Other complicating factors have also been identified, two of which are convergence and divergence (Table 7.3). Convergence, also called equifinality, refers to the case in which similar responses or geomorphic effects are produced by different types of disturbance; divergence is the occurrence of different effects by similar causes (Schumm and Brakenridge 1987; Downs and Gregory 2004).

Theoretically, fluvial adjustments to disturbance can be described in terms of the direction of the change (increasing or decreasing dimensions), the magnitude of the change, and the rates at which the alterations occur. It is also important to understand when and where the adjustments will occur. Given our current state of knowledge, conceptual and other modeling routines are rather good at determining the direction that a change in the hydrologic or sedimentologic regime will bring about, but quantifying the magnitudes and rates of adjustment is more problematic. Take, for example, the potential for climate change to cause a threshold crossing event. Predictive models of river adjustment to climate change are primarily based on observed alterations in channel form to past changes in temperature and precipitation that have occurred over varying temporal and spatial scales. Documentation

Table 7.3. Difficulties encountered when interpreting channel adjustments

---

Convergence: different processes and causes may cause similar effects; also referred to as equifinality
Divergence: Similar disturbances may produce different effects
Multiplicity: an observed response is produced by multiple, simultaneously acting disturbances
Singularity: randomness or unexplained variation in response to a disturbance; also called indeterminacy
Complexity: multiple and spatially-out-of phase responses to a disturbance
Sensitivity: the likelihood that a system will respond to a disturbance; denotes how close the system is to a threshold

---

After Schumm (1991) and Downs and Gregory (2003)

of channel change is typically based on a combination of historic measurements, morphologic data (e.g., terraces and abandoned channels), stratigraphic information that records the timing and extent of erosional and depositional events, and sedimentologic data that can be used to assess changes in flow regime. Almost without exception, the data sets are complex and fragmented, yielding a spatially and temporally incomplete record of channel change (Schumm and Brakenridge 1987). Depending on the time frame under consideration, changes in climate may be documented using historical data or a variety of climatic proxy indicators, including pollen, packrat middens, tree-rings, and stable isotopes. The climatic record is also temporally and spatially incomplete.

Documentation of past climate change and its impact on fluvial systems generally reveals the direction in which channel morphologic parameters have been altered, but information on the rate or magnitude of those changes is generally weak. Moreover, the responses may not be consistent across the landscape. Some rivers may absorb minor shifts in climate and remain in a state of equilibrium, while others go through a complete metamorphosis in process and form. It has even been shown that different segments of the same river have responded differently to similar alterations in sediment load and discharge (Schumm and Brakenridge 1987). These contrasting responses are at least partly attributed to variations in landscape (or landform) sensitivity.

Sensitivity has been defined in a number of different ways, depending on the objectives of the investigation (see, for example, Downs and Gregory 1993). For our purposes, it will be considered as the tendency of landforms to respond to an environmental disturbance by attaining a new equilibrium state (Schumm and Brakenridge 1987). Conceptually, sensitivity involves both the propensity for the system to change and the ability of the system to resist change by absorbing the forces of the disturbance. It is closely tied to the interaction and magnitude of the driving and resisting forces, and, therefore, is controlled by such factors as the erosional resistance of the underlying bedrock and the channel forming materials, basin relief and morphometry (e.g., drainage density and stream frequency), and watershed hydrology (Gerrard 1993). While sensitivity varies between basins, it also varies spatially within a basin. As a result, asynchronous geomorphic responses are likely to occur because thresholds are crossed under different magnitudes of disturbance. Perhaps the most significant variations have been observed between headwater and downstream reaches of a given catchment (Knighton 1998). Headwater areas, with their closer proximity to the sources of water and sediment production, their lower storage capacities, and their greater potential energy, tend to be more responsive than downstream reaches (see Balling and Wells 1990, for an example).

A conceptual model that is widely used for describing the direction of channel adjustment to a change in water and sediment discharge was presented by Schumm (1965, 1968, 1969) in a series of papers describing the concept of *river metamorphosis*. Schumm pieced together a host of empirical equations that had been developed to assess the relationships between river processes and form to produce a comprehensive, though qualitative, model of the possible river adjustments to altered hydrology and load. When change in both sediment and water discharge are considered, the possible adjustments can be described by four separate equations (Schumm 1969):

(1) $Q_w{}^+Q_t{}^+ = w^+, L^+, F^+, P^-, S^\pm, d^\pm$

(2) $Q_w{}^-Q_t{}^- = w^-, L^-, F^+, P^+, S^\pm, d^\pm$

(3) $Q_w{}^+Q_t{}^- = d^+, P^+, S^-, F^\pm, w^\pm, L^\pm$

(4) $Q_w{}^+Q_t{}^- = S^+, F^\pm, d^-, P^-, w^\pm, L^\pm$

where Q_t is the percentage of the total load transported as bedload (sand-sized or larger), and Q_w can be either the mean annual discharge or the mean annual flood. The other variables are width (w), depth (d), slope (S), meander wavelength (L), width-depth ratio (F), and sinuosity (P). The exponents, expressed as either a plus or minus, indicate whether the dimensions of the variables are increasing or decreasing.

As an example of how these equations may be used, let us assume that an area of forest is clear-cut. It is likely that Q_w will increase because infiltration rates will be lowered and direct runoff to the stream will be enhanced. Q_t will also increase because coarse sediment that was previously stabilized on the slopes by plant roots can now be eroded and make its way to the channel. Moreover, the coarse sediment can be moved more frequently because the peak flood discharge has increased. With positive changes in both Q_w and Q_t, Eq. 1 indicates that width, wavelength, and W/D ratio are likely to increase, but sinuosity will probably decrease. Depth and slope may vary in either direction. Slope will probably increase, however, because the channel becomes straighter.

It is important to recognize that Eqs. 1–4 do not indicate which variables will actually respond to a change in water or sediment discharge. They only suggest the direction of the change, if in fact, that particular parameter adjusts to the disturbance. Moreover, the equations provide no indication of the magnitude of the changes that may take place. Given our current access to high speed computers and the progress that has been made towards modeling climate change, rainfall-runoff processes, and other phenomena these types of predictive models may seem archaic. To be sure, a number of more quantitative, numerically intensive models have been put forth and appear promising (see, for example, Chang 1986; Wilcock and Iverson 2003), but their application to adjustments over long periods of time is often plagued by significant uncertainty.

One type of adjustment that is not included in Schumm's equations, but which was clearly recognized by him, is the potential for rivers to change their channel pattern in response to a disturbance. Perhaps the most precisely defined transition in pattern is between braided and meandering rivers. More specifically, a number of studies have delineated a threshold between braided and meandering rivers on the basis of slope and discharge (Lane 1957; Leopold and Wolman 1957; Ackers and Charlton 1971). In each case, an increase in discharge for a given slope (or an increase in slope for a given discharge) will convert a meandering channel to a braided channel (Fig. 7.11). Actually, the components that are being used to define the threshold (slope and discharge) describe changes in flow strength because both are inherent in the equation for stream power ($\omega = \gamma QS$).

The slope-discharge relation is a powerful tool to assess the potential for a river or river reach to change from meandering to braided, or vice versa. It is, nonetheless, a simplification of the controls on channel form in that a river's pattern is strongly influenced by other parameters as well. As might be expected from our earlier discussions of the controls on river systems, two of the more important parameters are sediment size and relative sediment supply (Ferguson 1987; Knighton and Nanson 1993). The latter is defined as the load supplied to a river relative to its ability to transport sediment downstream. Another important factor is the resistance of the banks to erosion (Ferguson 1987; Knighton and Nanson 1993). Channels formed in highly resistant banks are likely to be straight, whereas meandering streams require localized erosion at meander bends. Braided channels, in contrast, can only occur where extensive bank erosion is possible. In light of the above, channel pattern is probably delimited by a combination of factors including flow strength, bank resistance to erosion, and sediment type

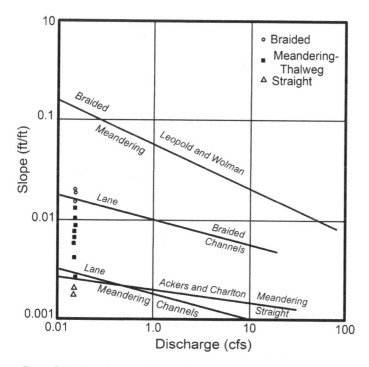

Figure 7.11. Channel pattern thresholds defined by channel slope and discharge by different studies
(From Schumm and Khan 1972)

and supply. Slope is probably not an independent variable, but adjusts along with pattern to the changing hydrologic and sedimentologic regime (Ritter et al. 2002).

7.5.   EFFECTS OF RIVER METAMORPHOSIS

It should be clear from the preceding section that natural and anthropogenic distur-bances can lead to a wide range of responses in the fluvial system. It follows then that readjustments to a new set of environmental conditions can affect the physical dispersal of sediment-borne trace metals in a variety of ways, each of which may vary in time and position along the valley floor. Nonetheless, the impacts of threshold-crossing events on the dispersal of contaminated particles fall within two broad categories which often work in concert with one another. First, disturbances may alter the rates of channel bed and bank erosion, and, thus, sediment transport rates. Second, channel modifications can significantly influence the location, quantity, and nature of contaminated sediment storage along the river valley. As an example, let us return to the laboratory studies conducted by Schumm and Parker (1973) in which the effects of base level lowering were investigated within a watershed. Their work suggested that a single lowering of base level resulted in multiple events. Initially, headcuts (knickpoints) migrated upstream rejuvenating the lower reaches of the drainage network. During this phase of readjustment, sediment transport rates increased by reworking channel bed and bank materials. Had these sediments been contaminated, one would suspect that the discharge of contaminated sediment from the basin would increase. As the erosion moved further up into the tributaries, an increasing quantity of sediment was deposited and stored along the axial channel, reducing sediment yields from the basin mouth. Reduced erosion in the headwaters eventually led to re-incision along the axial channel, thereby temporarily increasing sediment loads from the catchment. Re-incision, however, led to short-lived episodes of downstream cut and fill as the sediment was moved out the system episodically. Each cycle of cut and fill produced minor increases and decreases in sediment yields from the catchment as a result of changes in the rate of sediment reworking and storage along the axial channel (Fig. 7.12).

Few studies of natural systems have been able to document these types of variation in sediment transport rates in response to environmental disturbances. The lack of such data is probably not systemic. Rather it is more likely caused by the long periods of time over which such responses occur, and the absence of detailed, long-term monitoring records. In fact, stratigraphic studies have demonstrated that these types of complex responses have occurred within many basins. The laboratory data, then, demonstrate that it is extremely important to understand how short-term monitoring data fit into the sequence of geomorphic events that are likely to occur over longer periods of time for rivers in a state of disequilibrium. One might suspect, for example, that the transport of contaminated sediment may increase significantly if there is a likelihood for upstream channel incision during an ongoing phase of system readjustment.

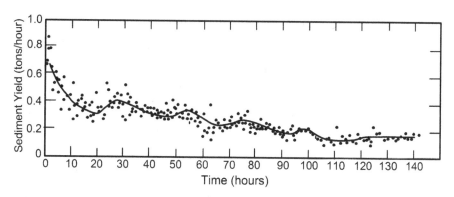

Figure 7.12. Variations in sediment-yield during drainage-basin evolution in an experimental facility. Solid mean line is based on a moving average (From Schumm 1977)

Perhaps one of the most significant influences of river metamorphosis on contaminant dispersal is related to its influence on sediment storage processes and volumes. The Carson River system of Nevada, for example, underwent a significant episode of channel bed aggradation during the late 1800s as a result of an increased influx of mine and mill tailings as well as sediment derived from clear-cut areas in the upper basin. The aggradational event resulted in the deposition of sediment-borne trace metals in alluvial deposits along the valley floor. Some of these deposits locally exceeded 1 to 2 m in thickness. Moreover, the disturbances led to lateral channel instability and rapid rates of meander cutoff and migration. This lateral channel instability, in turn, resulted in the storage of contaminated debris in abandon meander loops and produced a complexly structured stratigraphic sequence characterized by extreme variations in trace metal concentrations. Once mining had effectively ceased, reduced sediment loads led to channel incision that re-exposed the contaminated materials in the channel banks. The exposed sediments were then vulnerable to bank erosion processes, and their reworking now serves as primary source of Hg to the modern river. The significance of the above example is that the distribution of trace metals within the channel bed is a product of the geomorphic responses of the Carson River to changes in sediment influx, and the effects of those responses on sediment transport and storage processes (Miller et al. 1998).

The effects of channel change on metal mobility are not limited to the physical transport and storage of contaminated particles. Such changes may affect the chemical mobility of the trace metals as well. Within fluvial environments, physio-chemical changes are often driven by alterations in channel gradient which result in episodes of channel aggradation or entrenchment. Channel incision, for example, is likely to lower groundwater levels in the adjacent floodplain, thereby changing the local redox conditions from a reducing to an oxidizing state. Metal containing sulfide minerals, such as those derived from mine sites, may then be oxidized. The net result is an increase in metal mobility as trace metals are released from the sulfides and the pH of the pore waters is reduced (see, Macklin and Klimek 1992, for an example). Aggradation and increased water levels may also enhance metal mobility by reducing metal enriched Fe and Mn hydroxides (see Chapter 2).

The potential for such changes to increase metal mobility is rooted in the concept of a chemical time bomb. Stigliani (1991) defined a *chemical time bomb* as "a chain of events resulting in the delayed and sudden occurrence of harmful effects to ecosystem structure and function that result from the mobilization of chemicals stored in soils and sediments in response to prolonged changes in the environment." A key element of a chemical time bomb is that while the impacts of contaminants on the aquatic environment may be minimal at the present time, they may potentially produce adverse affects in the future. The potential consequences of chemical time bombs have been widely discussed (Lacerda and Salomons 1992; Appelgren and Barchi 1993; Smal 1993; Várallyay et al. 1993), particularly with regards to the degradation of contaminated soils. However, the "detonation" of chemical time bombs, in which the system is subjected to a delayed and sudden impact from chemical substances stored within alluvial deposits, has seldom been documented. Nonetheless, there is no question that changes in the physical and chemical environment can increase metal mobility and bioavailability and, therefore, must be analyzed as part of any sound assessment program. Examples of factors which may be considered include soil acidification, salinization or alkalinization, changes in redox conditions, and alterations in the cycling of organic matter (Várallyay et al. 1993).

It should be clear in the light of the above discussion that the environmental changes leading to a threshold crossing event, combined with the resulting geomorphic responses of the river system, may alter multiple parameters within the watershed. Thus, predicting the potential impacts of a disturbance on metal mobility can be an extremely complicated process. For example, an increase in flood frequency along the River Ouse in the U.K. during the past 20 years has enhanced the rates of contaminated sediment delivery to both the channel and the adjacent riparian farmlands (Macklin et al. 2006). These elevated levels of metal contamination are thought to present an increased risk to ecosystem and human health (Dennis et al. 2003). However, to understand the nature of the increased risk, or if it even exists, would require information on how the change in flood frequency, along with the associated change in rainfall duration and intensity, will alter such processes as leaching of the floodplain soils, mineral weathering, organic matter content, and soil pH and Eh. In addition, a prediction of the changes in channel morphology, which may further alter the chemical conditions within the floodplain deposits and metal mobility, is needed.

7.6.   TERRACES

7.6.1.   Definition and Formative Processes

A significant component of the geomorphological-geochemical approach is the ability to map trace metal distributions on the basis of the sedimentologic and stratigraphic nature of the alluvial deposits. In the case of relatively stable channels, for example, it was shown in Chapter 6 how floodplain deposits could be used to both predict the geographical distribution of sediment-borne trace metals within a river valley and to construct a history of trace metal loadings to a river system. The same basic principles used in the analysis of floodplains can be applied to rivers that have, or are undergoing, a phase of river metamorphosis. A common difference, however, is that rivers which have experienced a state of disequilibrium may have undergone an episode of channel bed aggradation or degradation. Where this is the case, part of the contaminant record will exist beyond the limits of the floodplain. In aggrading channels, the contaminated deposits will be buried beneath the modern alluvial deposits; in incising channels, trace metals will be contained within an alluvial terrace or suite of terraces.

Perhaps the simplest definition of a *terrace* is that it represents an abandoned floodplain that formed when the channel floor was at a higher level than at present. Topographically, it is composed of a flat-lying *terrace tread*, which represents the former floodplain, and a *scarp* that connects it to any surface which resides at a lower elevation (Fig. 7.13). In most cases, the lower surface is the floodplain or another (younger) terrace. Identifying a terrace in the field is not as easy you might think. Most geomorphologists, for example, would recognize a surface that is being constructed by vertical accretion as a floodplain in that the landform and the river are still hydrologically connected (Ritter et al. 2002). However, it is not uncommon for these topographically higher surfaces to be inundated during a flood event, although much less frequently. Thus, it can be highly problematic to define a terrace on the basis of inundation frequencies, or the occurrence of recent, vertically accreted deposits. The question, then, becomes one of how to distinguish between a floodplain and terrace when both may exhibit similar attributes. For our purposes,

Figure 7.13. Cross sectional view of a terrace located along the Rio Pilcomayo, southern Bolivia

the distinguishing factor in terrace formation is that the surface was created when the river was at a topographically higher elevation at some point in the past. From this perspective, a terrace is the product of channel bed degradation (downcutting) and can only exist where degradation has occurred. Three of the most common causes for downcutting include (1) a decrease in sediment load or material size to the channel, (2) a change in the sediment load to stream flow ratio, resulting in excess stream energy, and (3) a downstream change in base level, defined as the lowest level to which the channel can erode (Brown 1997).

At any give site, geomorphic surfaces including both the floodplain and the adjacent terraces are identified numerically, with "1" being the oldest and therefore topographically highest surface. Along some rivers, terraces are continuous and the observed staircase type progression from the floodplain to the highest flat-lying, alluvial surface changes little along the valley. More often terrace preservation is incomplete, and they occur as isolated remnants that may be kilometers apart. Moreover, the downstream slope of the tread may differ significantly from one terrace to another because degradation, by its very nature, produces a change in channel gradient. As a result, terraces often converge or diverge downstream, making the identification of a particular terrace on the basis of elevation above the channel bed problematic. Take, for instance, the case in which a terrace possesses a greater slope than the modern floodplain. It is entirely possible that further downstream the terrace surface will plunge beneath the floodplain where it will no longer exist as a terrace, but as a buried floodplain surface within the alluvial valley fill (Fig. 7.14) (Brown 1997).

Terraces are generally classified genetically or on the basis of topography. When classified genetically, they can be further categorized as either erosional or depositional (Fig. 7.15). Erosional terraces are primarily formed by processes of lateral

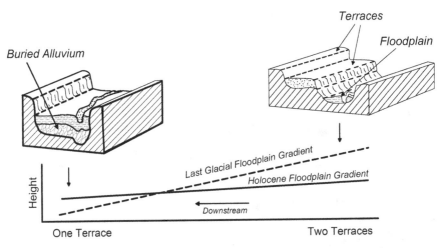

Figure 7.14. Terrace and fill characteristics in relation to changes in the longitudinal profile (Adapted from Brown 1997)

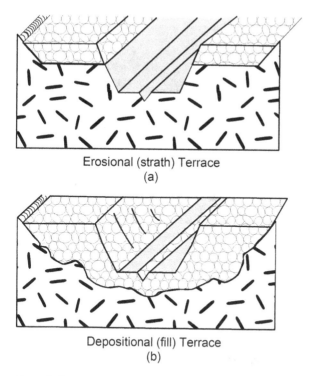

Erosional (strath) Terrace
(a)

Depositional (fill) Terrace
(b)

Figure 7.15. Erosional (A) and depositional (fill) terrace (B). Erosional terrace is characterized by thin
alluvial cover of a smooth truncated surface composed of bedrock or older alluvium. Depositional
terrace exhibits a tread underlain by alluvium that represents the highest level of fill deposited in
valley. Thickness of alluvium is irregular and may be substantial (Modified from Ritter et al.2002)

channel migration. Different types of erosional terraces can be distinguished on
the basis of the materials which have been eroded as the channel migrated across
the valley floor. If bedrock is truncated, the terrace is referred to as a *strath, a
rock-cut terrace*, or a *rock-cut strath* (Fig. 7.15). If the eroded material is unconsolidated sediment, they are commonly referred to as *fillstraths* or *fill-cuts*. Erosional
terraces, whether cut on bedrock or unconsolidated materials, possess a distinct
set of properties which allows them to be distinguished from depositional forms.
Of primary significance is that (1) they are capped by a relatively thin layer of
sediment, deposited during channel migration, whose thickness is dependent on
the scouring depth of the river channel, and (2) the eroded surface beneath this
layer closely reflects the topography of the terrace surface (Fig. 7.16). In contrast,
depositional terraces are underlain by a thick sequence of valley fill, and the terrace
tread represents the surface of the fill (Fig. 7.15). Although the terrace tread is
likely to be flat, the lower bounding surface beneath the fill may be highly irregular.
While these differences may seem of little importance, in reality they are critical to
the reconstruction of past geomorphic events in the basin because the two indicate

Figure 7.16. Rock-cut (strath) terrace located along a tributary to the Rio Pilcomayo, Bolivia

different historical pathways of formation. Depositional terraces require a period of significant valley filling over an irregular, underlying surface followed by an episode of channel incision. The development of the eroded surface beneath the terrace alluvium occurred at some time before the deposition of the overlying fill. Thus, the two events are separated by a distinct time interval. In contrast, the layer of sediment contained in an erosional terrace is deposited at the same time that the underlying rock or unconsolidated debris is being cut. In addition, erosional terraces are typically considered to be a manifestation of a previously stable river in which significant episodes of channel aggradation or degradation are absent.

From a topographic perspective, terraces are often classified as either *paired* or *unpaired terraces* (Fig. 7.17). Where terraces on both sides of the channel are at the same elevation and are presumable of the same age, they are considered paired. If the elevations of the terraces differ from one side of the river to the other, they are unpaired. The general assumption is that unpaired terraces are created by a channel that is downcutting as it migrates across the valley floor.

7.6.2.   Trace Metal Distributions

The factors controlling the distribution and concentration of sediment-associated trace metals on terraces has been less thoroughly investigated than for floodplains. Of course, there is some degree of overlap between the two landforms in that by definition, a terrace is an abandoned floodplain. Assuming that trace metals were available for incorporation into its deposits prior to terrace formation, its trace metal patterns will

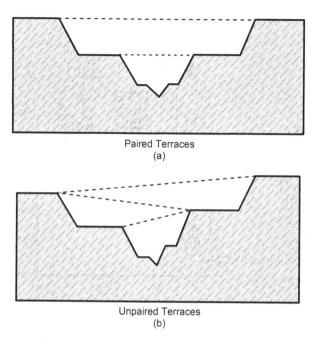

Paired Terraces
(a)

Unpaired Terraces
(b)

Figure 7.17. Terraces classified according to topographic relationships. (A) Pair terraces exhibit treads at the same level on both side of channel, (B) Unpaired terraces stand at different elevation from one side of the channel to the other (From Ritter et al. 2002)

reflect all of the controlling variables previously discussed in Chapter 6. However, this is not always the case; downcutting and terrace formation may occur before, during or after the influx of trace metals to the system. Moreover, once downcutting has occurred, deposition of alluvial sediments on the terrace may partially or completely cease, depending on the ability of flood flows to inundate the terrace tread. Thus, trace metal concentrations should, at least theoretically, be controlled by the age of the terrace deposits relative to the influx of the contaminants, and the height of the terrace above the channel floor. Actually, the importance of these two factors in controlling metal concentrations on terrace surfaces has been documented for a number of rivers. An example is the work by Brewer and Taylor (1997) on the upper Severn Basin in the U.K. They collected samples from the upper 15 cm of a large number of well-defined terraces within three different reaches positioned along the river. Their data illustrated that terraces positioned at intermediate elevations above the channel tended to possess the highest trace metal concentrations (Fig. 7.18). Sediments within these terraces were primarily deposited during mining operations, or they were inundated relatively frequently during the mining era allowing for the enhanced accumulation of trace metals. Terraces located at higher elevations were constructed prior to mining operations in the basin and, therefore, reflected natural (background) concentrations, which were not modified by overbank flooding. The lower set of terraces exhibited

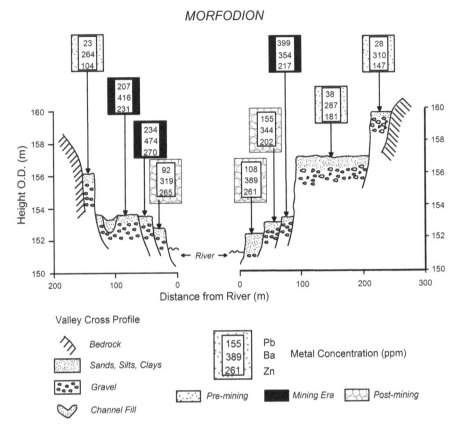

Figure 7.18. Cross valley profile at the Morfodion study reach on the upper Severn River, U.K (Adapted from Brewer and Taylor (1997)

intermediate concentrations because they were formed after mining operations had ceased and were composed of a higher percentage of clean sediment.

Comparison of data from the three study sites investigated along the upper Severn revealed that significant variations in the geographical patterns of trace metals existed along the river, even though the general trends were similar. Brewer and Taylor (1997) attributed these spatial differences to variations in the styles of channel instability and flood hydraulics which, when examined separately, could be used to develop a conceptual model of controls on the distribution of trace metals on alluvial terraces. Their model is presented in Fig. 7.19 to demonstrate the geographical patterns that would be expected before, during, and after the influx of trace metals from a point source for an erosional terrace sequence. In the case of a vertically and laterally stable channel or channel reach, the quantity and concentration of trace metals will decrease with terrace height. The distribution reflects a higher frequency of inundation of the lower terraces which allows for a great

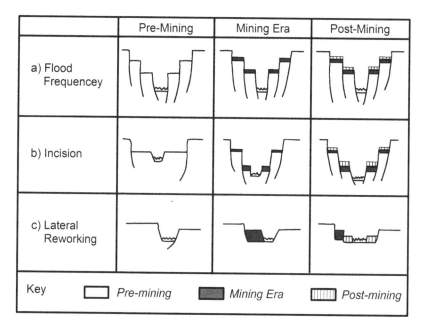

	Pre-Mining	Mining Era	Post-Mining
a) Flood Frequencey			
b) Incision			
c) Lateral Reworking			

Key ☐ *Pre-mining* ■ *Mining Era* ▥ *Post-mining*

Figure 7.19. Styles of terrace deposition prior to-, during, and following the influx of mining debris. Similar depositional styles can be associated with specific geomorphic processes during the influx of sediment-borne trace metals from any contaminant source. Differences in stratigraphy depend on whether floodplain development is controlled by flood frequency (a), channel incision (b), or lateral channel migration (From Brewer and Taylor 1997)

opportunity of trace metal accumulation. After contaminant influx has ended, cleaner sediment accumulates on the terraces and, again, sediments accumulate faster on the lower terrace treads. Assuming that the post-influx sediments moving along the channel exhibit lower metal concentrations, it is possible that the topographic order of most to least contaminated may reverse itself with time.

Along incising channels, those terraces formed during the period of contaminant influx will possess the highest concentrations. Thus, if terrace formation corresponded precisely with the onset and termination of metal inputs to the river, the highest terrace would be clean, the intermediate terrace would be highly contaminated, and the lowest terraces would exhibit intermediate concentrations as they would be formed of reworked sediments from both terraces. However, once formed the pattern will be modified by differences in the frequency with which each surface is inundated. Thus, the thickness and trace metal content of the post-mining deposits would decrease with height. Rivers which are vertically stable, but which are migrating laterally, may rework significant quantities of sediment and the concentration of the laterally accreted deposits will reflect the composition of the material that is available for remobilization (Brewer and Taylor 1997). In general, however, the latter changes in trace metal concentration will reflect deposit age relative to the influx of the contaminants.

The terraces along the Severn, and described in the above model, are assumed to be produced by episodes of downcutting that are separated by periods of floodplain formation. A different sequence exists along the Glengonnar Water in the Leadhills of Scottland. Here, an influx of mining debris led to significant aggradation and storage of trace metals in the valley fill. Once mining ceased, channel bed incision ensued, producing a depositional terrace. That is, the tread was underlain by fill composed in large part of mining debris. Multiple terraces were subsequently formed by continued, episodic downcutting and sediment reworking. In contrast to upper Severn, the highest concentrations of trace metals along the Glengonnar Water were located on the highest terrace which corresponded to the period of mining activity. Concentrations progressively decrease with declining elevation above the channel bed as contaminated materials eroded from the fill was mixed with "clean" sediment following the termination of mining operations. The contrasts between the Severn and Glengonnar illustrates the need to understand the mechanisms of terrace formation in order to determine the geographical patterns of trace metals that can be expected in a terrace sequence.

Determining whether a terrace is erosional or depositional can be a difficult process. After all, both types are characterized by a flat-lying surface underlain by alluvium, and if we were to walk across their surface, there would be nothing that could be used to tell the difference between them. The key, as noted above, is provided by stratigraphic exposures in the terrace scarp which allows for a determination of whether the fill is thin and uniform or variably thick. Even then the two may be difficult to separate without multiple, obliquely oriented views. The problem has, however, been overcome in a number of cases by using various types of geophysical methods to determine the thickness of the terrace alluvium and the configuration of the underlying surface. These characteristics reveal the terrace origin.

7.7.   QUANTIFYING EXTENT AND MAGNITUDE OF CONTAMINATION

An important component of nearly all site investigations is a determination of the type, extent and magnitude of contamination along the river valley. These data are required for a variety of purposes, a few of which include the identification of likely zones and pathways of exposure, an analysis of the locations and quantity of material potentially requiring remediation, and the assessment of current and future contaminant loadings to and from the aquatic environment. It is perhaps useful, then, to take a closer look at the use of geomorphic and stratigraphic principles covered in Chapters 5–7 to delineate the extent and magnitude of contamination in river valleys.

Unlike many other types of contaminated sites, it is common for the project boundaries of a contaminated river to include exceedingly large areas of complex terrain. In some cases, literally tens of kilometers of stream channel may require investigation. When combined with the need to cleanup sediment-borne trace metals stored within the adjacent floodplain and terraces, the site can be enormous, and determining the distribution of contaminants at a level required for sediment remediation or other purposes can seem overwhelming. Consider, for example, the costs and effort required to characterize a 30 km reach. If we were to randomly collect just one sample from the channel, the floodplain, and a terrace every 100 m along the river, a total of 900 samples would be taken for analyses, and the sampling program would be entirely inadequate for delineating those areas potentially requiring cleanup. In fact, if we were to increase the frequency of sampling by three-fold, so that there were 2,700 samples, we would still be unable to effectively determine contaminant distributions. Part of the problem is that a linear extrapolation of concentrations between sampling points in riverine environments often generates erroneous geographical patterns in trace metals because fluvial depositional processes can produce abrupt changes in sediment-borne trace metal concentrations.

Our earlier discussions have demonstrated that it is often possible to link sediment-borne trace metal concentrations within the channel, the floodplain, and the adjacent terraces to specific alluvial deposits. The linkage is based on their sedimentological character and age, the geomorphic history of the river system, the dispersal processes operating at the site, and the trace metal-sediment relationships. Once these linkages have been defined, the spatial distribution of sediment-borne trace metals can be estimated by mapping the alluvial deposits.

In practice, contaminant concentrations in landforms and their deposits will be determined by studying and sampling discrete reaches along the river system as it is simply impossible, in most cases, to conduct detailed analyses of the entire project area. It is therefore necessary to extrapolate data from these detailed study reaches to the entire river valley. The advantage of using this geomorphological-geochemical approach is that linking trace metal concentrations to specific alluvial deposits and landforms allows contaminant distributions to be predicted on the basis of the processes responsible for their dispersal. Moreover, the approach can identify and account for non-linear variations in concentrations between discrete sampling points. Thus, the approach is likely to produce a detailed understanding of the three-dimensional distribution of the contaminants over large areas.

Due to the high degree of variability in the sedimentological and geochemical nature of alluvial deposits, an iterative approach is typically required. For example, the information from a number of detailed study sites can be used to place the deposits or landforms into specific categories or groups classified by concentration. These deposits and/or landforms can then be mapped throughout the entire valley, after which a new phase of sampling can be used to verify and update predicted geographical patterns in trace metal concentrations determined on the basis of the site geomorphology.

The mapping of alluvial deposits and landforms typically relies on the integration of field data with data collected remotely over large areas by some form of air- or space-borne platform. Historically, the remotely sensed data included black and white or color aerial photographs. The procedures used for mapping of alluvial

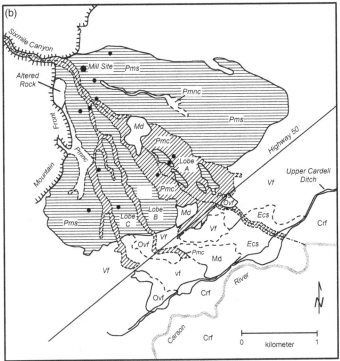

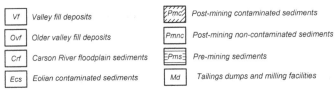

deposits on aerial photographs and other types of remotely collected data is beyond the scope our analysis, but has been discussed in a number of other documents (Lueder 1959; Wolf 1974; Paine 1981; Burnside 1985). Suffice it to say that, provided with photographs of an appropriate scale, distinct landforms and sedimentary units can be effectively mapped for most river systems. With regards to sediments, depositional units and landforms often exhibit differences in surface reflectance as a result of variations in the sediment's moisture, organic matter, and Fe content, grain size distribution, and surface structure, thereby allowing for their delineation (Fig. 7.20) (Gilvear and Bryant 2003).

While the mapping of alluvial landforms and deposits is still heavily dependent on aerial photographs, the use of other types of remotely sensed, space- and air-borne data (Table 7.4) has increased dramatically in recent years. The trend is driven by the recognition that these data possess several advantages over aerial photographs, including (1) enhanced spatial and temporal resolution, and (2) the analysis of the emitted or reflected electromagnetic radiation from sedimentary units and water that cannot be detected by the human eye (Gilvear and Bryant 2003). In addition, high resolution topographic data can be obtained by laser altimetry or Lidar technology. Not only do these methods allow for the distinction of subtle differences in elevation between landforms (e.g., terraces), but they provide for an understanding of the three-dimensional geometry of the features which is required to calculate the volume of sediment within it.

Although it has been mentioned previously, its importance justifies repeating here. Every river system exhibits its own unique characteristics and, therefore, the plan used to characterize the spatial distribution of trace metals in three-dimensions will need to be tailored to that particular system. Nonetheless, it is perhaps instructive to examine how the geomorphological-geochemical approach can be applied in a relatively simple hypothetical case. The goal is to determine the area, location, and quantity of contaminated material that may be remobilized by chemical or physical processes or that must be removed as part of the remediation process. The hypothetical river, Mud Creek, is located in a humid temperate climatic regime and is characterized by a low-gradient meandering channel (Fig. 7.21). Sediment within the channel bed is dominated by coarse sand and gravels, whereas floodplain and terrace deposits are composed of a combination of sand, silt, and clay-sized particles. Our hypothetical site was contaminated by the release of tailings from a Pb–Zn mine which began operation in the early 1900s. A detailed study of the geomorphic history of the river system demonstrated that instability during the late Holocene led to the formation of two alluvial terraces, both of which pre-date mining operations (Fig. 7.21). The first terrace is located approximately 8 m above the modern channel bed; the other terrace about 5 m

---

Figure 7.20. (A) Aerial photograph of an alluvial fan at the mouth of Sixmile Canyon, Nevada. Light colored materials are Hg contaminated mine and mill tailings. (B) Surficial geologic map of the Six mile Canyon fan produced using a combination of cartographic and field data (After Miller et al. 1996)

Table 7.4. Summary of high and medium-resolution satellite remote sensing platforms and sensors

Sensor and platform	Launch date	Spatial resolution	Temporal resolution
SPOT 4 (HRV)	1998	10–80	2–26 days depending on overlap coverage
Landsat 5 (SS)	1986	80	16–18 days
Landsat 4	1982	80 to 120 depending on bandwidth	16 days
Landsat 3	1978	80	16 days
Landsat 2	1975	80	16 days
Landsat 1	1972	80	16 days
Landsat 5 (T)	1984	30	16 days
AVHRR NOAA-15 (K)	1988	1100	12 hours
IRS-1C/D	1995 (IC) 1997 (ID)	5.8–23 depending on bandwidth	5–24 days depending on latitude
Ikonos-1/2	1999	1–4 depending on bandwidth	1–3 days
Landsat 7 (ET+)	1999	15–60 depending on bandwidth	16 days
SPIN-2	1999	1–10 depending on bandwidth	Variable
Terra (EOS-A1)	2000	15 to 90 depending on bandwidth	Variable
CERBS (CCD)	1999	20	26 days
ADEOS (AVNIR)	1996	8–16 depending on bandwidth	41 days (3-day sub-cycles)

Adapted from Gilvear and Bryant (2003)

above the channel bed. Differences in elevation between the lower terrace and the floodplain are subtle, measuring less than 1.5 m in most locations.

As part of the site assessment program, landforms and their associated deposits were delineated for a number of discrete reaches. Each unit was sampled, and their trace metal content was characterized in terms of their mean, median, and range of concentrations. The primary metal of concern was Pb as the deposits were observed to exceed the $1,000\,\mu g/g$ remediation standard for sediment in a number of locations. Data from the studied reaches revealed that because of the coarse grained nature of the channel bed, Pb concentrations were relative low, ranging from 50 to $275\,\mu g/g$. However, lateral channel migration during the past 100 years incorporated sediment-borne Pb into point bar and laterally accreted floodplain deposits. Pb concentrations generally exceeded $1,500\,\mu g/g$. The width of the laterally accreted deposits was difficult to discern stratigraphically because

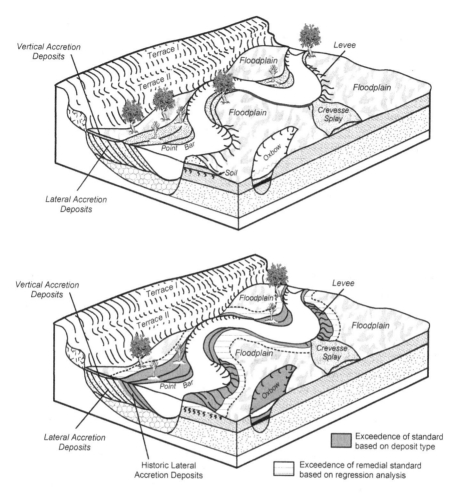

Figure 7.21. Highly idealized model of determining areas requiring cleanup using the geomorphological-geochemical approach. Assumes that the remedial standard is exceeded for specific depositional units, including sediment within oxbow lakes, natural levees, historic lateral accretion deposits, and vertically accreted floodplain deposits adjacent to the channel

there were few bank exposures and topographic variations which could be used to identify these deposits were lacking. However, core data revealed that their width could be approximated by the age and species of riparian vegetation.

On the floodplain, an oxbow lake deposit exhibited the highest Pb concentrations, owing to the clay-rich nature of its sediment and the relatively high levels of organic matter. The areal extent of the oxbow could be easily delineated on aerial photographs. The thickness of the oxbow deposits was variable, but marked by a basal sand and fine gravel layer of the old channel bed. Also on the floodplain, splay-deposits consisting predominately of fine- to medium-sand were delineated and mapped. Lead concentrations in these deposits ranged from 250 to 800 μg/g. The splay-deposits overlaid finer-grained floodplain materials. A wedge shaped body of sediment associated with a natural levee paralleled the channel. The remainder of the floodplain was characterized by a low-relief, rather non-descript surface. There was no means of visually subdividing these portions of the floodplain into smaller units of differing Pb concentration. However, geochemical data collected at uniform increments from the channel revealed that Pb concentrations decreased semi-systematically with distance from the river, as expected from our discussions in Chapter 6. The highest concentrations were associated with the levee and exceeded 1,200 μg/g. Beyond the limits of the levee, concentrations dropped to less than 750 μg/g. A series of cores taken along a transect perpendicular to the channel revealed that the thickness of the contaminated deposits also decreased with distance from the channel, and was a function of the levee topography. Concentrations beneath both the levee and other parts of the floodplain dropped abruptly below a dark colored, A-horizon buried beneath the surface.

Above the floodplain level, surface sediments of the upper terrace exhibited concentrations consistent with background values. Surface deposits of the lower terrace possessed concentrations above background levels (20 μg/g), but generally below 100 μg/g. These contaminated sediments were attributed to the deposition of fine sediment during recent floods.

Based on the geomorphic, stratigraphic, and geochemical relationships, the data collected from the studied reaches were used to map the entire valley floor within the project area. After the mapping was completed, a new round of sampling and analysis was conducted, and minor variations in unit boundaries were performed. These cartographic data could then be used to identify areas possessing "significant" sediment-borne Pb for the entire site, specific reaches along the channel, or with respect to their potential to be remobilized and enter the water column. Alternatively, areas could be delineated according to whether they were likely to exhibit Pb concentrations exceeding the action level and therefore requiring remediation. Once identified, the collection of additional field data could be used to verify the mapping results. For example, in the case of our hypothetical site, the channel bed and both terraces could be eliminated from further consideration as concentrations were well below the action level of 1,000 μg/g (Fig. 7.21). Sediments from the oxbow lake, the natural levees, and the laterally accreted deposits all exceeded 1,000 μg/g, and, thus, are potentially in need of remediation. The remaining floodplain deposits possessed concentrations that were both above and below the remediation standard. However, because concentrations tended to decrease with distance from the channel, a simple regression model could be used to approximate the region of the floodplain where concentrations were likely to exceed 1,000 μg/g.

Determining *areas* of contamination usually represents only part of the equation. The *volume* of contaminated sediment must also be known to estimate the mass of contaminants that may be exchanged with the water column, or that may require remediation as part of the cleanup process. While estimating the area possessing pollutants is a relatively straightforward process, determining the volume of sediment characterized by a specific range of concentrations is much more difficult because it requires an understanding of both the thickness and geometrical form of the depositional units. Unit thickness, of course, is not only dependent on surface topography, but the nature of the lower bounding surface. While variations in surface topography can be very accurately determined, the nature of the lower boundary is problematic. Thus, in most cases we must simplify the nature of the lower boundary to estimate sediment volumes.

Determining the geometrical form of the deposits can be even more difficult. In fact, we are usually left with using rather crude surrogates of the unit's geometric form based on field observations. For instance, the cross-sectional geometry of the natural levees in our hypothetical example could be envisioned as a wedge of sediment where the lower boundary is characterized by a flat, horizontal surface. The maximum thickness of the levee, in this case, can be determined according to height of the levee crest above other portions of the floodplain, on the basis of core data, or on a combination of the two. Unit thicknesses and geometry may also be determined in some areas using shallow geophysical techniques such ground penetrating radar.

7.8. SUMMARY

Rivers strive to develop a specific combination of cross-sectional dimensions, shape, gradient, planimetric configuration, and pattern that allow them to transport the available load as efficiently as possible given the existing hydrologic regime. As sediment load and discharge change, rivers will attempt to maintain this high degree of efficiency by adjusting the above properties. Because discharge and sediment load are continually changing, rivers cannot maintain equilibrium as a steady-state condition, except perhaps over very short periods of time. Rather the various properties of a river must perpetually adjust. Nevertheless, over long-periods of time (years to decades), a balanced or equilibrium condition is commonly maintained as the adjusting variables vary around a mean. Major changes in climate, tectonics, or land-use, however, may alter the hydrologic or sedimentologic regime to the point where adjustments around the mean can no longer maintain the most efficient system. When these limits to the equilibrium condition, called thresholds, are reached, major changes in channel form, planimetric configuration, or pattern will occur by altering the rates and magnitudes of erosion and deposition. The responses of the river to disturbance may involve adjustment in one or more of the above parameters and the rate, magnitude, direction of the adjustment can vary in both time and space. Moreover, the adjustments in river morphology may lag behind the disturbance by significant periods of time and require decades to complete.

Major readjustments in channel morphology can significantly alter (1) the rates of sediment-borne trace metal transport through time, (2) the quantity and location of contaminated sediment storage, and (3) the local physiochemical environment within the alluvial deposits, particularly with regards to the pH and Eh of the porewaters. It is therefore important to understand the past history of the river system as well as the current and future stability of the channel. Determining the age of fluvial terraces with respect to the rates of contaminant influx is particularly useful in predicting patterns in trace metal concentration with height above the channel bed and the location of contaminant hotspots along the river valley.

7.9.   SUGGESTED READINGS

Brewer PA, Taylor MP (1997) The spatial distribution of heavy metal contaminated sediment across terraced floodplains. Catena 30:229–249.

Ritter DF, Kochel RC, Miller JR (1999) The disruption of Grassy Creek: implications concerning catastrophic events and thresholds. Geomorphology 8:287–304.

Rowan JS, Barnes SJA, Hetherington SL, Lambers B, Parsons F (1995) Geomorphology and pollution: the environmental impacts of lead mining, Leadhills, Scotland. Journal of Geochemical Exploration 52:57–65.

Schumm SA (1969) River metamorphosis. Proceedings of the American Society of Engineers, Journal of the Hydrology Division 1:255–273.

Schumm SA, Brakenridge GR (1987) River response. In: Ruddiman WF, Wright HE Jr (eds) North American and adjacent oceans during the last deglaciation, Geological Society of America, Centennial Special Volume K-3, Boulder.

Stigliani WM (1991) Chemical time bombs: definition, concepts, and examples. Executive Report 16, International Institute of Applied System Analysis, Laxenburg Austria.

CHAPTER 8

REMEDIATION AND REMEDIATION STANDARDS

8.1.   INTRODUCTION

Remediation is a process conducted to prevent, minimize, or mitigate damage by a contaminant to human health or the environment (USEPA 1988). It can involve a wide range of techniques aimed at removing, containing, destroying, or reducing the exposure of humans and other biota to a pollutant. There are few environmental issues that are more arduous, complex, and controversial than the remediation of contaminated sites, including river basins, as the debate includes a number of highly sensitive societal concerns. These concerns range from the protection of human health and valuable ecosystems, to questions of liability and property valuation, to local economic development and faith that the applied remedial technologies will achieve their objectives (Soesilo and Wilson 1997). As a result, nearly everyone will complain about some aspect of a site remediation program. There is, in fact, a sort of continuum that develops between those who argue that remedial action of any kind is absurd and those who contend that numerous additional and often unreasonable steps are required to adequately protect human and ecosystem health.

The remediation of contaminated rivers can be an extremely expensive process. For example, Table 8.1, which provides estimated costs for the characterization, assessment, and remediation of selected contaminated rivers, illustrates that it is not uncommon for site investigation and cleanup to exceed tens of millions of dollars. The costs involved in site remediation are dependent in no small way on the technologies that are applied to cleanup the site. The selection of the proper technologies, in turn, depends on a number of factors, not least of which is the regulatory environment (Grasso 1993; Soesilo and Wilson 1997). Other factors to be considered include the technology's long-term effectiveness, permanence, and implementability, their capacity to reduce contaminant toxicity, mobility, and/or volume, their acceptance by the community, and various engineering considerations (see, for example, the criteria used to evaluate technologies in the RI/FS process used by the U.S. Superfund Program, USEPA 1988).

Significant discussion has occurred during the past several decades regarding the decision-making processes that are most effective for selecting remediation

Table 8.1. Estimated costs of investigation and remediation for selected contaminated rivers

Site name	Major contaminants	Cleanup costs (in millions of $)	Reference
St. Louis River, MN	Various Contaminants	0.25	2
St. Clair River	Pentachlorophenol	0.35	2
Lower Menominee River, WI/MI	Paint sludge	0.50	2
Rouge River, MI	PCBs	0.75	2
Davis Creek, MI	PAH, Hg, Pb	0.9	2
Rouge River, lower branch, MI	Zn	1	2
Lower Menominee River, WI/MI	Arsenic	1.3	2
Black River, OH	PAH	1.5	2
Welland River, Ontario	Various Contaminants	2.6	2
Monguagon Creek	Various Contaminants	3	2
Maumee River, OH	PCBs	5	2
River Raisin, MI	PCBs	6	2
Kalamazoo River, MI	PCBs	7.5	2
Gill Creek, NY	Various Contaminants	10	2
Shiawassee River, MI	PCBs	13.6	2
Black and Bergholtz Creeks, NY	Dioxin	14	2
Manistique River, MI	PCBs	25	2
Saginaw River, MI	PCBs	28	2
Ashatabula River, OH	PCBs, llrw,*heavy metals, chemicals	>50	2, 3
St. Lawrence River, NY	Various Contaminants	53.7	2, 6
Grand Calumet River, IN	PAHs, PCBs, heavy metals	55	2
St. Lawrence River, NY	PCB	78	2, 6
Hudson River, NY	PCBs	100–150	4
Alamosa River, CO	Cyanide, trace metals	>170	5
Rio Agrio, Spain	Pb, Zn, other metals; tailings from dam failure	100–200	7
Grasse River, NY	PCBs	250	2, 6
Fox River, WI	PCBs	>300	1
Coeur d'Alene	Zn, Pb, other trace metals	360	8

1 Fox River Watch (2002); 2 Zarull et al. (1999); 3 Environmental News Service (2006); 4 ENR (2006); 5 Radio free Europe (1995–1999); 6 EPA (2006); 7 WISE (2006); 8 NRC (2005)

alternatives (Corbett 1988; Hodges et al. 1997; Petts et al. 1997; Soesilo and Wilson 1997; Peterson et al. 1999; Younger et al. 2005). This has resulted in the creation of numerous guidance documents for specific cleanup programs (e.g., USEPA 1988, 1993, 2005). While significant variations in methodologies exist, most contain a series of closely related steps, including (1) the development of remediation objectives, (2) a characterization of the potential threats of a pollutant(s) to human or ecological health (as determined from the site assessment), (3) the identification of the potential remediation alternatives that may be applied to the site to meet those objectives, (4) an evaluation of the identified remediation alternatives, and the subsequent selection of one or more remedies to reduce risks at the site, and (5) the establishment of action or cleanup levels, where appropriate, that will protect human and ecological health from the contaminant of concern (USEPA 2005). It is important to recognize that a particular technology can be applied to the entire site, a specific geographical region of the site (sometimes called an operable unit), or a particular media (e.g., soil, sediment, or water). It is possible, then, that decision-makers will have multiple choices for specified regions or media, and the selection process must incorporate these technologies into a set of alternatives that are most effective for the remediation of the entire river basin (which we will refer to as the remedial strategy).

Many of the processes and data types examined in the preceding chapters form the foundation for completing the general steps involved in developing a remediation strategy (described above). Others, such as the methods and criteria for selecting remedial alternatives, include social, economic, and political factors which fall well outside of our focus on geochemical and geomorphological processes. Thus, the intent of the following two chapters is not to describe the processes involved in selecting remediation technologies or alternatives. Rather, the goal is to provide an overview of the techniques that are commonly applied to remediate metal contaminated alluvial sediments, and to highlight important aspects of fluvial systems that must be taken into account when developing a remediation strategy. First, however, we must briefly examine a number of important factors which are used to both define and prioritize areas or units within the site for remediation.

8.2.   SETTING REMEDIATION STANDARDS

For every contaminated site where some action is necessary, a concentration must be defined above which the risks are considered to be unacceptable. Development of these remediation standards or targets is one of the most difficult, controversial, and important aspects of the remediation process because it governs the time and cost of remediation, the area to which remediation will be applied, the remedial technologies which can be effectively used, and the overall reduction in the risks of the contaminant to human and ecosystem health.

It may seem prudent to take a conservative approach in which all of the contaminants released to the river basin are extracted, thereby returning the system to a pre-contaminated state. Such an approach, however, is now seen as impractical in many cases. Not only does it push the limits our technology, but it also adds significantly to the time and costs required for cleanup while providing only minimal benefits to human health and the environment (Wildavsky 1995).

The selection of a remediation standard for a given media is usually based on one or more criteria. The most frequently used criteria include: (1) geochemical background levels (which for some chemical substances is the level of analytical detection),

(2) health or risk-based standards, (3) site-specific human and/or ecosystem risk assessments, (4) the levels achievable using the best available technologies, and (5) for soils and sediment various extraction or leaching tests (e.g., the toxicity characteristic leaching procedure (TCLP), which is widely used in the U.S.). Exactly which of these criteria are applied to a site depends on the regulatory framework, the nature of the contaminants, and the characteristics of the site, particularly the potential for humans and other biota to come into contact with the contaminated media.

Geochemical background, as used above, refers to the concentration of a trace metal or other substance within a given media which has not been affected by a release from a contaminant source. For organic contaminants which are not found naturally in river systems, background is effectively zero. The use of background represents a conservative approach because contaminant concentrations are being returned to those that existed prior to the release of the substance to the river system.

In contrast to using background, *health* or *risk-based standards* represent a single value that may be applied uniformly from one site to another (although the established standards often vary between regulatory bodies). These standards essentially define a level of contamination that is acceptable, where acceptability is dependent on a specific risk level for which corrective action is warranted (Soesilo and Wilson 1997). A significant advantage of these pre-determined standards is that they can be used to immediately determine if a specific media at a site requires remediation. For example, if the alluvial sediment from a specific area contains 1,200 mg/kg of Pb, and the standard is 1,000 mg/kg, then the sediment must be addressed. The approach leaves little room for debate and because everyone understands the cleanup objectives, it greatly accelerates the remediation process (Soesilo and Wilson 1997). A difficulty in this approach, however, is that for some rivers, particularly those in mountainous terrains draining mineralized zones, the standard may be below local background values. The remediation of water and sediment to levels below background is prohibitively costly, may actually damage the ecosystem, rather than protect it, and usually makes little sense because the site will be "recontaminated" to background levels (Reimann and Garrett 2005). Perhaps a more important shortcoming of the approach is that it does not consider the unique aspects of the site which play an important role in controlling the effects of the contaminant(s) on human health or the environment. For these reasons, health or risk-based standards are usually used in combination with other criteria.

A recent trend in remediation is the use of *site-specific, risk assessments* to set safe, but less stringent remediation standards. This site-specific approach represents a process that determines the extent to which a contaminant threatens human and ecological health. In many, if not most instances, emphasis is placed on the human component. In this case, risk characterization consists of two linked but separate analyses referred to as a toxicity assessment and an exposure assessment. The *toxicity assessment* is intended to identify the potential health effects that may occur as a result of exposure to the contaminant of interest. The outcome is the establishment of a dose level above which the substance in question may affect human health and below which the contaminant poses no direct threat (Soesilo and

Wilson 1997). The *exposure assessment* determines the frequency, magnitude, and pathways of exposure of a substance to humans (typically of different age groups). The product of the assessment is given in terms of exposure (e.g., as daily intake as expressed in units of mg/kg/day). When the two assessments are combined, the results can be compared to a target risk level (e.g., 1:1,000,000) in order to determine what, if any, action should be taken.

One of the advantages of a site-specific risk assessment is that both current and future land-use considerations can be directly included in the analysis (i.e., whether the property is characterized by residential, industrial, or some other form of non-residential activity). Moreover, it tends to be highly cost effective. In fact, Millroy (1995) estimated that science-based risk assessments could reduce the costs of waste site remediation by 60%. In spite of its advantages, it has been criticized because (1) there is no readily accepted methodology, (2) the analytical methods are usually difficult to understand, (3) the approach often over emphasizes cancer risks, and (4) the toxicity of a specific contaminant is often in question. Nonetheless, its cost effective nature, and its focus on the site of concern, has led to increased use of risk based assessments in recent years, a trend that is likely to continue in the future.

The TCLP was designed to assess the potential mobility of both organic and inorganic contaminants from waste disposed in landfills. The actual procedure is rather complicated and involves the mixing of a solid, which may be filtered from a fluid, with a slightly acidic solution, after which the leachate is analyzed. The TCLP test has become an important and costly industrial analysis in the U.S. because potential waste generators must use it to determine if they are producing hazardous waste. More specifically, if concentration of the leachate exceeds the threshold (maximum) concentration for one of 40 different substances, it is considered a hazardous waste under the Resource Conservation and Recovery Act (RECRA) (40 CFR Part 261).

In the case of metal contaminated rivers, the TCLP may be used to determine at what total concentration sediment will likely be labeled as a hazardous waste (Table 8.2). A hazardous waste, in this case, is defined as (1) a waste that may cause an increase in mortality, or an increase in a serious illness, or (2) a waste

Table 8.2. Summary of TCLP maximum concentration of selected metals and metalloids for toxicity characteristics

Metal	Regulated Level (mg/l)
Arsenic	5.0
Barium	100.0
Cadmium	1.0
Chromium	5.0
Lead	5.0
Mercury	0.2
Selenium	1.0
Silver	5.0

that could pose a threat to human health and the environment when improperly managed. Clearly there will be resistance to leaving sediment in place without some form of treatment when the materials exceed the TCLP threshold values. The TCLP test may also influence how the contaminated sediment is treated and disposed of if, in fact, it is removed from the site.

Establishing standards based on the contaminant levels which are achievable using the best available technology is rarely used for sediment contaminated solely by metals. It usually applies to certain types of organic compounds or mixed wastes which can be difficult to treat. The logic behind its use is that there would be no need to set a cleanup limit below that which can be effectively achieved using the most sophisticated techniques currently available.

A common problem inherent in setting an effective cleanup level for many contaminated rivers is that the remediation goals are imprecisely stated. Typical examples drawn from actual projects are to "eliminate adverse ecological impacts" to "reduce human health risks from direct contact or from fish consumption", or to "reduce contaminant levels in fish" (see, Cushing 1999 for additional discussion and other examples). None of these statements provide a clear indication of how the attainment of these objectives can be measured (i.e., what constitutes success). Moreover, quantitative data linking the concentration of sediment-borne trace metals and other contaminants to human health is limited in comparison to other media such as air, water, and food (IJC 1997). The definition, then, of a remediation standard is left open for debate often allowing the final remedial standard to be inconsistent with the level of risk reduction that is warranted for the site.

Another unfortunate, but perhaps not surprising, discovery in recent years is that the action level as well as the remediation strategy is often influenced more by stakeholder perceptions than by sound scientific data. For example, in 1992 approximately 25,000 to 50,000 m^3 of acidic, metal enriched mine waters were released into the Carnon River in the U.K. from the Nangiles Adit (Younger et al. 2005). In spite of the enormity of the release, detailed scientific investigations demonstrated that the ecological health of the Fal estuary, into which the Carnon River drained, remained intact. Moreover, the scientific data indicated that even if the continuing flow of waters from the mine were left untreated, they would result in negligible impacts to the estuary (Younger et al. 2005). However, there was a strong public perception that any discoloration of the water entering the estuary by mine effluent would degrade the quality of maerl (a calcified seaweed which provided significant income for the local community). Although the perception was not based on any scientific data, it resulted in the continued and costly application of pump and treatment operations at the mine site (Younger et al. 2005).

8.2.1. Background Concentrations

In the previous section, we indicated that background concentrations of trace metals in water and sediment can be used as a remediation standard (i.e., the level to which cleanup will occur). Even where background is not selected as the remediation standard, it will likely be necessary to determine background values for specific media. For example, where risk-based standards are used to set action levels, it will be necessary to insure that the standards do not exhibit concentrations below the local background values for a specific element. With this in mind, we need to take a closer look at what constitutes background, and the difficulties that can be encountered while attempting to define it for trace metals.

The term background has its roots in economic geology. It was developed to separate those areas characterized by "normal" elemental concentrations from regions of the landscape exhibiting anomalously high values and which may therefore contain ore bodies (Hawkes and Webb 1962). These early studies made it clear that background could not be described by a single value. The elemental concentrations observed in non-mineralized regions possessed a range of values resulting from minor differences in particle size, organic carbon content, mineralogy, the abundances of Fe and Mn oxides and hydroxides, and a host of other parameters. If, for example, 20 samples were collected from the channel bed of a river draining an area devoid of mineralization and analyzed for Pb, the measured concentrations would fall within a definable range, which often exhibits a bell-shaped distribution (Fig. 8.1). It follows, then, that in order to identify an anomalous

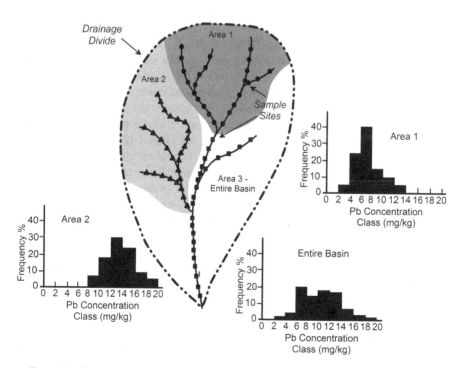

Figure 8.1. Hypothetical differences in the distribution of Pb concentrations collected from three separate areas (subcatchments) of the basin. Distribution assumes no anthropogenic contamination; differences are due to variations inherent in the underlying bedrock

area it is necessary to quantitatively define an upper boundary to these "normal" variations in elemental concentrations. This upper limit is referred to as a *threshold*. Values exceeding the threshold are considered by definition to be anomalous.

In the case of contaminant studies, the anomaly that is being identified is that of contamination by anthropogenic substances. As such, *background* can be defined as the range of elemental concentrations that occur in a given media which have not been affected by human activities. A significant problem that emerged from this idealized definition, however, is that there are few, if any, sites which have not been affected by the input of trace metals from anthropogenic sources. The analyses of ice cores from the poles have even shown that metals released from industrial activities have been transported to, and redeposited in, these remote localities by atmospheric processes. The recognition that truly non-contaminated materials are extremely rare led to a plethora of definitions for background (Reimann et al. 2005), none of which have been universally accepted. The term *ambient background*, for example, has been used by some to describe materials in which the natural (or true) background has been affected by the input of minor (and unmeasurable) quantities of the element of interest, whereas *pre-industrial background* is often used to describe materials representative of the pre-industrialized condition (Reimann et al. 2005). Others, particularly regulatory bodies focused on the remediation of specific sites, have elected to define background as a function of the contaminant source. In the case of the USEPA, background is defined "as substances or locations that are not influenced by the release from a site (USEPA 2002)." From this perspective, the anomaly is defined according to whether a geographical location or media exhibits constituent levels that are elevated with respect to the surrounding area because of onsite releases of the substance. Moreover, background concentrations, in this case, are the product of both natural and anthropogenic sources (Fig. 8.2).

No matter how it is defined, one of the difficulties of determining background is that it is scale dependent. In fact, Reimann and Garrett (2005) argue that no single range of background values exist for any element except for very localized and specific cases. Rather, background will change between project areas and even from one area to another within the boundaries of the contaminated site. As illustrated in Fig. 8.1, their conclusions are particularly relevant to contaminated rivers. Background concentrations determined for one tributary will often differ from that determined for another, both of which will differ from the basin as a whole. Differences reflect spatial changes in the underlying bedrock or soil types as well as differences in various physiochemical processes. It may be necessary, then, to define background for specific subbasins or regions of the watershed and, even when this approach is undertaken, the defined range will possess a degree of uncertainty (Matschullat et al. 2000).

Numerous methods have been developed for defining background and for comparing background data to sites where contamination is suspected (Matschullat et al. 2000; USEPA 2002; Reimann and Garrett 2005; Reimann et al. 2005). In most cases, the comparison is not based on the range of background concentrations per se, but on the *threshold* (i.e., the concentration that defines the upper limit

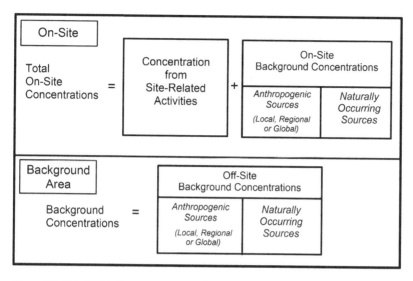

Figure 8.2. Relationship between on- and off-site concentrations and anthropogenic sources as used by the USEPA when defining background at hazardous waste sites (From Breckenridge and Crockett 1998)

of the background values). Most quantitative approaches are based on a statistical analysis of the data. Many of these techniques define background as the mean plus or minus two standard deviations (mean $+/-2\sigma$) which roughly includes 95% of the samples. The use of the mean $+/-2\sigma$ approach has been criticized because it is based on the assumption that the data exhibit a normal or log-normal distribution, and, thus, the data are part of a single population. In the search for a contaminant anomaly, however, trace metal data are likely to be associated with different populations because they were derived from different sources. In addition, both the mean and standard deviation are significantly influenced by extreme values. Thus, Reimann et al. (2005) argue that the mean $+/-2\sigma$ rule is invalid for many sites, and that the use of techniques based on median $+/-2$ median absolute deviation values are better suited to characterizing background.

One of the oldest and most simplistic methods available to determine background values is through the use of cumulative distribution plots. The approach involves the graphical evaluation of cumulative frequency and elemental concentration data on either arithmetic or logarithmic scales (Fig. 8.3). Deviations from a straight line are assumed to represent anomalous values, and therefore, inflections in the trend can be used to establish both the range of background values in an area and the threshold concentration. Specific line segments may also be used to identify the range of concentrations associated with individual contaminant sources or populations.

Although cumulative distribution plots are based on normally distributed data, Reimann et al. (2005) point out that geochemical data often exhibit a symmetrical distribution, or can be transformed to approximate one. Moreover, they argue that the method has several advantages over other methods including (1) the portrayal

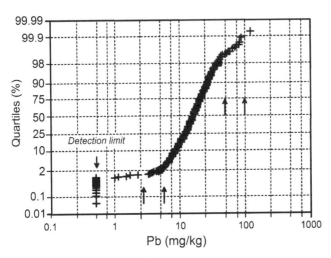

Figure 8.3. Cumulative probability plot for Pb within the Ap-horizon of agricultural soils in northern Europe. Arrows correspond to inflection points that may indicate thresholds between individual populations with a distinct range of Pb values (Adapted from Reimann et al. 2005)

of individual data points on the graph, (2) the identification of extreme outliers as a single value, (3) the easy determination of outliers and their degree of deviation from the rest of the data, and (4) the display of data in a widely used format.

There are several rules-of-thumb that should be followed in any attempt to compare background concentrations to those derived from a potentially contaminated site. First, the background sediment must be derived from areas which have not been impacted by the suspected contaminant source. Second, samples must be collected and analyzed using the same analytical procedures as the materials suspected to be contaminated, and preferably, at the same time. Finally, sediment from the background and project site should be similar in terms of their grain size distribution, organic matter content, mineralogy, etc.

In the case of rivers, potential background sites include reaches located upstream of the source, tributaries, terraces located above the historic flood regime, and which pre-date site contamination, and buried alluvial deposits that are known to pre-date the contaminant release (Fig. 8.4). It is probably best to avoid the use of buried sediment located downstream from the suspected source unless the vertical translocation of metals can be ruled out.

In most cases, it is difficult to find background sediment having the same or similar characteristics to that sampled along the contaminated reach. A commonly used method to compensate for sedimentological differences is the development of normalization models (Loring and Rantala 1992). These models typically involve the uni- or multivariate regression of selected independent variables (e.g., percent of silt and clay, organic matter content, or the concentration of Fe and Mn oxides and hydroxide) with metal concentrations. The "true" background concentration of

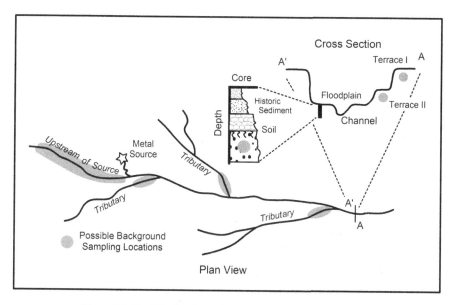

Figure 8.4. Possible sites from which to sample background sediment

a sample from the contaminated site can then be determined from the regression equations on the basis of similar parameter characteristics (Fig. 8.5). In other words, if a sample contains 35% silt and clay, the background concentration of the sediment containing that quantity of fine material can be determined using a regression equation between metal concentration and the percent of silt and clay in the background samples. It is then possible to estimate the degree to which the contaminated sample has been enriched in a given trace metal by dividing the measured concentration of the sample ($M_{observed}$) by its predicted concentration ($M_{predicted}$) thereby yielding a non-dimensional enrichment factor (EF), or:

$$(1) \qquad EF = M_{observed}/M_{predicted}$$

When only a single normalizing parameter is used, enrichment factors can be determined directly (i.e., without utilizing a regression model), using the following equation:

$$(2) \qquad EF = (X/Y)_{sample}/(X/Y)_{background}$$

where, X is the concentration of the trace metal, Y is the value of the normalizing parameter (Covelli and Fontolan 1997).

The selection of the parameter(s) to use within the regression model is dependent on the trace metal under consideration, the characteristics of the site, and the preferences of the investigator. It is important, however, for the normalizing variable(s) to reflect processes that strongly influence sediment-borne trace metal concentrations

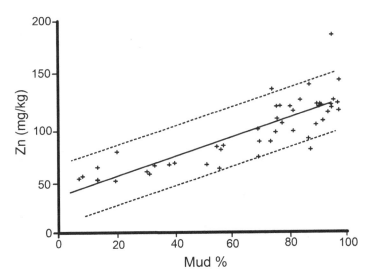

Figure 8.5. Scatter plot of Zn versus percent silt and clay for sample from the St. Lawrence estuary. Solid line represents regression ($r = 0.82$); dashed lines define the 95% confidence interval (From Loring and Rantala 1992)

at the site (Covelli and Fontolan (1997). A commonly used proxy to account for variations in grain size is Al because of its abundance in aluminosilicate minerals (Windom et al. 1989; Covelli and Fontolan 1997). Loring (1990), however, found that Li was equal to or superior to Al for the normalization of sediment derived from many types of crystalline and non-crystalline rocks.

A final note on background concentrations concerns data obtained from a number of surveys conducted since the 1960s that are used to provide insights into the typical ranges of trace metals and other elements that occur naturally in soils (Kabata-Pendias and Pendias 1984) and specific rock types (e.g., shale; Turekian andWedepohl 1961; Taylor 1964) (Table 8.3). These survey data should not be used as background concentrations (Breckenridge and Crockett 1998; Matschullat et al. 2000) because they do not consider the specific characteristics of the drainage system. Nonetheless, they may provide valuable insights into whether elevated levels of sediment-borne contaminants exist along the investigated reach, particularly where background data from the site are lacking.

8.2.2.   Health and Risk Based Standards

The establishment of a remediation standard for river bed sediment is likely to depend, at least in part, on guidelines that have been established to determine at what concentration the contaminant poses a threat to sediment-dwelling organisms, wildlife, or humans. For alluvial sediments, these standards have been referred to by a variety of names such as sediment quality criteria, sediment quality objectives,

Table 8.3. Typical concentrations of selected elements rocks and soils; all values reported in µg/g

Element	Granite	Basalt	Shale	Sandstone	Soils[1] (average)	Soils[1](Range)
Lithium	30	17	66	151		
Chromium	10	170	90	35	54	1.0–2000
Manganese	450	1500	850	50	560	<1–7000
Nickel	10	130	68	2	19	<–700
Copper	20	87	45	2	25	<1–700
Zinc	50	105	95	16	60	<5–2900
Arsenic	2	2	13	1	7.2	<0.1–97
Selenium	0.05	0.05	0.6	0.05	0.39	<0.1–4.3
Molybdenum	1	1.5	2.6	0.2	–	<3–7.0
Silver	0.04	0.1	0.07	–	–	–
Gold	0.002	0.002	0.005	0.006	–	–
Cadmium	0.13	0.2	0.3	–	0.06	0.01–0.7
Antimony	0.2	0.1	1	0.4	0.67	<1–8.8
Mercury	0.03	0.01	0.4	0.03	0.089	<0.01–4.6
Lead	17	6	20	7	19	<10–700
Uranium	3	1	4	2		

Adapted from Turekian (1971), Martin and Meybeck (1979), and Buonicore (1996)
[1] Data from Buonicore (1996)

sediment quality standards, or sediment quality guidelines. The specifics of how they are used in the cleanup process can vary significantly from one regulatory body to another. In many, if not most, instances, two sediment quality guidelines are presented which subdivide contaminant concentrations into three distinct ranges. The lower guideline concentration is intended to represent a level below which the contaminant will cause no adverse effects. The upper concentration represents a value above which adverse effects are expected with a high degree of certainty. Unfortunately, the nomenclature used to delineate these ranges has been inconsistent as the upper and lower guideline concentrations have been given a host of names (Table 8.4). Here we will follow the lead of MacDonald et al. (2000) and refer to the lower guideline as the threshold effects concentration (TEC), and the upper guideline as the probable effect concentration (PEC).

Selected TEC and PEC concentrations, based on their potential to harm sediment-dwelling organisms, are shown in Table 8.5 for a number of trace metals and metalloids. The listed sediment quality guidelines can differ by several fold as a result of differences in the procedures and assumptions that were used to develop the guidelines as well as their intended use. Exactly which set of guidelines will be applied by a specific regulatory body depends on the receptors which will be considered (e.g., benthic organisms, wildlife, or humans), the region to be considered, the purpose for which they are applied, and the degree of protection that is deemed necessary (MacDonald et al. 2000). For some contaminated rivers, however, there are no pre-described set of guidelines that will be applicable to the site and the program manager will need to select an appropriate sediment quality guideline for the river system. This can be a daunting task as no one method may be best suited for the full range of biophysical conditions

Table 8.4. Description of selected freshwater sediment quality guidelines

Sediment Quality Guideline	Acronym	Description	Reference
Threshold Effect Concentration (TEC)			
Lowest effect level	LEL	Sediments are clean to marginally polluted; no effects on majority of sediment-dwelling organisms below this concentration	Persaud et al. (1993)
Threshold effect level	TEL	Concentration below which adverse effects are expected to only rarely occur	Smith et al. (1996)
Effect range-low	ERL	Represents concentration below which adverse effects are rare	Long and Morgan (1991)
Threshold effect level for *Hyalella azteca* in 28-day tests	TEL-HA28	Concentration below which adverse effects on *Hyalella azteca*, an amphipod, are expected to occur only rarely in 28-day test	USEPA (1996a)
Minimal effect threshold	MET	No effects are expected below this concentration; sediments are considered to be clean to marginally contaminated.	EC and MENVIQ (1992)
Chronic equilibrium partitioning threshold	SQAL	Concentration in sediments predicted to be associated with concentrations in pore waters below the chronic water quality criteria; adverse effects are thought to be rare below this concentration.	Bolton et al. (1985; Zarba 1992); USEPA (1997a)
Probable Effect Concentration (PEC)			
Severe effect level	SEL	Sediments are heavily polluted; adverse effects on majority of sediment-dwelling organisms are expected.	Persaud et al. (1993)
Probable effect level	PEL	Concentration above which adverse effects are expected to occur.	Smith et al. (1996)
Effect range-median	ERM	Concentration above which adverse effects are expected to occur	Long and Morgan (1991)
Probably effect level for *Hyalella azteca* in 28-day tests	PEL-HA28	Concentration above which adverse effects on *Hyalella azteca*, an amphipod, are expected to occur frequently in 28-day test	USEPA (1996a)
Toxic effect threshold	TET	Sediments are heavily contaminated; concentration above which adverse effects on sediment-dwelling organisms are expected.	EC and MENVIQ (1992)

Modified from MacDonald et al. (2000)

Table 8.5. Selected sediment quality guidelines for metals in freshwater ecosystems (in mg/kg); see Table 8.4 for guideline description

	Sediment Quality Guideline					
Threshold Effect Concentration						
Element	**TEL**	**LEL**	**MET**	**ERL**	**TEL-HA28**	**Consensus Based TEC**
Arsenic (As)	5.9	6	7	33	11	9.79
Cadmium (Cd)	0.596	0.6	0.9	5	0.58	0.99
Chromium (Cr)	37.3	26	55	80	36	43.4
Copper (Cu)	35.7	16	28	70	28	31.6
Lead (Pb)	35	31	42	35	37	35.8
Mercury (Hg)	0.174	0.2	0.2	0.15	–	0.18
Nickel (Ni)	18	16	35	30	20	22.7
Zinc (Zn)	123	120	150	120	98	121
Probable Effect Concentration						
Element	**PEL**	**SEL**	**TET**	**ERM**	**PEL-HA28**	**Consensus-Based PEC**
Arsenic (As)	17	33	17	85	48	33.0
Cadmium (Cd)	3.53	10	3	9	3.2	4.98
Chromium (Cr)	90	110	100	145	120	111
Copper (Cu)	197	110	86	390	100	149
Lead (Pb)	91.3	250	170	110	82	128
Mercury (Hg)	0.486	2	1	1.3	–	1.06
Nickel (Ni)	36	75	61	50	33	48.6
Zinc (Zn)	315	820	540	270	540	459

From MacDonald et al. (2000)

that exist. Moreover, the causes of uncertainty inherent in each approach may differ significantly. To compensate for these shortcomings, MacDonald et al. (2000) have developed consensus-based sediment quality guidelines in which the geometric mean of TEC and PEC obtained from multiple methods are combined to set guideline concentrations. These consensus-based guidelines are likely to be applicable to a relative wide range of environments (Table 8.5).

8.3.   PRIORITIZING CLEANUP

Large contaminated river basins are often referred to as megasites by the USEPA as they cost more than $50 million to remediate and may require decades to complete. Cleanup of the Cour d'Alene River basin in Idaho, for example, is expected to exceed 30 years (NRC 2005). When dealing with such large and complex sites, it is important to separate areas or media which require immediate attention from those of lower priority which can be put on hold. A significant consideration in developing the prioritized list will undoubtedly be the ability to rapidly reduce the risk to human health. As such, attention will focus on the removal or containment of contaminants in soils and sediment from residential properties, parks, or other areas where humans may come into contact with contaminated materials. The removal of point sources of contamination is also likely to be viewed as a high priority action because of the need to terminate further contamination of the aquatic system. Other priority areas often include valuable habitats (e.g., riparian wetlands) which serve a wide range of species.

A criterion for prioritizing areas of remediation along contaminated rivers which is usually recognized, but which is often given inadequate attention, is the area's potential to be re-contaminated. Clearly, it makes no sense to restore a riparian wetland if it is likely to be flooded and receive contaminated sediment on a regular basis. While this may seem obvious, it often occurs because discrete regions of the basin are delineated and addressed separately, rather than using a systems approach (see, for example, NRC 2005). As a result, remedial actions in one part of the basin are not integrated with those in other parts of the basin. It is also related to the misconception that if concentrations of the upstream deposits are below the action level, remobilized materials cannot re-contaminant downstream areas to levels exceeding the remediation standards. This is not necessarily true. It is quite possible, for example, for the upstream deposits to consist of a mixture of sand, silt and clay. Once eroded, the fine-grained sediment can be partitioned and concentrated during deposition in low-energy environments, such as floodplains (as was shown in Chapter 6). If partitioning is significant, these newly formed deposits may possess trace metal concentrations exceeding those found in the sedimentary units from which they were derived. It follows, then, that for areas where recontamination is a potential issue, it may be necessary to (1) examine the current rates of sedimentation within the area or unit proposed for remediation (as determined by short-lived radionuclides or some other method; Chapter 6) to determine how quickly sediment-borne trace metals are likely to accumulate, and (2) determine the textural characteristics of the deposits within the area relative to those of the upstream sources to gain an understanding of the potential for sediment partitioning.

For relatively small sites, remediation strategies are typically intended to produce a final remedy. For complex river systems, however, developing an all encompassing remediation strategy may not be realistic because the high degree of uncertainty involved in describing fluvial processes may lead to the application of remedial actions that are either ineffective or produce unexpected consequences. Thus, the NRC (2005) has argued that an adaptive management strategy should be utilized. Adaptive management has been widely used in natural resource management (Holling 1978; Lee 1993; NRC 2003), particularly in the fields of river and riparian restoration. As applied to contaminated rivers, the adaptive management approach can be viewed as a six step process, organized in the form of a feedback loop in which the original expectations of the remedy are evaluated and used to modify and improve future actions (Fig. 8.6). It is important to recognize that simply monitoring the system for selected parameters does not constitute an adaptive management plan. Rather, adaptive management is a process where monitoring data are used in a pre-determined manner to assess if the implemented strategy has been successful. It necessarily requires the establishment of quantitative indicators

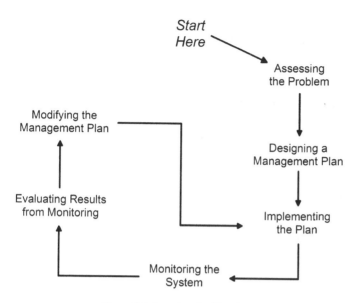

Figure 8.6. Steps involved in adaptive management approach

that can be monitored over the available time-frame to determine if implemented remedial action(s) should be modified (NRC 2005).

8.4.   SUMMARY

Remediation is a process conducted to prevent, minimize, or mitigate damage by a contaminant to human health or the environment. The development of remediation strategies generally revolves around (1) a characterization of the potential threats of a pollutant(s) to human or ecological health (as determined from the site assessment), (2) the identification of the potential remediation alternatives that may be applied to the site, (3) an evaluation of the potential impacts of the identified remediation alternatives on the site, and the subsequent selection of one or more remedies, and (4) the establishment of action or cleanup levels, where appropriate, that will protect human and ecological health from the contaminant of concern (USEPA 2005).

One of the most controversial aspects of the remediation process is the development of a remediation standard or target above which remediation will be required. These standards are set for specific contaminants of concern in specific media (e.g., water, sediment, and soil) and for specific areas or land-use types. In most cases, the standards are based on one or more of the following criteria: (1) background concentrations, (2) health or risk-based standards, (3) site-specific human and/or ecosystem risk assessments, (4) the levels of cleanup achievable using the best available technologies, and, for soils and sediments, (5) various extraction or leaching tests (e.g., TCLP).

Many river systems identified for site assessment and possible remediation cover large areas of diverse terrain. It is therefore necessary to prioritize remediation by focusing on areas which require immediate attention. Prioritization should include the potential for the area to be re-contaminated by sediment-borne trace metals as well as the usually considered concerns regarding the need to immediately reduce risks to both humans and the environment. An adaptive management approach, in which the original expectations of a remedy are evaluated and used to modify and improve future actions, may also be needed when addressing large, complex areas.

8.5.   SUGGESTED READINGS

MacDonald DD, Ingersoll CG, Berger TA (2000) Development and evaluation of consensus-based sediment quality guidelines for freshwater ecosystems. Archives of Environmental Contamination and Toxicology 39:20–31.

Matschullat J, Ottenstein R, Reimann C (2000) Geochemical background – can we calculate it? Environmental Geology 39:990–1000.

Reimann C, Garrett RG (2005) Geochemical background-concept and reality. Science of the Total Environment 350:12–27.

Soesilo JA, Wilson SR (1997) Site remediation: planning and management. Lewis Publishers, CRC Press Inc, Boca Raton.

Younger PL, Coulton RH, Froggatt EC (2005) The contribution of science to risk-based decision-making: lessons from the development of full-scale treatment measures for acidic mine waters at Wheal Jane, U.K. Science of the Total Environment 338:137–154.

CHAPTER 9

EX SITU REMEDIATION AND CHANNEL RESTORATION

9.1. INTRODUCTION

A remedial option considered for almost every polluted river system is the removal of contaminated sediment by means of excavation or dredging. In fact, in the U.S. excavation and dredging have been the preferred methods of remediation for contaminated sediment, having been used at more than 100 Superfund sites across the country (USEPA 2005). The wide spread use of sediment removal as a remedial strategy can be attributed to two factors. First, it has been shown to be an effective remedial technology in a wide range of riverine and marine environments. Second, it is consistent with recent legislation which favors remedial technologies that *permanently* reduce the volume, toxicity or mobility of the contaminant of concern.

Whether sediment removal is conducted by means of excavation or dredging depends in large part on where the contaminated materials are located. *Excavation* is defined as the subareal extraction of sediment using earthmoving equipment (e.g., backhoes and front-end loaders). As such, it is primarily applied to the removal of contaminants from floodplains and terraces, although it has also been used along shallow channels where flow can be diverted to other areas, or where sediment can be removed by equipment located along the channel margins. Dredging refers to the extraction of sediment from an underwater environment (NRC 1997). Dredging, however, is not only conducted to remove contaminated sediment (called *environmental dredging*), but for such purposes as the maintenance of navigable channels and other facilities (*navigational dredging*). The objectives of environmental and navigational dredging programs are quite different. In the case of environmental dredging, the intent is to remove all of the contaminated sediment from the riverine environment which is above the specified action level or target, while minimizing the spread of contaminants to the surrounding ecosystem (USEPA 1993; NRC 1997; USEPA 2005). In contrast, navigational dredging is designed to remove large volumes of subaqueous sediment as efficiently as possible. Here we will only examine environmental dredging and the term dredging will be limited to this form of subaqueous sediment extraction.

The removal of contaminants from alluvial deposits is not limited to the use of excavation or dredging techniques. Other methods, such as in situ soil flushing, electrokinetics, or phytoremediation may also be applied (Table 9.1). Moreover, a combination of remediation options may be combined as part of the remedial action plan to both remove and contain contaminated sediment at a site. The latter methods (containment) are aimed at reducing the exposure of humans and biota to the pollutant, without removing them from the riverine environment.

In this chapter, we will briefly examine the most significant remedial technologies that are available for dredging, excavating, and the ex situ treatment of the removed sediment. Other in situ methods of extraction and/or containment (Table 9.1) that can be applied to trace metals will be examined in the following chapter. In keeping with our focus on fluvial processes, a significant component of our analysis is related to the potential impacts of the various techniques on the geomorphic stability of the river and, when the system has been significantly modified, the methods that may be used to re-create stable and ecologically productive channel systems.

Table 9.1. Selected remediation alternatives for metal contaminated alluvial deposits

Method	Comments
Ex Situ Treatments	*(applied to excavated or dredged sediments)*
Dewatering	Extraction of water from removed sediment
Particle separation	Selective removal of sediments (e.g., fine particles) that contain relatively high concentrations of trace metals
Soil washing	Extraction of metals from the sediments using a water-based solvent which may or may not be combined with other reagents
Vitrification	Heating of contaminated materials to high temperatures to produce a glass-like non-leachable material with low-permeability
Solidification-stabilization	Addition of binding agents to produce a hardened material of low-permeability
In Situ Treatments	
Phytoremediation	Use of plants to extract trace metals from the soils and sediments
In situ vitrification	Same as above, but heat source is typical produced by an electrical current delivered through electrodes
In situ soil washing (soil flushing)	Same as above, with the exception that solutions are applied to and extracted from in situ materials
In situ solidification, stabilization	Encasing of contaminated material with a low-permeability substance to reduce mobility
Electrokinetics	Use of an electric current to concentrate and remove ions
In situ (subaqueous) capping	The placement of a clean, isolating material over contaminated sediment in a subaqueous environment without relocating or causing a major disruption to the original channel bed material
Soil and sediment capping	The placement of clean material over contaminated sediment in a subareal environment

9.2. EX SITU REMEDIATION TECHNIQUES

Ex situ methods include those techniques in which the contaminated sediment is removed for disposal or treatment. Once a decision has been made to remove contaminated sediment at a site, significant attention must clearly be devoted to determining the extraction processes that will accomplish the remediation objectives, while minimizing the physical and chemical effects of the process on the aquatic environment. Determining the methods of removal, however, represents only one aspect of the remedial process. Attention must also be given to such factors as the treatment and disposal of the sediment, the location and availability of treatment or pre-treatment areas, and the means of transporting materials between the excavation, treatment, and disposal sites. Figure 9.1 provides a typical flow diagram of the excavation/dredging process and the primary components that need to be evaluated when considering sediment removal as a remedial alternative. The most simplistic operation may consist of as few as three components, such as sediment removal, transport, and disposal. These relatively simplistic operations, composed of a limited number of steps, tend to be effective at lowering both costs and the risk of contaminant loss to the surroundings (USEPA 2005). Cost, however, is not only dependent on the number of steps in the process, but the volume of material to be disposed of. Thus, some form of treatment may be required to reduce the volume and, therefore, cost of sediment disposal. This prompted the National Research Council to suggest that excavation and dredging operations should be developed using a systems approach which integrates each step of the process (NRC 1997). In other words, the products generated during one step of the process must be compatible with those that follow, while the entire program must reduce the costs of remediation and environmental risk. For example, the costs of environmental dredging are often only a fraction of the costs of treatment (NRC 1997). It follows,

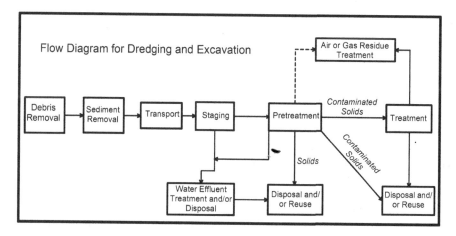

Figure 9.1. Common steps involved in environmental dredging and excavation operations (From USEPA 2005)

then, that the extra cost and effort required for the development of an efficient dredging or excavation program may be offset by reducing the quantity of material requiring treatment and disposal.

9.2.1. Excavation

The excavation of geological materials is one of the oldest and most widely used methods of remediating contaminated sites (Lambert et al. 2000). With respect to river systems, it is primarily applied to the removal of material contained within floodplain and terrace deposits, the bed of small streams and rivers, and contaminant sources such as mine and mill tailings. The volume of material removed during these operations can vary dramatically, ranging from 10^1 to $10^6\,m^3$. Following the Aznalcóllar tailings dam failure in Spain, for example, more than $4.7 \times 10^6\,m^3$ of tailings and contaminated sediment were removed from along the Rios Agrio and Guadiamar in 1998–1999 as part of an emergency cleanup operation (Hudson-Edwards et al. 2003). In the U.S., Cleland (2000) found in a study of 89 completed, ongoing, or planned sediment cleanup projects in the U.S., that approximately 1.4 million cubic yards of material had been removed by dry excavation methods.

Excavation is not a complicated technique, although when combined with the costs of treatment and disposal, it can be one of the more expensive. Costs in the U.S. are on the order of $300 to $510 per metric ton, including excavation, transportation, and disposal at a RCRA permitted site (FRTR 2006). The variations in cost reflect, in part, the methods used in excavation, the nature of the contaminant, the ability to efficiently access the site, and distance to the disposal facility.

Excavation is generally conducted using standard earthmoving equipment such as backhoes or front-end loaders, and can be performed by a wide-range of companies that specialize in the excavation of contaminated sediment. The typical approach requires (1) the area of excavation to be defined on the basis of geochemical data, and (2) an understanding of the remediation standard or target. The horizontal dimensions of the area are typically defined using data collected from a gridded sampling program, often without regard to the depositional units that exist. Determining the depth of excavation is accomplished by following one of several approaches. An approach commonly used for emergency cleanup operations is the excavation of contaminated soils to a uniform depth across the entire site or operable unit without regard to vertical variations in contaminant concentrations. After excavation, the area is usually backfilled with clean soils, thereby limiting the exposure of humans and other biota to any contaminated sediment that remains. The technique limits the number of geochemical samples that must be collected and analyzed, and speeds the cleanup process.

A more commonly applied method involves the characterization of contaminant concentrations using surficial sampling and core data, followed by the delineation of the area of excavation (and the volume of material to be removed) on the basis of the remediation standard. An assumption inherent in this approach is that excavation of the defined area will protect human and ecosystem health while limiting removal costs. The degree to which this assumption is met depends, in part, on the potential for any remaining contaminated sediment to be remobilized by physical or geochemical processes. In the case of floodplains or terraces, remobilization is a function of such factors as channel stability, the rates of lateral channel migration, distance of the excavated site from the river, the frequency and magnitude of inundation during floods, and the nature of the physiochemical environment. It is also dependent on the direction and magnitude of groundwater flow, a factor which is commonly overlooked.

In practice, the area and depth of excavation, defined on the basis of core data, provides only a rough guide to the volume of material to be removed. The actual volume and distribution of sediment extracted from a site is usually based on a iterative process in which a small quantity of material is removed, samples are collected and analyzed to determine if the action levels of the newly exposed sediment is above or below the target concentration, and then a new round of excavation is conducted, depending on the analytical results.

A potentially more effective, but underutilized geomorphological-geochemical method for defining the limits of excavation takes advantage of the relationships between contaminant concentrations and sedimentary deposits. More specifically, where contaminant concentrations vary systematically between depositional units, it may be possible to set the boundaries of excavation according to the distribution of mapped alluvial units in the area, or on the basis of a specific subsurface horizon which denotes a change in deposit age or type (e.g., a buried, organic rich A-horizon). Although sampling during excavation will still be necessary, the number of samples required to define the excavated zone may be dramatically reduced. In addition, the volume of sediment extracted may be considerably lower because the equipment operator has a visible marker that can be followed during sediment removal.

Excavation can also be applied to the removal of channel bed sediment. In these instances, some program managers have classified excavation as being either wet or dry. Wet excavation refers to the removal of sediment from the channel using a backhoe that is located on the channel banks. Sediment is removed from beneath the water and, thus, wet excavation is similar to dredging that uses a mechanical dredge (as we will see later in the chapter). Dry excavation includes the removal of sediment from ephemeral channels during the dry season, or from sections of the channel that have been partitioned off from the rest of the river's flow and dewatered. Isolation of a specific segment of the river can be accomplished using a variety materials and techniques including sheet pilings, earthen dams, cofferdams, and geotubes (USEPA 2005). Alternatively, the river can be temporarily or permanently routed around the reach to be excavated by the use of dams or pipes.

The primary advantage of dry excavation over dredging is that the channel bed deposits are visible and, thus, the contaminated materials can be more precisely removed. Thus, dry excavation reduces the volume of material requiring treatment and/or disposal, and generally eliminates problems associated with sediment resuspension. In addition, sediment removal is not as significantly affected by debris or large clasts (cobbles, boulders, etc) as it is during dredging. The isolation and dewatering of the channel bed, however, can drive the costs of sediment removal up considerably. Thus, any cost savings achieved by dry excavation (versus dredging or wet excavation) may be offset by the additional costs of site preparation.

9.2.2.   Dredging

Dredging is typically applied to large river systems characterized by relatively deep waters which allow for the floatation of a barge and which discourages the isolation of a section of the river bed. A number of other factors, however, will influence the decision of whether to use dredging or some other alternative such as dry excavation. The most significant of these considerations are listed in Table 9.2.

In a review of 89 sediment remediation projects (including both river and marine environments) Cleland (2000) found that 72% used dredging or wet excavation, and had removed over 1.7 million cubic yards of contaminated material. The majority

Table 9.2. Constraints and conditions conducive to dredging and excavation

Conducive site conditions	Site constraints to consider
Suitable disposal site(s) are available and close	Occurrence of sensitive or important habitats and species
Sediment staging and handling areas are available	Navigational considerations
Water depth is adequate to accommodate dredge, but not too great to be infeasible; for excavation, dry conditions occur during the year	Residential, commercial, or military anchorage sites (e.g., marinas)
Water diversion is practical, or flow velocity is low, thereby reducing the potential for resuspension and downstream transport of sediment-borne trace metals	Flood control and river training structures
Contaminated sediment overlies clean or much cleaner sediment	Utility and road crossings, water supply intakes, storm water or effluent systems and discharge points, and other forms of infrastructure
Sediment contains low quantities of debris or is amenable to effective debris removal prior to dredging or excavation	Recreational facilities and uses (e.g., boating, fishing, canoeing, kayaking, etc.)
Sediment does not overlie bedrock and contains a low proportion of coarse sediment (gravel/boulders)	Existing dredge disposal sites
High-contamination concentrations cover discrete areas of channel bed or floodplain	Waterfront developments or industries
Contaminant concentrations are highly correlated with sediment grain size to facilitate separation, reduce disposal volumes and reduce costs	Narrow valley widths or limited access as a result of geomorphic/geological factors
Reduction in long-term risk by sediment removal outweighs impact(s) of contaminant remobilization and habitat disruption	Socio-political constraints, including community acceptance

Modified from USEPA (2005)

of the identified river projects targeted less than approximately $100,000 \, m^3$ of sediment. Costs associated with the eleven ongoing projects in 2000 ranged from $1 to $44 million, or $37 to $1,793 per cubic meter (Cleland 2000). The variability in per volume costs primarily depends on disposal costs and the rates of dredge production. Lower rates of sediment removal are usually related to the occurrence of debris in the channel, shallow water, floating oil, the need to control sediment resuspension, and limited water holding and treatment capacity. Other factors that influence dredging costs are the volume of material to be removed, the type of river to which it is applied, and the distribution of contaminants on the channel bed (i.e., whether removal focuses on isolated hotspots or occurs over contiguous areas).

Dredges operate by dislodging sediment from the channel bed and subsequently removing it from the bed to the surface. This two step process can actually be performed by a wide-variety of platforms that fall into three categories: mechanical, hydraulic, and pneumatic. Mechanical dredges utilize some form of bucket suspended on a cable or an articulated arm that scoops-up sediment from the channel floor. Three types of mechanical dredges have been frequently used in the U.S. (Table 9.3) (USEPA 2005). The most important advantage of using a mechanical dredge is that sediment is removed without adding significant quantities of water to the material. The volume of the removed sediment, then, is similar to that of the in situ deposits, minimizing the quantity of both water and sediment that needs to be treated and/or disposed of. Mechanical dredges are generally used where the volume of sediment to be removed is small, or where a temporary or permanent sediment handling facility is not within pumping distance of the site (NRC 1997).

Table 9.3. Types of mechanical and hydraulic dredges

Mechanical dredges	Hydraulic dredges
Clamshell	*Cutterhead*
Wire supported, open clam bucket; circular shaped cutting action	Conventional hydraulic pipeline dredge with convential cutterhead
Enclosed Bucket	*Horizontal Auger*
Wire supported; near watertight or sealed bucket; either circular or level cutting action	Hydraulic pipeline dredge with horizontal auger dredgehead
Articulated Mechanical	*Plain Suction*
Backhoe designs, supported by articulated fixed-arm; clam-type enclosed buckets, with hydraulic closing mechanisms	Hydraulic pipeline dredge using dredgehead design with no cutting action (only suction)
	Specialty Dredgehead
	Other hydraulic pipeline dredges with specialty dredgeheads or pumping systems
	Diver Assisted
	Hand-held hydraulic suction dredges with pipeline transport of sediment

Adapted from USEPA (2005); Palermo et al. (2004)

Hydraulic dredges dislodge the material from the bed using some form of rotating blades, augers or high-pressure water jets, and then transport the sediment to the surface through a pipeline in the form of a slurry created by the addition of large volumes of water (USEPA 2005) (Table 9.3). The total volume of the water and sediment mixture is significantly greater than that of the in situ material. Thus, dewatering of the sediment at a treatment or temporary handling facility is required prior to disposal. Depending on the nature of the contamination, it may be necessary to treat the excess water before it is released back into the river (USEPA 2005), a process that can be expensive. Hydraulic dredges are most commonly used when large volumes of sediment must be removed, when a sediment handling facility is within pumping distance, and where a pipeline (which carries the material from the dredge to the sediment staging area) can be constructed without obstructing transportation on the water (NRC 1997).

Pneumatic dredges rely on some form of air operated pumping system. They have not been extensively used for environmental dredging and, therefore, will not be discussed here.

Selection of the appropriate dredging equipment depends on several factors. The most important of these are (1) water depth, (2) the total volume of sediment to be removed, (3) the characteristics of the sediment (e.g., grain size and quantity of debris), (4) the precision and accuracy of sediment removal that is required, and (5) the availability of sediment staging areas. A number of new, innovative dredges have been designed specifically for environmental dredging and are now widely used in Europe. They have not been extensively used in the U.S., due in part to their lack of availability, and to the fact that conventional dredges have been shown to be effective in the removal of contaminated sediment.

The need to precisely remove contaminated sediment during dredging is driven by the high costs of sediment handling, treatment, and disposal. In other words, a cost-effective program must avoid overdredging to the extent possible, while ensuring that as much of the contaminated sediment targeted for removal is collected. Recent technological advances (e.g., differential global positioning systems, DGPS) have greatly improved the horizontal positioning of the dredging equipment, while the use of video cameras and acoustical instrumentation can be used to monitor the depth of sediment removal. Nonetheless, the USEPA (2005) notes that "contaminated sediment cannot be removed with surgical accuracy even with the most sophisticated equipment." In practice there may be no need for high resolution dredging because the geographical mapping of contaminated sediment on the channel bed is unlikely to be better than a meter horizontally, and perhaps a few to a few tens of centimeters vertically (depending on the characteristics of the river and its deposits).

In addition to influencing the overall costs of cleanup, accuracy of the dredging process influences the amount of residual contamination at the site (i.e., the concentration of the sediment-borne contaminants which are not recovered during the process). Even after several passes some contaminated particles will remain because (1) dredging did not occur to sufficient depth or lateral extent, (2) particles dislodged by the dredge were not subsequently captured by it, (3) particles on the margins of the dredged zone fall into the dredged depression, or (4) particles which are resuspended during the dredging process settle over the area (USEPA 2005).

The amount of contamination that remains following dredging or wet excavation has been a topic of intense and ongoing debate. It is an important issue because it ultimately dictates the effectiveness of the dredging or remediation program, including the potential for continued contaminant uptake by biota. Most residual contaminant data indicate that contaminant concentrations are usually reduced by more than 85% (Table 9.4). In many instances, the reduction can be much higher, approaching 99%. At some sites, however, investigators have reported that dredging has resulted in only minimal decreases in contaminant concentrations. For example, at a site referred to as Deposit N on the Lower Fox River in Wisconsin dredging was conducted in 1998 and 1999 using a cutterhead hydraulic dredge to remove PCB contaminated sediment. A total of $5,474\,m^3$ of material was removed, along with 5.4 kg of PCBs. The PCB concentration in surface sediment was reduced from an average of 16 mg/kg prior to dredging to 14 mg/kg after dredging had ceased (Romagnoli et al. 2002).

In the case of Deposit N, the reduction in concentrations appears to have been limited. The environmental significance of dredging at Deposit N may not, however, be adequately illustrated by examining only PCB concentrations because while

Table 9.4. Changes in contaminant concentrations in bed sediment following remediation at selected sites where sediment removal was conducted

Site	Change in concentration ($\mu g/g$)			
	Contaminant	Pre-remediation	Post-remediation	Percent change
Grasse River, NY	PCBs			
	– Average	518	75	86
	– Maximum	1,780	260	85
Sheboygan River, WI	PCBs			
	– Average	640	39	94
	– Maximum	4,500	295	93
River Raisin, MI	PCBs			
	– Maximum	1,400	136	90
St. Lawrence River, NY	PCBs			
	– Average	200	9.2	95
	– Maximum	8,800	<100	99
Lower Fox River, WI (Deposit N)	PCBs			
	– Average	16–130[1]	14	12–89[1]
	– Maximum	61–186[1]	130	0–30[1]
Lower Fox River, WI SMU 56/57	PCBs			
	— Maximum	710	17	98
Manistique River, MI	PCBs			
	— Maximum	4,200	1300	69

Adapted from Cleland (2000); RETEC (2002)
[1] Reflects multiple pre-dredging sampling programs

dredging resulted in only limited declines in concentration, it resulted in an 89% reduction in the total mass of PCBs in the deposit (FRRAT 2000). Moreover, the 2.59 kg of PCBs remaining in the dredged area represented only about half of the total annual PCB load from the area to the Lower Fox River before dredging (Cleland 2000; FRRAT 2000). Thus, dredging appears from the perspective of total contaminant mass to have been effective at reducing contaminant loads. In addition, the data indicate that a mass-balancing approach which focuses on the total mass of contaminants remaining in alluvial deposits (and transported downstream) may be a better indicator of dredging effectiveness than are concentration data alone.

Cushing (1999) also argues that from a risk perspective the amount of contaminated material left in the deposit is an important measure of dredging success. For example, if 99% of a given trace metal is removed, the remaining 1% may still pose a risk to the aquatic environment (NRC 1997), depending on the total mass of contaminant that remains. It follows, then, that another measure of project success to consider is the potential reduction in risk that is, or is expected to be, achieved by dredging.

Part of the difficulty in quantifying the degree of dredging effectiveness at a site is that changes in contaminant levels in sediment and biota can be due to multiple factors in addition to dredging, such as the elimination of the source materials, natural recovery processes, or the influx of contaminated particles from upstream areas or other sources. Figure 9.2, for example, shows the changes in PCB concentration in white perch following the dredging of 830 m^3 of contaminated sediment from the South Branch of Shiawassee River in Michigan. PCB concentrations in the fish clearly decrease following the termination of PCB discharge to the river in 1976. They then continue to decrease until 1984, following the dredging operation in 1982. However, the degree to which the decline in PCB concentrations in the fish after 1982 can be attributed to dredging, rather than source control and natural attenuation, is difficult to determine. Nonetheless, the data illustrate that when used as part of an overall remediation strategy, dredging can be an effective remedy.

In cases where dredging is less efficient in removing the contaminated sediment, it is often possible to reduce contaminant remobilization and biotic exposure by

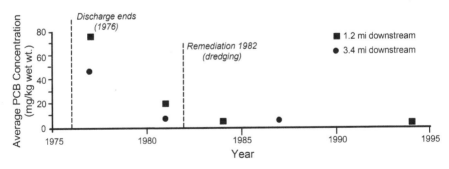

Figure 9.2. Changes in average PCB concentrations in white perch within the South Branch of the Shiawassee River before and after dredging (Figure from Cleland 2000; data from FRG 1999)

placing a layer of uncontaminated sediment over the recently dredged channel bed (USEPA 2005). The effectiveness of this process over long periods of time will undoubtedly depend on the river system, including the nature of the fill in comparison to the original channel bed sediment, the contaminant of concern, and the geochemical conditions (see chapter 10). For example, along rivers where scour and fill during floods is significant, the cap sediment may be eroded and transported downstream, while deeper contaminated sediments may be brought to the surface. Thus, a layer of clean sediment or cap would provide little benefit to the site.

Currently there is no accepted method of estimating what the residual concentrations will be during or following a dredging project. Existing data demonstrate, however, that lower residual contaminations may be expected when contaminated sediment is located at the surface and can be removed during an initial pass. Subsequent passes can then remove materials of lower concentration. Following each pass it is likely that uncontaminated material will mix with the contaminated sediment, reducing the overall concentration of the channel bed materials. The degree of residual contamination is also directly related to quantity of debris and coarse-(cobble or boulder) sized sediment, and indirectly related to the depth to bedrock or other hard material (e.g., a layer of dense clay). In fact, the poor recovery observed for Deposit N on the Fox River was due in part to these two factors.

Another controversial issue associated with dredging is the degree to which dredging and its related activities (e.g., boat traffic and the installation of sediment containment barriers) resuspend contaminated sediment. Of particular concern is the possible resuspension of sediments which are deeply buried and are therefore effectively removed, at least temporarily, from the aquatic environment. Once the sediment-borne contaminants are resuspended, they may be transported downstream beyond the borders of the affected site, and negatively impact aquatic biota.

The degree of resuspension depends on the complex interplay between a host of variables including the physical properties of the bed sediment, the vertical distribution of the contaminated sediment, the velocity and turbulence of the water, the type of dredge and the methods employed in the dredge operation, the quantity of woody and other types of debris, and the magnitude of boat traffic (USEPA 2005). In most cases, an attempt is made to limit the magnitude of resuspension by selecting an appropriate (and perhaps specialized) dredge and through the use of containment barriers. These barriers may consist of sheet-pile walls, silt curtains, or silt screens, to mention just a few. Silt curtains and screens have probably been used most extensively. Although the terms are frequently used synonymously, the USEPA (2005) note that they represent different types of containment barriers. Silt curtains consist of an impervious material (e.g., coated nylon) and are used to re-direct the flow of water around the dredged area. Silt screens, on the other hand, are composed of synthetic geotextiles which allow water to flow through them while removing a large portion of the suspended solids (Averett et al. 1990; USEPA 2005). An unfortunate disadvantage of using containment barriers is they tend to reduce dredging efficiency and increase overall project costs.

Contaminant barriers are usually effective at retarding the downstream migration of contaminated particles. Nonetheless, there is some evidence to suggest that minor increases in concentration downstream of the dredged site may occur. Steuer (2000), for example, showed that for a site along the Fox River called SMU56/57, samples collected and analyzed up- and downstream of the dredged area exhibited a relative percent difference of 59%, the downstream concentrations being higher. Interestingly, little of the increase was associated with the movement of contaminated sediment. In fact, measurements taken up- and downstream of the site demonstrated that the downstream particulate load was not significantly affected by dredging. Rather, the higher downstream PCB concentrations were largely due to the movement of contaminants in the dissolved phase, which accounted for 35% of the total PCB load in the water column.

An increase in the dissolved load can primarily be attributed to two processes. First, many contaminants, including trace metals, will tend to exhibit higher concentrations in waters located in the spaces between the channel bed sediment (i.e., in pore waters) than in the overlying water column. The increased concentration results in part from the longer contact time between the water and the contaminated particles. When the channel bed sediment is disturbed by dredging, the pore waters are released and mixed with the overlying river water, thereby increasing the dissolved concentrations of the contaminants. Second, changes in the physio-chemical conditions to which the particles are exposed during dredging can led to a release of contaminants to the river water as a result of various sorption-desorption reactions. Adams and Darby (1980), for example, found that concentrations of Pb, Zn, Cd, and Cu were higher in dredge effluent than in sediment pore waters, presumably because of the geochemical remobilization of the elements as the sediment was removed from the channel bed.

It is important to recognize that the degree to which the total contaminant load is increased downstream of dredging by changes in dissolved concentrations is usually minor. In the case of the Fox River, for example, Steuer demonstrated that the resuspended PCB load was $< 2.5\%$ of what was dredged from the deposit, and approximately 9% of the total annual load transported in 1994–1995 (Fig. 9.3).

A number of studies have attempted to develop methods to predict the amount of resuspension that will occur for a given site (Bohlen 1978; Adams and Darby 1980; Cundy and Bohlen 1982; Herbich and DeVries 1986; Crockett 1993; Collins 1995). Currently, however, there are no fully accepted models available to determine the degree of resupension that is likely to occur (USEPA 2005). The development of such models is made difficult by the number of variations in dredging operations that can be undertaken at a site, and the wide range of variables that control sediment resuspension (NRC 1997). The limited data that are available, however, suggest that the quantity of sediment lost to resuspension is on the order to 2–5% of the in situ dredged volume (NRC 1997).

The degree to which resuspension increases the short-term risks to the system varies from river to river as well as between sites on any given river. For example, the analysis of caged fished located up- and downstream of the dredged area on the

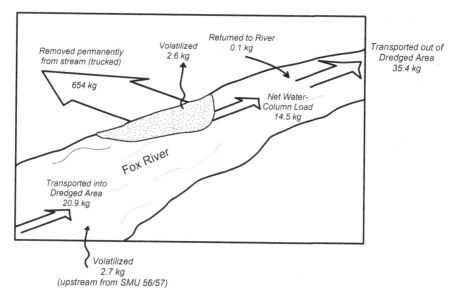

Figure 9.3. Estimates of polychlorinated biphenyl (PCB) mass associated with the primary transport pathways operating during the September 1–December 15, 1999 environmental dredging project at SMU 56/57 on the Fox River, Wisconsin (From Steuer 2000)

Fox River did not reveal any significant differences suggesting that PCB bioavailability was not significantly affected by dredging. However, dredging along the Grasse River, located in New York, appears to have caused a significant increase in PCB levels in local fish communities (Fig. 9.4). The differences in response may have been related to the presence of cold water during dredging of the Fox River (Romagnoli et al. 2002). Other studies have also observed or predicted increases in contaminant concentrations in biota to short-term rises in contaminant concentration including Rice and White (1987), Otto et al. (1996), and Su et al. (2002). Importantly, however, the impact of dredging over periods of a few years (as opposed to a few months) appears in nearly all cases to have produced beneficial results, such as a decline in contaminant concentrations within biota (Cleland 2000).

Perhaps one of the most significant ecological effects of both dredging and excavation is the disruption of stream and riparian habitats. In fact, depending on the nature of the project, it is unlikely that rehabilitation or reconstruction of the site will be able to completely return the area to its previous condition. It is therefore important to insure that the benefits of removing the contaminated sediment outweigh the potential short- and long-term ecological impacts to the area. Such decisions can only be undertaken when a sound understanding of the source of contaminants to aquatic and other biota have been determined, and the potential for future contaminant remobilization has been thoroughly analyzed.

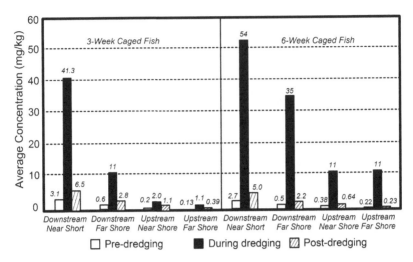

Figure 9.4. Polychlorinated biphenyl concentration in caged fish before, during, and after dredging operations on the Grasse River, New York (From Romagnoli et al. 2002)

9.3.   EX SITU TREATMENTS

The *treatment* of excavated or dredged sediments is performed to reduce toxicity, mobility, or volume of the contaminated materials as a treatment (USEPA 2005). The process may consist of a single step, or a series of integrated steps, occasionally referred to as a *treatment train*. Some investigators have attempted to subdivide the components of the process into pre-treatments and treatments, although the distinction between the two is often vague. What may be considered a pre-treatment in one case is often thought of as a treatment in another. The primary difference between the two is that pre-treatments include activities that pre-condition the material for further manipulations. Here we will consider any action intended to modify the removed sediments as a treatment, although it should be recognized that it may also be used as a pre-treatment in some instances.

The most common treatments applied to excavated or dredged sediments include dewatering and particle size separations. Both of these processes are described in more detail below. The point to be made here is that they are almost always applied, either solely or in combination, to dredged and excavated sediments prior to disposal. In contrast, more sophisticated methods (e.g., soil washing, solidification/stabilization, or vitrification) are rarely used in the U.S. at the present time because of their high costs, the uncertainties surrounding their effectiveness, and the concern by local communities that the treatments will result in the release of the contaminants to the local environment. In contrast, these more sophisticated treatments have been more widely utilized in Europe. In the Netherlands, for example, there are more than ten treatment plants that possess soil washing capabilities. They have a combined capacity to treat about one million tons of contaminated sediment and soil per year (Rem et al. 2002). The extensive use of these treatment processes in the Netherlands is driven by legislation that prohibits the cost-effective disposal of material unless a certificate is issued stating that the materials are unsuitable for cleaning. It is likely that as the costs of material transport and disposal increase in the coming years the use of selected treatment processes will rise in most developed countries (Raghavan et al. 1991; Peters 1999).

Many organic contaminants can be broken down into less or non-toxic substances by a wide range of physical, chemical, or biological processes (e.g., thermal volatilization or biodegradation). Although their chemical speciation can be altered, metals cannot be transformed into other substances. Thus, most of the cost-effective options that can be applied to organic contaminants are unsuitable for the remediation of metals (Vangronsveld and Cunningham 1998). The treatment of trace metals, then, is limited to those that reduce the volume of the contaminated sediment that must be disposed of, reduce the mobility and/or bioavailability of the constituents, or do both. The following sections briefly describe treatments that are commonly applied to excavated and dredged sediment contaminated by metals and metalloids.

9.3.1. Dewatering and Particle Separation

The separation of sediment and other materials into specific classes is often conducted on excavated or dredged sediment on the basis of size, density or surface chemistry. The logic behind particle separation is that for many contaminated rivers, most of the sediment-borne trace metals or other hydrophobic substances will be associated with particles of a specific size or composition (see Chapter 2). Isolation of these contaminant-enriched particles, which may comprise only a small percentage of the alluvial deposits, may therefore allow the remaining sediment to be returned to the river system, or to be used for beneficial purposes (e.g., building materials) without further treatment. The method is most often applied to sediment in which the contaminants are concentrated in the silt- and clay-sized fraction. Once removed, the silt and clay-sized material will possess a higher concentration than was observed for the bulk sediment, and will probably require disposal in a landfill approved for hazardous substances, or a designated contaminant disposal facility (CDF) constructed for the site. However, the volume of material requiring disposal will have been considerably reduced; in fact, it is not uncommon to be able to discard 75–80% of the total sediment without further treatment (Rem et al. 2002), depending on the grain size distribution of the alluvial deposits and the sediment-contaminant relationships.

Depending on the nature of the sediment removed and the methods of removal, it may also be necessary to reduce the water content of the material prior to disposal or further treatment. Sediment collected from the channel bed by hydraulic dredges will almost always require dewatering, but even alluvial materials excavated from floodplain and terraces may require dewatering when the water table is near or at the ground surface. The methods used for dewatering can vary considerably from site to site. The process may include dewatering in ponds or a CDF through the combined actions of seepage, drainage, consolidation, and evaporation (USACE 1987). Alternatively, dewatering may involve application of more rapid methods including centrifugation, filtration and filter pressing, and gravity thickening. Most of the more rapid dewatering methods cannot be effectively applied to sediment containing significant quantities of silt and clay (NRC 1997).

It is not uncommon for dewatering and particle size separation to be conducted together in the form of a treatment train. Figure 9.5 shows the key components that are commonly used in the process including settling basins, debris and particle size separation equipment, filter presses or other forms of dewatering equipment, and sediment handling or staging areas (Cleland 2000).

Depending on the type, nature, and concentration of the contaminant(s), it may be necessary to treat the effluent produced from dewatering prior to its release back into the river system. For example, sediment dredged from the Fox River in Wisconsin was dewatered by means of a filter-press using a stepped process roughly similar to that shown in Fig. 9.5. The effluent was then passed through a sand and carbon filtration system before being discharged to the river (Steuer 2000).

It is not uncommon for the solid materials produced by the process to be dealt with differently. Coarse sediment containing very minor amounts of the contaminant may, in some cases, be returned to the river or used for beneficial purposes. Finer-grained sediment exhibiting higher levels of contamination are most often disposed of in an upland landfill. In the U.S., several options for upland disposal exist,

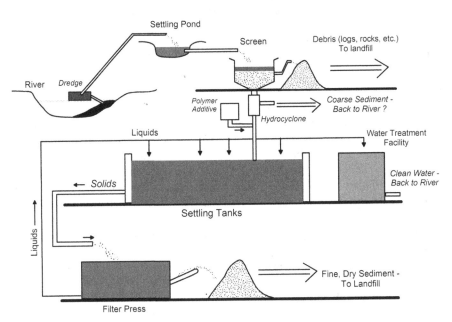

Figure 9.5. Common components of an environmental dredging operation (Modified from Foth and Van Dyke 2000)

including Resource Conservation and Recovery Act (RCRA), non-hazardous solid waste landfills, RCRA hazardous waste landfills, and Toxic Substances Control Act (TSCA) toxic waste landfills (Cleland 2000). To reduce costs, the coarse- and fine-grained materials may be disposed of in different landfill types, according to their degree of contamination. The most contaminated materials are usually sent to a TSCA facility.

9.3.2. Soil Washing

Soil washing is a general term used for the extraction of a wide range of organic and inorganic contaminants from soils and sediment using a water-based fluid as a solvent (NRC 1997). In many parts of Europe, including the Netherlands, it became an important remedial process in the 1980s; in the U.S. its use has been more limited, but is currently increasing. The goal of soil washing is to separate as much of the contaminants from the bulk sediment as possible to reduce the volume or weight of material that requires further disposal and/or treatment. Soil washing concentrates the contaminants into a small portion of the total sediment, or strips the contaminants from the sediment which can then be disposed or treated using another technique. It has been shown to be a cost effective alternative to landfilling alone, the costs (including excavation) averaging approximately $170 per ton in the U.S., depending on site characteristics (FRTR 2006).

The basic mechanics involved in soil washing were derived from the mineral processing industry. The method can involve both physical and chemical processes, and can be applied to in situ sediment and, more commonly, sediment which has been removed by excavation or dredging. The primary steps inherent in soil washing include the formation of a slurry in a water-based solvent, the physical separation of the material on the basis of size, density or surface chemistry, and, when dealing with insoluble metals, the washing of the sediment with chemical extractants to remove the contaminants. Depending on the nature of the alluvial deposits and contaminants of interest, washing of the sediment may need to be repeated several times to achieve the desired levels of removal (Peters 1999). The process generates several waste streams that require further attention (Fig. 9.6). These include (1) contaminated solids from the washing process, (2) leachate, (3) sludge, residue, and other materials removed from the treated water, and, in some cases, (4) vapors released during processing (USEPA 1998). In the case of sediment contaminated solely by metals, the gases released during the process generally do not require treatment.

The physical separation of the sediment rests on the assumption that the contaminants are associated with specific materials as defined by grain size or composition. In contrast to the use of simply dry screening, however, multiple size or density fractions may be segregated and treated separately. For example, mechanical screens and hydrocyclones are often used to partition the bulk sediment into specific grain-size fractions (Peters 1999). The silt- and clay-sized fraction, which is usually enriched in trace metals, will typically be hauled off and disposed of or treated using

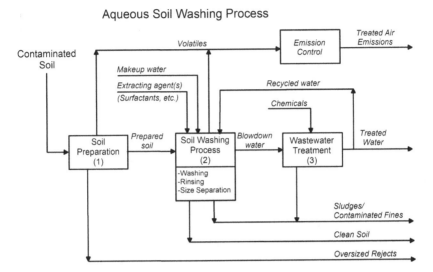

Figure 9.6. Flow diagram of soil washing process (From USEPA 1990)

other techniques after which it is landfilled. The sand-sized fraction can be further manipulated to isolate materials characterized by a specific density range or surface chemistry into fractions that possess relatively high- and low-metal concentrations. Those exhibiting high-concentrations can subsequently be washed using extractive solutions to remove the trace metals.

The soil washing of excavated or dredged sediment has relied on a wide variety of reagents including acids, bases, chelating agents, surfactants, and alcohols. In the case of metals, the extracting agents are typically strong acids or chelating agents (Raghavan et al. 1991). The use of acids (e.g., hydrochloric, nitric, sulfuric, fluorosilicic, and citric) is based on the observed increase in metal solubility with a decrease in pH as described in Chapter 2. However, there are several limitations to the use of acids as an extracting agent. First, soils treated with acids may require further treatments, such as neutralization, before the materials can be returned to the environment. Second, acids tend to destroy or alter the basic characteristics of the treated materials. In some cases, these alterations will be so significant that the material is no longer suitable for revegetation (Peters 1999). Finally, acids may be ineffective where the sediment contains neutralizing minerals such as calcite or dolomite (Neale et al. 1997).

A large number of chelators have been examined as the extracting agent, including ethylenediaminetetraacetic acid (EDTA), nitrilotriacetic acid (NTA), and diethylenetriaminepentaacetate (DTPA). Chelation was defined by Lehman (1963) as "the equilibrium reaction between a metal ion and a complexing agent, characterized by the formation of more than one bond between the metal and a molecule of the complexing agent and resulting in the formation of a ring structure incorporating the metal ion." The net result of chelation is that metallic ions which were immobile under pre-existing conditions are mobilized by the complexing agents, allowing them to be removed from the contaminated particles. The most extensively studied chelating agent for soil/sediment remediation has been by far EDTA (Fig. 9.7). The removal of many trace metals using EDTA and other chelators has been shown to be quite effective (Allen and Chen 1993; Abumaizar and Khan 1996; Steele and Pichtel 1998; Peters 1999). However, their use thus far has been limited by (1) their high cost, (2) the difficulties of removing the trace metals from the leachate following treatment of the contaminated materials, and (3) the high potential for the leachate to contaminate other surface and subsurface water bodies as the chelated metals are highly mobile and the chelators (particularly EDTA) do not biodegrade very rapidly under natural conditions (Juang and Wang 2000).

The removal efficiency of the soil washing process is highly variable and depends largely on the physical characteristics of the sediment (e.g., its particle size distribution), the trace metal of interest and its chemical form within the sediment mass, the chemistry of the extracting agent, and the processing conditions, including the number of repetitive washing cycles (Peters 1999). The particle size distribution of the material is particularly important as the difficulty and effectiveness of soil washing decreases as the percentage of silt- and clay-sized particles increases (Fig. 9.8). Krishman et al. (1996) have argued, however, that soil washing of materials consisting of high percentages of fine particles is possible using an electrode assisted soil washing process.

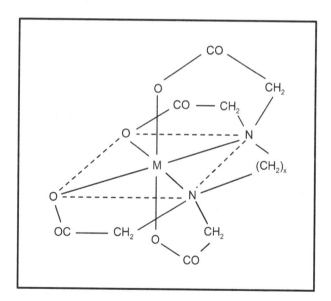

Figure 9.7. Structure of EDTA complexing agent. Metallic ion (M) is bonded within the structure (From Lehman 1963)

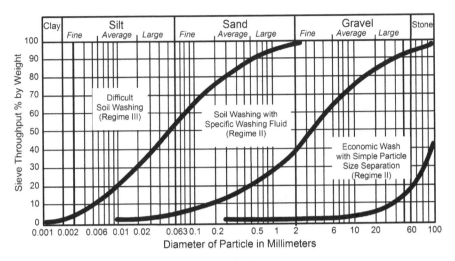

Figure 9.8. Influence of grain size distribution on the soil washing process (From Krishman et al. 1996)

9.3.3.   Other Treatment Alternatives

In addition to the techniques described above, there is a large number of additional methods that have been proposed for treating contaminated sediment (Table 9.1). These methods are often categorized as being thermal, chemical, or biological in nature. Biological methods are rarely used to treat metal contaminated sediment and will not be discussed here. Thermal treatments are also most often applied to sediment contaminated by organic pollutants. In this case, the thermal techniques are used either to separate organic substances from the material (e.g., in thermal desorption), or to permanently destroy the organic contaminants (e.g., through incineration) (Soesilo and Wilson 1997; USEPA 1998). The applicability of these thermal treatments to metals is limited because metals cannot be broken down into other constituents, and because their volatility is extremely low even at high temperatures (Vangronsveld and Cunningham 1998). A form of thermal treatment that is applicable to metal contaminated sediment is called vitrification. *Vitrification* refers to the melting of the soil mass at extremely high temperatures, ranging from about $1,300°$ to more than $2,000°C$. Upon cooling, the metals are contained within a highly durable, glass-like material which is chemically inert, of low permeability and from which metals can only be leached with extreme difficulty. Organic contaminants are usually vaporized and/or destroyed in the process.

Vitrification can be performed both in situ and on excavated or dredged sediment. The exact nature of the ex situ treatments vary, primarily as a function of the heat source and containment system used in the process. Heat sources include natural gas-fired burners, electrical current, plasma torches, and radioactive reactors (see, Iskandar and Adriano 1997, for a description of several different methods). There are several limitations, however, to the use of either in situ and ex situ forms of vitrification; of most importance is that both methods are extremely expensive and require highly specialized equipment and personnel. Thus, vitrification probably cannot be justified other than in the case of treating sediment contaminated by highly toxic substances such as radioactive materials.

There are several chemical treatments that are used to reduce the mobility of trace metals in excavated or dredged sediment. The most widely applied process is solidification/stabilization. *Solidification/stabilization* involves (1) the mixing of sediment with some form of binding agent which encases the materials in a stabilized mass (solidification), and (2) the introduction of reagents which react with the material and any remaining water to reduce the solubility and mobility of the trace metals (stabilization) (USEPA 1994; Iskandar and Adriano 1997). The solidification of small particles of contaminated sediment with some form of low-permeability material is referred to as *microencapsulation; macroencapsulation* encloses a large block of contaminated material in a casing of low-permeability. Solidification by means of either micro- or macro-encapsulation retards contaminant migration by changing the physical characteristics of the material, such as its surface area, water content, and permeability. In contrast, stabilization may not necessarily alter the physical properties of the material, but reduces the contaminant's solubility through chemical reactions. Solidification/stabilization is most often used for inorganic contaminants, including radionuclides. Costs, including excavation, are typically on the order of $100 per metric ton (FRTR 2006).

Both inorganic and organic binding agents may be used in the solidification/stabilization process. Typical inorganic agents include cement, lime, and siliceous materials such as fly ash. Examples of commonly utilized organic binders include asphalt, epoxide, polyethylene, and polyesters. Inorganic binders are almost always used except for radioactive wastes or specific hazardous organic substances because of their lower costs (USEPA 2000). In the case of inorganic binders, one or more of the binding agents may be combined with other additives and mixed with the contaminated sediment to form a hard, rock-like block (USEPA 1993; Iskandar and Andriano 1997). Less often, the treated sediment may produce an end product that is friable and that maintains many of the physical properties of the original material (Brown 1997).

The effectiveness of solidification/stabilization for metals was analyzed by Erickson (1992) using data collected from several hundred solidification studies (Table 9.5). He found that cement combined with kiln dust resulted in the greatest decreases in leachability for As, Cd, Cr, and Pb. In addition, the studies showed that one or more binding agents had led to a 99.9% reduction in leachate concentrations between the treated and untreated materials for As, Ba, Cd, Cr, Pb, Hg, and Se. There is no question, then, that solidification/stabilization can be highly effective in reducing metal mobility. Interestingly, however, some treatments resulted in an increase in metal concentrations within the leachate, rather than a reduction (Table 9.5). This was most common for materials containing Pb that had been treated with a combination of lime and fly ash (Brown 1997). Thus, it is important

Table 9.5. Percent reduction in leachable metal concentrations following solidification with selected binder types. Negative values indicate an increase in leachable metal concentrations. Presented data represent range of values from several hundred studies

Binder type	As	Cd	Cr	Pb
# of Studies	65	92	44	280
Cement	> 99.9	94– > 99.9	99.9	79– > 99.9
+ kiln dust	> 99.9	98– > 99.9	> 99.9	> 99.9
+silicates	57–87.8	88–99.8	57–82	11–99.9
Pozzolan				
+ silicates	−85–89.8	68– > 99.9	89– > 99.9	−160–98
Lime				
+ fly ash	−85–89.9	68– > 99.9	89– > 99.9	−160–98
+ silicates	24–72.9	—	91	66– > 99.9
Silicates	—	0–71	< 0	−67–99.9
Fly ash	—	81–99.0	99.7	—
Kiln dust	−79–97.2	92– > 99.9	—	8–99.5

Data from Erickson (1992)

to insure that attention is given to the type of binders used and their effects on the leachability of the metals from the contaminated materials.

A significant advantage of solidification/stabilization is that the mixing process can be performed using readily available equipment which is capable of processing large volumes of material quickly. Thus, it tends to be a relatively cost effective method (USEPA 1994). Its most significant limitation is that it can increase the volume of contaminated material by 10% to more than 30%, which, depending on the nature of the contaminant, may still require disposal as a hazardous waste. In addition, there is some question as to whether changes in the physiochemical environment will affect long-term stability of the stabilized mass (USEPA 1994; Iskandar and Adriano 1997; FRTR 2006). For example, a majority of the inorganic binding agents produce an end product possessing a high pH (> 10). Waters with lower pH values may attack the material and result in the leaching of metals from the solidified mass.

9.4.   RIVER RESTORATION AND REHABILITATION

9.4.1.   Goals and Objectives

The removal of contaminated sediment from a river valley may include all of the alluvial materials lying within and adjacent to the channel. This is a significant and costly undertaking that requires the complete reconstruction of the river channel as well as the structure and function of the riverine and riparian ecosystems. Even for less ambitious remediation projects, however, disruption of these ecosystems can be significant. Thus, excavation and dredging operations are likely to be accompanied by an attempt to recreate and/or improve aquatic and riparian habitats. Such a process is often referred to as river restoration.

In its purest form, *river restoration* is defined as the structural and functional return of a degraded riverine ecosystem to its pre-disturbance condition (Berger 1990; NRC 1992). In the U.S., the pre-disturbance condition is generally taken as the state of the river prior to European settlement; in Europe, it is often assumed to be the bronze-age. In either case the idea is to return the degraded ecosystem to its previously held, natural state. In reality this ideal objective can be rarely, if ever, achieved because we lack the necessary information on the structure, function, composition, or dynamics of the ecosystem prior to significant human disturbance (NRC 1992; Hobbs and Norton 1996). Moreover, even if the natural characteristics of the river system were known, the pre-disturbance condition may not fit the modern stable state(s) because of alterations in climate, land-use, and system hydrology (Wade et al. 1998).

In light of our inability to return the system to its original natural condition, the historical meaning of *river restoration* has often been redefined as the return of a degraded ecosystem to a close approximation of its remaining natural potential (USEPA 2000). Ecosystem potential is typically determined on the basis of biological indicators such as the quantity and quality of existing habitat or the diversity of riparian and/or aquatic species (Shields et al. 2003). The intent, then, is to return the system to the most natural state allowed by the existing conditions within the watershed. Hobbs and Norton (1996) argue that this includes a return of a riverine ecosystem to a more natural working order that is not only sustainable over the long-term, but is more productive, aesthetically appealing, and valuable from a conservation perspective.

It may seem that defining river restoration is simply an issue of semantics. Unfortunately, its varied definition has often led to confusion as to what is actually being done and for what purpose. For example, the majority of "restoration" projects in the U.S. utilize hardened, in-stream structures composed of rock and other materials to limit bank erosion and to stabilize the channel in a specific location on the valley floor. Bank erosion and channel migration, however, are natural processes that are required for properly functioning riparian and riverine ecosystems (Ward and Stanford 1995; Kondolf et al. 2001; Palmer et al. 2005). Thus, while projects that utilize these structures may improve the existing habitat for one or more species, they do not return the system to a naturally functioning, sustainable state. This had led some to argue that these kinds of efforts are more properly termed *rehabilitation* projects, a term widely used in the U.K. (see, for example, Waal et al. 1998).

A significant outcome of the discussion of what constitutes river restoration is the recognition that ecological success of a project is not necessarily the same as restoration success (Fig. 9.9) (Palmer et al. 2005). Restoration successes, for example, can be measured by evaluating whether the project was cost-effective, protected important infrastructure, led to increased recreational opportunities and community education, or improved system aesthetics. In contrast, Palmer et al. (2005) argue that true ecologically successful restoration must meet five criteria. First, ecological success requires that the system is moved toward the least degraded

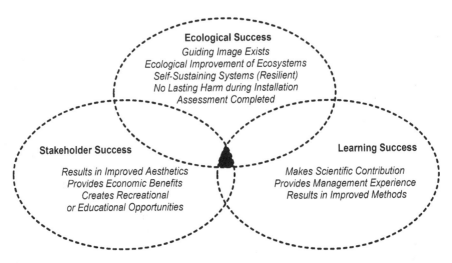

Figure 9.9. Various ways of defining restoration success. Stakeholder success is based on human satisfaction of restoration project, while learning success consists of advances that will benefit future restoration efforts. Ecological success is defined according to five criteria described in text. The most effective restoration projects are found at the intersection of the three ovals (After Palmer et al. 2005)

and most ecologically dynamic state possible, given the regional context in which the river is located. An ecologically dynamic state, as used here, is one in which the biotic composition and abundance vary through time and space as they would in a relatively undisturbed area. Moreover, it assumes that channel morphology and planimetric configuration can change in response to variations in water and sediment discharge. Second, ecological success requires measurable changes in the physiochemical and biological conditions of the river toward that of an envisioned target (or reference) system. Third, the riverine ecosystem that is produced must be more resilient and self-sustaining than before restoration. Creating resilient and self-sustaining systems is one of the most difficult tasks because it requires the restoration of natural fluvial processes including channel migration, which is inconsistent with the use of hardened engineering structures. Fourth, ecological success occurs only when no harm is done while implementing the restoration project. This includes the resuspension of potentially hazardous contaminated particles. Finally, ecological success requires an evaluation of the positive and negative outcomes of the project on the basis of the pre-project objectives. In other words, the project must provide information as to what works and what does not in order to improve future project designs and objectives.

In the case of contaminated rivers, the general objectives of the restoration project are likely to include: (1) the integration of fluvial processes in such a fashion as to develop a stable channel, (2) the development of high quality and sustainable habitats, and (3) the minimization of the risks associated with the remobilization of contaminated sediment (see, for example, Boyd and Skidmore

1999). It is important to recognize that the objectives should be precisely defined on the basis of quantitative criteria prior to project design and implementation. Moreover, the defined objectives are likely to vary along the river as a function of the remediation methods to be implemented, the geomorphic processes that predominate within the reach and the existing physical and social constraints. For example, where all of the contaminated materials forming the channel perimeter have been removed, it may be possible to reconstruct a highly dynamic channel capable of migrating across the valley floor without the risk of remobilizing contaminated sediment. In contrast, where in situ remediation has been performed the need to reduce erosion and contaminant remobilization rates may not allow for channel migration. In these areas, then, it will be necessary to design restoration programs that limit the erosion of contaminant bound materials.

9.4.2. Channel Reconstruction

Attempts at restoring both contaminated and non-contaminated rivers have increased dramatically during the past two decades in most developed countries. The price tag for these initiatives is unknown, but in the U.S. alone, restoration costs associated with small- to medium-scale projects are thought to exceed, on average, $1 billion per year (Bernhardt et al. 2005). The global explosion in river restoration has been accompanied by an exponential increase in the restoration literature (Bernhardt et al. 2005). It would be impossible for us to adequately summarize all of these investigations in the space that is available here. Thus, we will primarily focus on the methods which are used to design geomorphically stable channel systems. Information on other aspects of restoration can be found in NRC (1992), FISRWB (1998), Waal et al. (1998), Wissmar and Bisson (2003), Downs and Gregory (2004), and Roni (2005), to mention just of few.

The re-establishment of channel morphology and the physical characteristics of the aquatic habitat do not insure that ecologically successful river restoration has occurred (Hobbs and Norton 1996). Nonetheless, the approach is consistent with the widely held belief that geomorphic factors are a primary determinant of the spatial and successional patterns of stream and riparian communities (Statzner et al. 1988; Gregory et al. 1991), and that the geomorphic and hydrologic characteristics of the river provide the framework for riverine and riparian ecosystems (Harper et al. 1995; Kondolf and Micheli 1995; Hupp and Osterkamp 1996; Montgomery 1997). In other words, it is unlikely that river restoration will be successful if geomorphic stability cannot be achieved.

One of first and most important steps in channel reconstruction is the development of design criteria. *Design criteria* represent quantifiable benchmarks for specific components of a project to insure that the objectives of the restoration activity are met. With respect to channel reconstruction, design criteria may include measurable limits to lateral and vertical erosion, bank stability, areal extent of aquatic habitat, and flood conveyance (Skidmore et al. 2001). As an example, a project may include three different discharge conveyance criteria: a bankfull channel capable of transmitting a flood with a 2-year recurrence interval, a low-flow channel that allows for adequate water depths, temperatures, etc., between floods in order to sustain important aquatic species, and a floodplain that allows for conveyance of a 100-year event. Criteria may also be established to insure the integrity of individual components (e.g., in-stream structures) during large floods (e.g., 50- or 100-year flood). Separate criteria will need to be developed for each reach or operable unit, and they must be consistent with the objectives of the area's remediation and restoration strategies.

Project success is often dependent on the effort which goes into developing the design criteria. After all, the design criteria should serve as the primary guide as what needs to be accomplished for each component of the restoration project. Thus, care must be used to insure that the design criteria are the driving force behind channel design, rather than a product of the methods used to design the channel.

At the present time there are no standardized methods of designing stable river channels. There are multiple reasons for this lack of standardization including a high degree of variability in the hydrologic and geomorphic characteristics of the project sites. It is also related the fact that the field of river restoration in still in its infancy. Thus, there has been a strong tendency to try new methods, and to view the design process as one involving as much art as science.

Although approaches to channel design are highly variable, most attempts at reconstructing river channels have relied on the use of four general methodologies, either alone or in combination. These include: (1) the "carbon-copy" method, (2) the stream classification and reference reach method, (3) the empirical method, and (4) various analytical methods.

With respect to contaminated rivers, the carbon-copy method is most applicable where reconstruction will occur after the channel bounding materials have been removed from a geomorphically *stable* reach. As the name implies, the carbon-copy method involves the collection of detailed information about the river's cross-sectional form, gradient, planimetric configuration, and bed/bank material composition. Then, after the contaminated materials have been removed, a new channel is replicated using newly introduced fill. The process is not as easy as it might first appear because it is impossible to precisely reproduce the previously existing, stable channel. The grain size distribution of the bed materials, the cohesion and texture of the bank sediments, and the type and abundance of riparian vegetation will all be different. It is therefore likely that some form of temporary stabilization method will be required to maintain the integrity of the constructed channel until the riparian vegetation is established. Nevertheless, allowing the bed and banks to deform over time appears to work well in many locations where it is possible to do so because the river can adjust on its own to the existing sedimentologic and hydrologic regime.

A widely used and closely related method to the carbon-copy approach is referred to as the reference reach method. It is based on the collection of data from other stable rivers in the area, or from stable reaches along the same river, and their use to define the cross-sectional dimensions and planform of the reconstructed river. Data collection typically revolves around the dimensions of the bankfull channel. Where the reference reach method is used, it is usually necessary to scale the reference data to fit that which is appropriate to the restored reach. One of several methods can be used for scaling purposes (Hey 2006). Perhaps the most common method utilizes the concept of regional curves. Regional curves are empirical relationships that have been constructed between basin area and channel width, depth, and cross-sectional area using data from reference sites (Fig. 9.10).

The primary advantage of this approach is in its simplicity. The time and effort required for design and data collection is kept to a minimum, thereby reducing design costs (Skidmore et al. 2001). In some situations, particularly along small, low-gradient streams, the approach has proven to be effective. However, a significant limitation of the approach is that the chosen reference reaches may not be an appropriate representation of the channel to be reconstructed because of differences in the geology, climate, bed and bank material composition, hydrology, and biogeography. More importantly, reaches selected to serve as stable references are usually located in basins which have been only minimally affected by land-use changes or other human activities. Thus, their application to sites affected by anthropogenic disturbances is often questionable.

One of the earliest methods of channel design has its roots in regime theory which developed out of an attempt to create canals that were non-scouring and non-filling (Kennedy 1894). Regime theory is based on empirical relationships which generally relate width, depth, slope, and other morphological parameters to discharge and bed material size. The relations are often developed using data from a large number of

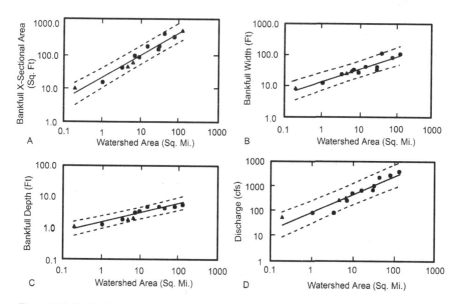

Figure 9.10. Regional curves developed for the rural piedmont of North Carolina. Triangles represent ungaged stream; circles represent gaged streams (From Harman et al. 1999)

rivers which are thought to be stable over a period of many years (Bray 1982; Hey and Thorne 1986; Williams 1986). The developed relationships are used in channel design to estimate the dimensions and form of the reconstructed channel provided that the discharge and bed material composition of the restored reach are known. In most instances, designers recommend that the empirical relationships used in channel design are collected from specific stream types and regions that are similar to the restored river. As is the case of reference reach approach, extrapolation of the empirical relationships to the restored reach assumes that similarities exist in channel characteristics (Skidmore et al. 2001). In addition, the relations generally cannot be applied to channels which are undergoing aggradation or entrenchment, or that are otherwise unstable because the channels are not in "regime" (non-scouring and non-filling), a basic assumption of the approach.

Analytical methods contrast significantly with the other approaches in that channel width, depth, slope, and meander characteristics are directly computed from an understanding of the channel forming discharge and bedload characteristics (White et al. 1981; Chang 1988; Copeland 1994). As such, analytical methods include some form of sediment transport analysis to insure that sediment coming into the restored reach from upstream areas can be transmitted through the reach. The analyses are usually based on hydraulic models and one of a variety of sediment transport functions to determine equilibrium channel dimensions (Skidmore et al. 2001). A significant advantage of the approach is that it does not rely on the assumption that data from other areas can be applied to the study site. Perhaps more importantly, the approach can be applied to unstable rivers which are adjusting to anthropogenic disturbances (Shields et al. 2003). The methods, however, are usually data intensive.

The above discussion is intended to show that none of the frequently used methods of channel design are completely satisfactory. Each possesses strengths and weaknesses, and the method that is most appropriate for a give situation will vary as a function of the site characteristics and the objectives of the restoration program. As a result, channel design is often based on more than one of the above techniques. For example, the most widely used approach in the U.S. is often referred to as Natural Channel Design. Natural channel design is heavily depended on the reference reach approach, although it may also include analyses based on both the empirical and analytical methods described above. The argument is that "by designing with nature, rather than imposing a solution on the river," the produced channel is likely to be more sustainable and cost effective than traditional engineering approaches (Hey 2006). Many engineers and ecologists have referred to natural channel design as a fluvial geomorphic method for designing stable channels (see, for example, Hey 2006). It is surprising, then, that the approach falls well short of the geomorphological analyses conducted and advocated by most practicing geomorphologists. Sear (1994), for example, notes that for many restoration projects geomorphology is used only for locating or scaling river features; "real" geomorphic investigations required to design restoration schemes that are sustainable over long periods of time are typically missing.

Data are now rapidly accruing to demonstrate that where natural channel design is applied without a sound understanding of the geomorphic processes that occur within the entire basin, the approach is likely to fail (Sear 1994; Kondolf et al. 2001; Shields et al. 2003). One such case was clearly documented by Kondolf et al. (2001). In November, 1995, a 0.9 km reach of Uvas Creek, California was reconstructed under the assumption that a C4 channel, classified according to the Rosgen stream classification system, was stable. A C4 channel denotes "a slightly entrenched, meandering, gravel-bed dominated, riffle-pool channel with a well-developed floodplain (Rosgen 1996)." The newly constructed channel washed out during a flood in February, 1996, only 3 months after construction. Interestingly, the channel did not suffer significant erosion or deposition, but rather was abandoned as the river adopted a new course through the constructed floodplain. Prior to the project, no historical geomorphic analyses had been undertaken to determine if a C4 channel was suitable for the reach, nor were any analyses performed to determine the possible effects of human activities on channel processes and form. After the project's failure, however, Kondolf et al. (2001) showed that the river had historically exhibited a braided channel pattern common for rivers draining the easily erodible Franciscan formation that underlies the watershed. With minor differences, the flood generated channel exhibited similar characteristics to its historic configuration. Thus, failure resulted from a lack of understanding of the predominant geomorphic processes operating along the restored reach, and the nature of stable channel system, both in the past and at the time of restoration.

The investigation by Kondolf et al. (2001), and others like it, demonstrates that the odds of success are greatly improved (1) where geomorphologic, hydrologic, and sediment transport data are combined, and (2) when the analysis focuses on the entire watershed, rather than only on the reach to be restored. One way through which this may be accomplished has been proposed by Shields et al. (2003). In this case, channel design combines an assessment of watershed geomorphology, empirical tools (e.g., hydraulic geometry, critical bed shear stress and velocity), hydraulic analyses, and sediment transport computations to develop a stable channel (Fig. 9.11). While this type of methodology is likely to require highly trained personnel as well as more time and money, it will undoubtedly reduce the uncertainty involved in producing stable river channels. Moreover, the higher initial costs may be more than offset because the restored (or reconstructed) rivers are less likely to require expensive repairs.

9.4.3.   In-stream Structures, Bank Protection, and Habitat Formation

Rivers in their natural state are dynamic systems. Not only do they perpetually change their morphology in response to fluctuations in sediment load and discharge, but rivers are continually altering their position on the valley floor. Meandering streams, for example, swing across the valley as cut banks are eroded and sediment is added to the adjacent point bar. Individual meanders may also sweep down valley

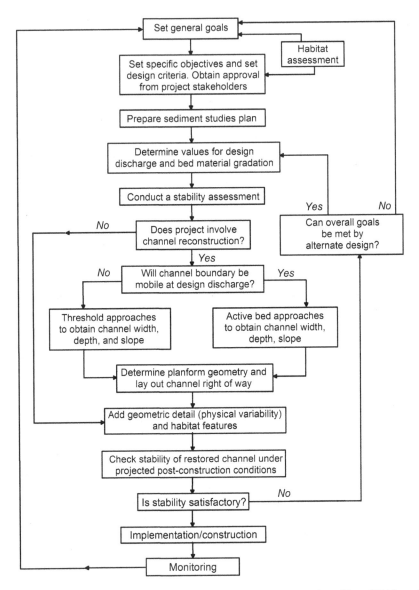

Figure 9.11. Flow diagram of channel design for stream restoration projects (From Shields et al. 2003)

as a result of enhanced erosion immediately downstream of the meander apex, a process which leads to meander cutoff and abandonment. These processes are of significant ecological importance as they create new surfaces for colonization of riparian communities and lead to the development of oxbow lakes, wetlands, and other forms of habitat.

The maintenance of native cottonwood forests provides an example of how channel migration may affect riparian ecosystems. Seedling establishment, in this case, depends in large part on the formation of bare, surfaces composed of unconsolidated sediment which is created during channel migration. In contrast, non-native species such as Russian olive can become established on previously vegetated surfaces. Thus, a reduction in channel migration as a result of the use of hardened structures may lead to a shift from cottonwood to non-native Russian olive forests (Howe and Knopf 1991; Miller et al. 2001).

In light of the above, it follows that the formation of dynamic stream systems are required for ecologically successful river restoration. Nonetheless, a stabilized channel may be required in some areas such as in urban or suburban environments where bank erosion may damage infrastructure (e.g., bridges, roads, pipelines, and buildings), or in the case of contaminated rivers, where it is necessary to inhibit the remobilization of sediment-borne contaminants within the channel banks. Stabilization is most often accomplished at the present time using a combination of bank treatments and rigid, in-stream structures which are designed to prevent bank erosion, control channel grade, and create habitat. Although the gross nature of these structures has changed little over the past several decades (Thompson 2002), the details of their construction and placement have changed. The most commonly used in-stream structures in the U.S. at the time of this writing include cross-vanes, rock and log vanes and weirs, and J-hooks (Fig. 9.12). The utilized rock and wood structures are thought to indirectly protect the channel banks from erosion by refracting the downstream flow of water away from the banks. In the process, the topographic complexity of the channel bed, and therefore aquatic habitat, is enhanced by creating a zone of flow convergence that produces bed scour and pool formation. In addition, the placement of the structures is often governed by an attempt to create pools, riffles, and other morphologic features as they would be found on the channel bed of an undisturbed river.

The extensive use of in-stream structures in river restoration has led to two topics of debate. First, the long-term integrity and performance of hardened structures (including rock and cross-vanes) has begun to be questioned as data on their durability and effectiveness increase (Ehlers 1956; Babcock 1986; Hamilton 1989; Frissell and Nawa 1992; Bestcha et al. 1993; Thompson 2002; Kochel et al. 2005). In one of the most commonly cited papers, Frissel and Nawa (1992) evaluated wood and stone structures in Oregon and Washington. They found that more than 50% of the log weirs, log deflectors, multiple-log structures, and boulder clusters had failed or exhibited significant impairment of function. The other studies cited above reported similarly high rates of failure. In contrast, Heller et al. (2000) found that the majority of the structures that they evaluated in Washington and Oregon withstood significant floods. Similarly, Slaney et al. (2000) determined that of 13 structures installed in a large river in British Colombia, all remained mostly intact. Moreover, rainbow trout populations studied by Slaney et al. (2000) were four times higher in the restored reach than in an untreated control reach.

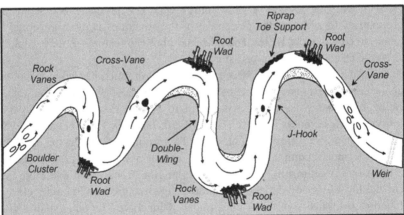

Figure 9.12. Commonly used in-stream structures: (A) cross-vane, (B) rock vane, (C) J-hook, (D) root wad. Plan view showing flow patterns during low flow at bottom

Undoubtedly, a significant part of the contrasting results from these studies is related to size and frequency of major storms to which they have been subjected, and the geomorphic and hydrologic characteristics of the river to which they are applied. For example, Kochel et al. (2005) showed that in North Carolina flow diversion structures (including rock and cross-vanes) performed well when flood flows were below bankfull. However, at higher discharges their effects are diminished or drowned out by relatively deep waters, a process also observed by Thorne et al. (1997). As a result, between 35 and 100% of the structures were found to be damaged or to have failed outright along streams that had experienced floods exceeding approximately the 10-year event. Although some structures were physically damaged or removed by the events, most failures were associated with bank erosion. Many of the structures ended up in the center of the channel as the river eroded around one side of the structure (Fig. 9.13). In these situations, the structures actually enhanced, rather than reduced, bank erosion, a conclusion reached by Thompson (2002) for the Blackledge and Salmon Rivers of Connecticut as well. Kochel et al. (2005) found that the geomorphic setting played a significant role as to whether the structures would fail over relatively short-time periods. More specifically, rivers that were in the process of adjusting to land-use changes in the basin experienced exceptionally high failure rates, either because they were buried by excessive sediment produced in upstream areas, or because they were subjected to intense channel bed and bank erosion.

Undoubtedly, more study is required to define what types of structures are most suitable for any particular river or river reach. It is clear, however, that currently utilized in-stream structures are most suited to rivers which are geomorphically stable (i.e., in a state of equilibrium), or are characterized by low rates of bedload transport. Even in these areas it is likely that their use will require continued maintenance. Thus, as pointed out by Skidmore et al. (2001), they should probably be used as a short-term tool to improve stream conditions until the primary drivers of degradation can be addressed. It should also be recognized that other types of approaches may be more effective in protecting the banks from erosion and in improving habitat. In Denmark, for instance, rigid in-stream structures are now prohibited because of their artificial nature (Iversen et al. 1993). They have largely been replaced by the use of constructed riffles which serve as important habitat and as a means of retarding channel bed erosion. Similarly, a large variety of bioengineering techniques is available which not only improve riparian habitat, but protect the channel banks from erosion (FISRWB 1998; Eubanks 2006). Live cribbing, a commonly used bioengineering treatment, was found to be particularly effective in North Carolina in resisting bank erosion during large floods, although its installation is typically more difficult, time consuming, and expensive (Fig. 9.14). Nonetheless, the extra expense may be justified where it is necessary to inhibit the remobilization of sediment-borne trace metals along contaminated rivers.

The second topic of recent debate rests on whether the use of rigid, in-stream structures is justified for all restoration projects. There is little question that some form of bed and/or bank treatment may be required where there is a need to

Figure 9.13. Cross-vane installed in 2002 along Beaver Creek, North Carolina. Bank erosion rapidly rendered the structure (outlined by black line) ineffective. Most of the more than 80 structures installed along the river had suffered significant damage by 2006 as a result of channel instability (Photo courtesy of R.C. Kochel)

protect infrastructure or contaminated sediment. Moreover, the currently used in-stream structures and bank treatments represent a significant improvement over the engineering practices of the past which relied heavily on riprap and concrete-lined channels because the latter provide little value in terms of aquatic or riparian habitat. What is less clear is whether these rigid structures should be used along stream reaches located in open areas devoid of infrastructure. Nonetheless, in many regions of the U.S., their application varies little between areas with or without infrastructure, in spite of the fact that they inhibit natural river processes.

The use of rigid structures is driven in part by the perceived need to reduce the risk of the designed channel to fail after construction. However, Kondolf et al. (2001) point out that if the designed channel was inherently stable, there would be no need for structures to stabilize the banks of the constructed river. In response to such assertions, it is typically argued that some form of stabilization needs to be used until the riparian vegetation is sufficient to prevent excessive bank erosion. While this is a valid argument, it is not clear that stabilization requires the installation of significant numbers of in-stream structures and other forms of bank treatments. A case in point is the restoration of Silver Bow Creek, Montana. The channel and floodplain of Silver Bow Creek were contaminated by historic metal

Figure 9.14. (A) Live (log) cribbing being installed along a the French Broad River, North Carolina in October, 2003; (B) same section of river in September, 2004 (Photo courtesy of R. Towmey)

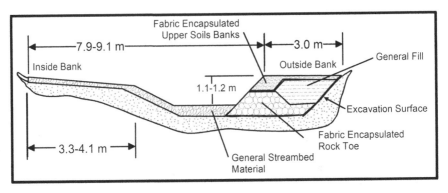

Figure 9.15. Deformable bank design for Silver Bow Creek, Montana (From Bucher et al. 2000)

mining activities to such an extent that the entire river valley required remediation. The utilized remediation strategy included the large scale removal of channel bed and floodplain deposits, creating a fresh template in which to construct a new river channel. The primary goal put forth as part of the remediation and restoration strategy was to "develop a naturally functioning, geomorphically stable channel that could transport the imposed water and sediment supply, protect infrastructure, and provide appropriate hydraulic conditions for aquatic habitat" (Bucher et al. 2000). In doing so, the main challenge to channel design was to create a channel that allowed for acceptable rates of bank erosion. Boyd and Skidmore (1999) and Miller (1999) suggested that the objective could be accomplished by producing deformable (i.e., erodible) banks which were not protected by rigid structures. The bank design that was utilized consisted of the placement of graded rocks that could be mobilized by flows ranging from the 2- to 50-year flood at the bank toe (Fig. 9.15) (Miller 1999). The rocks were wrapped in a fabric that while new provided resistance to erosion, but through time decayed, thereby exposing the materials to transport well after the establishment of the riparian vegetation. The basal rocks were then overlain by fine-grained soils which were also wrapped in the fabric and which were seeded to establish a vegetated riparian zone.

The success of such approaches as the deformable bank concept has yet to be adequately evaluated. There is no question, however, that designs advocating erodible banks, and which allow the river greater flexibility to self-adjust, are required to achieve ecologically successful restoration projects. Thus, this is an area that is greatly in need of further study.

9.5.   SUMMARY

The removal of sediment by means of environmental dredging or excavation is a remedial option considered for almost every contaminated river system. Excavation differs from dredging in that the former refers to the subareal extraction of sediment using earthmoving equipment (e.g., backhoes and front-end loaders), whereas environmental dredging extracts sediment from an underwater environment such as the channel bed. Sediment removal has been the method of choice for the remediation of contaminated sediment, and has been shown to be an effective remedial alternative in a wide range of riverine and marine environments. In addition, sediment removal can be achieved relatively quickly in comparison to many other remedial alternatives, and at a cost that can be accurately estimated. It is, however, an expensive option that can be highly disruptive to riparian and aquatic ecosystems, depending on the site characteristics.

Excavated or dredged sediment may be treated ex situ to reduce toxicity, mobility, or volume of the contaminated materials. The process may consist of a single step, or a series of integrated steps, that should be developed using a systems approach. Simple treatment processes such as sediment dewatering and particle size separation are generally used to reduce the volume of sediment that must be disposed of and to reduce contaminant mobility by means of chemical processes. More sophisticated methods, include soil washing, vitrification, or solidification and stabilization, are rarely used in the U.S. because of their high cost, the uncertainties surrounding their effectiveness, and concern by local communities that the treatments will release contaminants into the environment. Treatments, particularly soil washing, are more frequently used in Europe where space for disposal is more limited and recovered sediment is more valuable.

Where sediment forming the channel bed and banks have been partly or completely removed, it may be necessary to reconstruct the river channel. Designing a geomorphically stable and ecologically dynamic system is an extremely difficult task. At the present time, channel design generally utilizes one, or a combination of, four techniques including the carbon-copy method, the reference reach method, an empirical approach, and various analytical approaches. Recent data suggest that to be effective, channel design and restoration must include a detailed analysis of (1) reach- and basin-scale geomorphologic processes, and (2) river hydraulics and sediment transport processes.

Floodplain and terrace deposits exposed in the channel banks where sediment-borne trace metals have not be removed, or where they have been treated in situ, can be protected using in-stream structures (e.g., rock vanes, J-hooks, or weirs), bioengineering methods (e.g., live cribbing), or a combination of the two. Studies in North Carolina and elsewhere suggest that ridge in-stream structures are ineffective for long-term bank stabilization without continued maintenance; bioengineering methods appear more effective, although they may be more expensive and time consuming to implement. The higher initial costs may be offset, however, by the lower costs of repairs associated with the more detailed, upfront analyses.

9.6.   SUGGESTED READINGS

Kondolf GM, Smeltzer MW, Railsback SF (2001) Design and performance of a channel reconstruction project in coastal California gravel-bed stream. Environmental Management 28:761–776.
USEPA (1993) Selecting remediation techniques for contaminated sediment. EPA-823-B93-001.

Palmer MA, Bernhardt ES, Allan JD, Lake PS, Alexander G, Brooks S, Carr J, Clayton S, Dahm CN, Follstad SJ, Galat DL, Loss SG, Kondolf GM, Lave R, Meyer JL, O'Donnell TK, Pagano L, Sudduth E (2005) Standards for ecologically successful river restoration. Journal of Applied Ecology 42:208–217.

Steuer JJ (2000) A mass-balance approach for assessing PCB movement during remediation of a PCB-contaminated deposit on the Fox river, Wisconsin. USGS Water-Resources Investigation Report 00-4245.

Vangronsveld J, Cunningham SD (1998) Introduction to the concepts. In: Vangronsveld J, Cunningham SD (eds) Metal-contaminated soils: in situ inactivation and phytorestoration, Springer-Verlag and R.G. Landes Company, Berlin pp 1–15.

CHAPTER 10

IN SITU REMEDIATION

10.1. INTRODUCTION

Sediment removal by means of excavation and dredging can be an effective means of remediating contaminated rivers. The process, however, is not without its limitation, one of the most significant of which is the potential for sediment removal to negatively impact the short- to medium-term health of aquatic and riparian ecosystems. Biotic communities may be significantly disturbed or even completely erased during the removal process. Local communities are also subjected to the often annoying transport of contaminated sediment from the riverine environment, to sediment handling or treatment facilities, and finally to the disposal site. Compared to excavation and dredging, in situ remediation techniques offer alternatives to sediment removal as they are generally less invasive and are less likely to impact the local community. Costs of in situ treatments are highly variable, but in some cases, can be similar to or less than those associated with ex situ remedies. It should come as no surprise, then, that the application of in situ methods has increased over the past decade, and will continue to do so in the future with the development of several promising methods.

In this chapter, we will examine the most extensively utilized in situ remediation techniques for metal contaminated sediment and soils. In doing so, the various methods have been grouped according to whether they extract particulate-borne trace metals from the sediment or contain them in place. This classification system is not perfect in that some general techniques (e.g., phytoremediation) are used for both extraction (phytoextraction, rhizofiltration) and containment (phytostabilization). Nonetheless, it provides a framework that facilitates our discussion while describing the primary approach to remediation that is used by the in situ method.

While reading the following text, you will find that all of the in situ remedies have strengths and weaknesses. It follows, then, that the selection of an appropriate remedy represents a balancing act where the program manager, in cooperation with other stakeholders, must weigh the benefits and limitations of one method against those of another. Once more, what may serve as an entirely appropriate remedial alternative at one site may be entirely inappropriate at another, depending on the river's characteristics, the physiochemical environment, the contaminant(s) of concern, and various social and political constraints, among a host of other factors. It is therefore impossible to precisely define where and how an in situ alterative should be used. Thus, we focus on providing a brief understanding of how the various methods work, along with their uses, limitations, and costs, all of which represent important considerations in method selection.

Because of the variations observed between remediated sites, comparable cost data are difficult to obtain (NRC 1997b). Here we use data which are being compiled by the Federal Remediation Technologies Roundtable for remediated sites in the U.S. The advantage of using these data is that there has been an attempt to compile the information using similarly reported metrics and methods. It is important to recognize, however, that these data are provided only as a comparison of the costs associated with the discussed technologies as their actual costs will undoubtedly vary from location to location.

10.2.   IN SITU EXTRACTION

10.2.1.   Soil Flushing

In Chapter 9, we examined the process of soil washing as applied to dredged or excavated materials. *Soil flushing* is a similar process with the exception that the contaminated materials are not excavated from the site. Rather, contaminants are flushed from the sediment in place using water or some other aqueous solution. In general, soil flushing, which is sometimes referred to as *in situ soil washing*, usually involves a multi-step process (Fig. 10.1). The first step involves the application of an extracting solution consisting of water, acids, or chelating agents to the site. The solution can be applied using either injection or infiltration systems such as sprinkler networks, horizontal or vertical injection wells, or leach fields and basins. If applied to the vadose zone, the solution will migrate downward through the

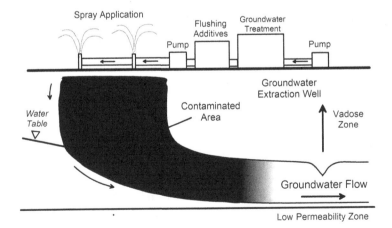

Figure 10.1. Schematic of the in situ soil flushing process (From USEPA 1991b)

unsaturated materials until reaching the groundwater table upon which it flows laterally through the subsurface flow system. As the solution moves through the subsurface, it solubilizes the contaminants which are then transported with the moving fluids. The fluid must subsequently be recovered, a process that is usually conducted using vertical or obliquely oriented wells positioned along the front of the migrating contaminant plume, or through the use of vacuum extraction systems when the fluids are within the vadose zone. In most cases, the recovered solution containing the desorbed metals and other contaminants will need to be treated to meet established discharge standards before it is released to the river, or re-used in the process (Fig. 10.1). To reduce costs, the extracting solution is most often applied, recovered, and treated in a cyclical manner until the remediation objective is achieved, or until the solution can no longer solubilize and remove the remaining contaminants from the soils or sediment (Vangronsveld and Cunningham 1998). Once completed, it may be necessary to reduce the potential mobility of the remaining metals by controlling infiltration over the site, or by applying additional reagents that produce immobile metal species that inhibit trace metal migration. The former is usually accomplished by capping the area with a material of low permeability.

Soil flushing is most often used for the extraction of inorganic pollutants, particularly relatively soluble metals (e.g., hexavalent Cr). It can also be used for the extraction of pesticides, fuels, and volatile organic contaminants, although more cost effective methods are likely to be available for these substances. The effectiveness of the method is closely related to the hydrologic conditions at the site. It is most effective where the extracting fluids can be continually cycled through the subsurface with little, if any, loss to the surrounding environment. Effective circulation generally requires sediment that exhibits a permeability in excess of 1×10^{-3} cm/sec.

The most significant advantage of soil flushing is that the contaminants can be permanently removed from the sediment without needing to remove the surficial materials, or other wise significantly disturb the site's surface. Perhaps the most important limitation is that if the hydrologic system is not adequately understood, or if the extraction methods fail to recover all of the circulating fluids, solubilized contaminants can be leached into the groundwater or river system, thereby increasing the area of contamination. Iskandar and Adriano (1997) suggest that in some situations the loss of contaminated fluid may be minimized by artificially freezing the sediment beneath and around the treated area to create an impermeable barrier (Fig. 10.2). It may also be possible in some cases to create an impermeable barrier beneath the flushed zone by means of injecting grout through directionally drilled boreholes. Some additional limitations of soil flushing are that treatment times are often lengthy, and where multiple contaminants are present, it may be difficult to formulate an effective extracting agent (Evans 1997).

The application of soil flushing in the U.S. has thus far been limited, in part because it can be expensive. Estimated costs of the method vary greatly, but range

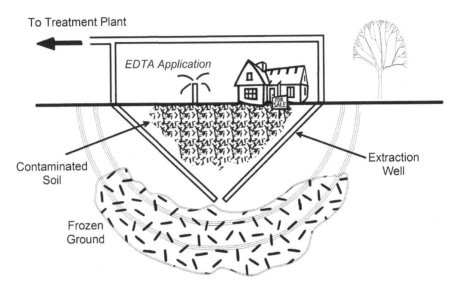

Figure 10.2. Potential design for in situ soil flushing using an artificially frozen barrier (From Iskandar and Adriano 1997)

from about \$30 to \$325 per m³, depending on the need to create impermeable barriers, and the type and concentration of the extracting fluids (Table 10.1).

10.2.2. Electrokinetic Separation

Electrokinetic Separation is a treatment process in which one or more electrode pairs are inserted into the ground and a low-intensity direct electrical current is applied between them to mobilize contaminants in the form of charged particles (Fig. 10.3). The method has also been referred to as *electroreclamation, electromigration*, and *electrokinetic soil processing*. It can be used to separate and remove a wide range of contaminants from sediment and soils including metals, radionuclides, and some forms of polar organic pollutants (Evans 1997).

The transport of the charged species occurs within pore waters of the geological materials. The electrical current mobilizes the charged particles toward an electrode depending on the nature of their charge; positively charged particles move toward the cathode, whereas negatively charged particles migrate to the anode (Fig. 10.3). Particle transport actually occurs by means of several mechanisms including (1) electromigration (the movement of *charged species* under an electric current), (2) electroosmosis (the migration of *pore fluids* in response to an electrical current), (3) electrophoresis (the transport of *charged particles* under a direct current), and (4) electrolysis (the initiation of chemical reactions in response to an electric field) (Evans 1997). Electromigration and, to a lesser degree, electroosmosis are the most important of the four transport processes (FRTR 2006).

Table 10.1. Summary of uses, limitations, and costs of selected in situ remediation methods

Method	Uses	Limitations	Costs
Soil Flushing	• Most applicable to metals & radioactivity contaminants • Can be applied to fuels, volatile organic contaminants (VOC), and pesticides, but may be less cost effective than other methods for these substances	• May be inappropriate for low permeability or heterogeneous materials • Surfactants can reduce material porosity • Potential for leaching of contaminant offsite • Treatability tests are generally required	• Range from about $30 to $325/m^3
Electrokinetic Separation	• Generally applied to metals, but may also be used for polar organic compounds • Most applicable to fine-grained, saturated or partially saturated sediment and soils • Contaminant concentrations range from a few ppm to 10s of thousands of ppm.	• Effectiveness is sharply reduced in materials where the moisture content is $< \sim 10\%$ • Can be ineffective in materials with high electrical conductivity • Oxidation-reduction reactions can produce undesirable products	• Direct costs of $15 per m^3 where energy costs are $0.3 per Kwh. • With enhancement ~$50–117/m^3
In situ Solidification & Stabilization	• Primarily used for metals and radionuclides • Limited effectiveness for SVOCs and pesticides	• Weathering may affect mobility of contaminants • May result in significant (2x) volume increase • Reagent mixing and delivery can be difficult • Solidified material may influence future site use • Application below water table may require dewatering • Treatability studies are generally required	• $50–$80/m^3 for shallow soil mixing & auger method • $190–$330/m^3 for deep, soil mixing method • $100 to $225 for injection method

(Continued)

Table 10.1. (Continued)

Method	Uses	Limitations	Costs
In Situ Vitrification	• Applicable to broad range of organic pollutants as well as metals and radionuclides • Most applicable to highly contaminated sites	• Solidified material may reduce site use • Dewatering of saturated materials may be required • Limited depth of application • Treatability studies are generally required	• $375–$525 per ton, plus $200–300k for equipment transport
In situ subaqueous capping	• Applicable to most waste forms • Can be applied to areas where sediment removal is difficult or ineffective	• Does not reduce toxicity or volume of contaminants • Reduces water depth • Alters habitat • May be affected by periodic floods or human activities • Long-term monitoring and maintenance required	• $300–$1235k per hectare[1]
Soil and Sediment capping	• Applicable to most waste forms, but small areas • May be used as final or interim remedy • Most effective where contaminant is above water table	• Does not reduce toxicity or volume of contaminants • Deep root penetration by plants may impact cap integrity • Influences future site use • Cap maintenance usually required	• Roughly $50 to several hundred dollars per hectare

Adapted from FRTR 2006

[1] Data from Naval Facilities Engineering Command (2006)

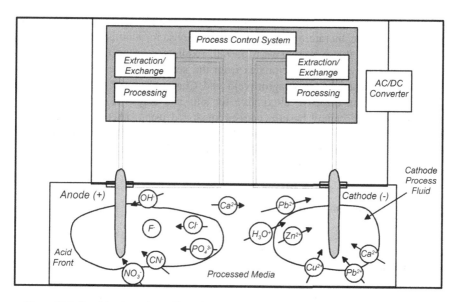

Figure 10.3. Possible electrode configuration used for in situ electrokinetic remediation (Modified from USEPA 1991a)

Probstein and Hicks (1993) found that the degree to which trace metals were moved through the subsurface during electrokinetic separation is related in part to metal desorption from particulate matter in response to hydrogen ion production at the anode. The produced H^+ ions lower the pH of the pore waters which subsequently migrate toward the cathode (Fig. 10.3). In the process, metallic cations are desorbed and solubilized until encountering hydroxide ions that have formed around the cathode. Metals which accumulate around the electrodes can then be removed by one of several processes. These include (1) the pumping of the pore waters from a well placed adjacent to the electrode, (2) electroplating or precipitation at the electrode, and (3) complexing with ion exchange resins (Evans 1997).

Given that the transport of the charged species occurs through pore waters, the technique requires saturated or partially saturated conditions. A commonly encountered problem when applied to partially saturated materials is the depletion of moisture in pore spaces adjacent to the anode as waters migrate toward the cathode. Electrokinetics, in fact, is a commonly used engineering technique for dewatering sediment and soils in place. When drying occurs it may be necessary to apply water or other solutions to the media, the latter of which may assist in the extraction of the contaminants from the in situ deposits. In some cases, the solution may be recirculated by extracting the fluids from one electrode, treating it, and then re-applying it at the other electrode. The flow of water through the subsurface as a result of pumping associated with recirculation will also enhance the transport of ions toward the well (Brown 1997).

Opinions regarding the universal applicability of electrokinetic separation differ significantly in the literature. Its primary limitations, in addition to those listed in Table 10.1, are that is expensive to install, requires constant maintenance, and tends not to remove metal precipitates from the sediments unless some form of extracting agents are used (Vangronsveld and Cunningham 1998). Nonetheless, it may represent an effective alternative for silt- and clay-rich deposits of low-permeability which cannot be effectively remediated using other in situ methods (USEPA 1995; Brown 1997). To date, its application has been limited, although the method has been successfully applied at number of demonstration sites (see, for example, USEPA 1995, 1997).

10.2.3.   Phytoremediation

Bioremediation, or the use of any living organism to transform or remove contaminants from the environment, has proven to be a particularly cost-effective method of remediating sites contaminated by a host of *organic* constituents. In general, the organic substances are broken down into non- or less-toxic forms by microorganisms in the soil, sediment, or water. The use of microbiota as a remedial technology for metal contaminated sites is still under development (Vangronsveld and Cunningham 1998), and is primarily limited to changing the chemical form (species) of the metal to reduce its mobility or bioavailability. Microbes, for example, have been identified that can reduce chromium (VI), selenate, and uranium (VI) to their less toxic forms represented by chromium (III), selenite, and uranium (IV), respectively (Litchfield 2005). Microbially mediated processes have also been proposed for use in suspension and solid-bed leaching processes to promote metal extraction from contaminated sediment (Seidel et al. 2004). These processes, however, have yet to be used in the remediation of metal contaminated rivers. This is not the case for *phytoremediation*, or the use of plants to remove or contain contaminants in soil, sediment, and water (Evans 1997; Kamnev 2003). Phytoremediation has been successfully tested at the field scale at a number of locations, and its use appears almost certain to increase in the future.

Phytoremediation actually involves three separate and distinct approaches: (1) *phytoextraction*, which is intended to remove contaminants from sediment or soil, (2) *phytostabilization*, a process that retards the mobility of the contaminants in sediment and soil, and (3) *rhizofiltration*, a technique aimed at removing contaminants from surface- and groundwaters. The current interest in these methods is primarily related to their ability to be applied to large areas with low-levels of contamination at a relatively low-cost. In addition, phytoremediation is an in situ treatment that does not require extensive disruption of the site. It can, in fact, be used as a means of ecological restoration and to prevent erosion, depending on the method(s) that are applied, while transforming the area into an aesthetically pleasing site (Vangronsveld and Cunningham 1998; Marmiroli and McCutcheon 2003). Phytoremediation, however, is not without its limitations, the most significant being that the in situ processes are relatively slow in comparison to most other technologies (Blaylock and Huang 2000), and the depth of treatment is generally limited to the root zone (often $<\sim 30$ cm).

The use of phytoremediation may be particularly applicable to the remediation of fluvial landforms such as floodplains and terraces which have been affected by low-levels of contamination as a result of overbank deposition. This follows because the contaminants on floodplains and terraces are often concentrated near the ground surface, and may be distributed over large areas that could not otherwise be addressed except at an enormous cost. Ultimately, the effectiveness of the processes depends on site characteristics, the nature of the contaminants, and the type of phytoremediation that is being used. The two forms of phytoremediation which extract contaminants from the environment are described in more detail below; the remaining method (phytostabilization) is described later in the chapter in the section on in situ containment methods.

10.2.3.1. Phytoextraction

Many trace metals are toxic to plants, a fact that has led to significant difficulties in rehabilitating abandoned mine and smelter sites. It is long been recognized, however, that there are a number of plant species (about 400 world-wide) that can not only tolerate high-levels of trace metals in the soil, but accumulate significant concentrations of metals in their tissues. Such plants are often referred to as *metal hyperaccumulators*. Hyperaccumulators are most often defined as plants which when grown in their natural habitats, may store more 1,000 mg/kg of Pb, Co, Cu, Ni, and Cr in their tissues, or 10,000 mg/kg of Mn or Zn on a dry weight basis (NRC 1997b; Blaylock and Huang 2000). The recognition of hyperaccumulators gave rise to the postulate that plants could be used to remove metals from contaminated soils and sediment, and concentrate them in the above ground biomass. Once in the tissues, the biomass can be harvested and their volume reduced by means of compositing, compaction, or incineration before their disposal in a landfill. More recent investigations have even suggested that the metals could be recovered from the biomass and reused for beneficial purposes.

The initial laboratory and field tests demonstrated that while hyperaccumulators were effective at removing metals from soils, the rate at which the metals could be recovered was limited because only small quantities of biomass were produced at a relatively slow rate (Baker et al. 1994; Brown et al. 1994). As a result, more recent investigations have primarily focused on other plant species that remove metals from the soil more quickly by optimizing the relationship between metal accumulation and biomass production. Blaylock and Huang (2000), for example, suggest that to be effective, the plant must be able to accumulate more than one percent of the metal of interest, while generating at least 20 metric tons of above ground biomass per hectare per year. Many of the currently investigated species represent high biomass agronomic crops (Blaylock et al. 1997; Huang et al. 1997) that can be planted, grown, and harvested using conventional agricultural practices.

The effective use of phytoextraction is dependent in large part on the abundance of metals within the soil and soil solution that can be readily accumulated by the plant roots. This is generally limited to metals that exist in solution within the pore waters, or that can be readily desorbed or solubilized by root exudates (Blaylock and Huang 2000). Root exudates are inorganic and organic compounds that are released by the roots to the surrounding soils. Depending on the source of contamination and the targeted metal, the readily available fraction may represent only a small portion of the total metal concentration within the contaminated materials. Thus, the plants would only be able remove part of the soil's total metal content. To help overcome this problem, investigators have examined the use of chelating agents (e.g., EDTA) to help desorb the metals from the soil, thereby making them more bioavailable. The chelators, in essence, complex the free metal ions in the pore waters which promotes further desorption and/or dissolution of the metals until an equilibrium is reached between the free ions, the complexed metals in solution, and the soil phase (Norvel 1991; Blaylock and Huang 2000). The results from these studies have shown that in addition to increasing the abundance of the metal that is available for uptake, the ability of the plant to translocate the metal from the roots to the shoots is dramatically increased. The chelators, then, can actually produce hyperaccumulation within a number of agronomic crops which are characterized by high biomass production (Blaylock and Huang 2000).

A potential shortcoming of using chelating agents is that they break down slowly in the subsurface environment, and once the metals are complexed by the chelators, they are highly mobile. It is possible, then, for the metals to migrate offsite and contaminate other surface- and groundwater resources. Thus, additional site management practices may be required to prevent inadvertent contamination of other areas or media (Vangronsveld and Cunningham 1998).

Even where chelating agents are used, phytoextraction is a relatively slow process. Its applicability, then, depends heavily on the time required for contaminant extraction by the plants, which in turn depends on the difference between the metal concentration in the sediment and the targeted level of cleanup (i.e., the remedial standard). For example, Blaylock and Huang (2000) calculated by using typical soil and phytoextraction data that the removal of Ni from a soil possessing a concentration of 2,800 mg/kg of the metal would require 50 crops to meet a residential cleanup standard of 250 mg/kg. Even assuming two crops per year, the application of phytoremediation to this site is likely to prohibitively long. In contrast, only eight crops would be required to meet a nonresidential goal of 2,400 mg/kg, and thus the strategy may prove highly valuable. Given the above example, it follows that phytoextraction is likely to be most applicable to sites where it is not necessary to remove large quantities of metals from the soil or sediment, including sites covering vast areas.

10.2.3.2.   *Rhizofiltration*

Rhizofiltration, alternatively called *phytofiltration*, is a relatively new method for removing toxic metals from surface or groundwater. It is based on the realization that hydroponically grown plants with extensive root systems are particularly effective at accumulating and precipitating metals from polluted effluents (Dushenkov et al.

1995). Hydroponically grown seedlings of various terrestrial plants also possess these characteristics (Salt et al. 1997). The actual mechanisms of metal removal are still rather vague and may differ between trace metals (Dushenkov and Kapulnik 2000); however, possible mechanisms are likely to include precipitation and adsorption along the cell wall, extracellular precipitation, and intracellular accumulation (Fig. 10.4).

Dushenkov and Kapulnik (2000) suggest that the most suitable plants for rhizofiltration are characterized by the ability to produce large quantities of root biomass under hydroponic conditions. Moreover, they should possess a high tolerance to the targeted metal, be able to accumulate large amounts of the metal, have a high root to shoot ratio, and be capable growing in controlled environments. Some of the

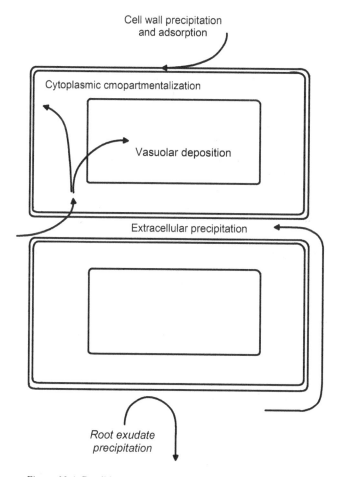

Figure 10.4. Possible mechanisms of metal removal from pore waters by plant roots or seedlings (From Dushenkov and Kapulnik 2000)

plant types that are particularly well-suited for rhizofiltration include sunflowers, Indian mustard, and tobacco (Dushenkov et al. 1995).

A number of field studies have been conducted to test the applicability of rhizofiltration to river and lake waters. The experiments usually involve the growth of selected plant species in special hydroponic cells, either in situ or adjacent to the water body, followed by an estimation of the quantity of metal(s) extracted from the water. The test results have shown that rhizofiltration holds considerable promise, and may to be comparable with many currently used pump and treat technologies (Dushenkov et al. 1995).

10.3.   IN SITU CONTAINMENT

10.3.1.   In Situ (Subaqueous) Capping

The National Research Council defines *in situ capping* as "the controlled, accurate placement of a clean, isolating material cover, or cap, over contaminated sediments without relocating or causing a major disruption to the original [channel] bed" (NRC 1997). In theory, the cap is intended to (1) physically isolate the contaminated materials from direct contact with biota, including burrowing organisms, (2) stabilize the contaminants in place to inhibit their remobilization by processes of erosion, and (3) limit the potential remobilization of the contaminants in the dissolved phase by geochemical processes (USEPA 2005). As of 2004, in situ capping had been selected as part of the remedial strategy at approximately 15 marine and freshwater Superfund sites in the U.S. (USEPA 2005), and had been applied to about the same number of sites elsewhere.

Selection of in situ capping is based on several perceived advantages over sediment removal. First, unlike dredging or excavation activities, resuspension is limited, and there is no residual contamination left along the water-sediment interface. Bottom-dwelling organisms are therefore immediately supplied with a clean substrate to recolonize, provided that the cap will not be re-contaminated by the transport and deposition of sediment-borne contaminants from upstream areas. Second, it limits the potential impacts on local communities as there is no need to develop sediment staging or treatment facilities, nor is there a need to transport contaminated materials through residential environments. Finally, it can be implemented quickly using conventional equipment and locally available materials. Thus, capping tends to be one of the least expensive remedial technologies for addressing contaminated channel bed sediment (Table 10.1) (NRC 1997).

Although in situ capping is usually considered as a potential remedial strategy for almost all contaminated rivers, sediment removal remains the method of choice for most sites. In fact, where capping has been applied, it is often used as a further risk reducing practice following the removal of highly contaminated sediment, or at sites where dredging activities had difficulties in meeting the cleanup standard (see, for example, Cushing 1999). The primary limitation of in situ capping is that the contaminants remain in the aquatic environment, and if the cap is breached, they could become exposed and dispersed. Moreover, it is possible that contaminants in the dissolved form will migrate through the cap in significant quantities. Thus, the method lends itself to questions about its ability to provide a permanent solution to the problem, particularly in highly dynamic rivers. Cushing (1999) notes, however, that capping should be thoroughly considered as a remedial alternative because in many cases it may exhibit more favorable characteristics than sediment removal in terms of implementation time, cost-effectiveness, difficulties of implementation, and degree of likely and realistic success.

The effective use of in situ capping ultimately depends on whether it can be appropriately applied at the remediated site as well as on the design, construction, and long-term maintenance of the cap (Palermo 1991). Perhaps the most frequently used materials in cap design include sand, silt, and clay, along with gravel and various geotextiles. Natural occurring geological materials are often preferred over processed (washed and size sorted) sediment because they are cheaper to obtain, and they possess larger quantities of reactive substances. The reactive substances can sequester contaminants before they are released to the overlying water column.

At sites where contaminant migration through the sediment is a significant concern, reactive substances are often added to the materials to increase the sorption properties of the cap. The most commonly used additives including apatite, activated carbon, clay minerals, zero-valent iron, zeolites, and various proprietary materials, such as AquaBlok™ (USEPA 2005).

Cap designs vary significantly depending on the site of application and the nature of the targeted contaminants. In order to stabilize and physically isolate the underlying sediment, the cap must be able to withstand the erosive forces which are applied to it over a wide range of flow conditions. Thus, caps are often designed to withstand at least the 100-year event by sizing the outer capping materials to resist entrainment by the associated discharge.

Another important factor to be considered during cap design is the geomorphic stability of the river system. Channel bed degradation, for example, is likely to result in cap failure by means of undercutting of the cap margins. It follows, then, that if the river is currently in a state of disequilibrium, or is likely to become unstable in the near future as a result of changing land-use conditions, or some other disturbance within the watershed, in situ capping should be avoided. In situ caps may also be inappropriate along highly dynamic rivers characterized by high rates of lateral channel migration, or significant scour and fill during flood events.

Cap design and performance should not only be evaluated on its ability to physically isolate sediment-borne contaminants over a long-period of time, but the ability of the system to retard the chemical flux of dissolved constituents through the cap to the overlying water column. As described in Chapter 2, trace metals are often associated with fine-grained, reactive channel bed sediment. Most fine-grained materials possess relatively low-shear strength, particularly when saturated. The weight of the cap on such fine-grained sediment may therefore result in consolidation and settling of the contaminated material over time. As settling occurs, cracks or fractures can be produced in the cap through which contaminated waters can migrate into the overlying water. In addition, pore waters and gasses within the channel bed sediment may be squeezed out of pore spaces as consolidation occurs. The expelled water and gas, along with any dissolved contaminants, may then be transported through the cap by advection (Cushing 1999; USEPA 2005). Other mechanisms through which contaminants may escape through the cap involve the upward flow of groundwater beneath the cap, and the process of molecular diffusion caused by differences in the chemical composition of the pore waters within the cap and channel bed sediment.

In situ caps can consist of a single layer or of several layers in which each horizon performs a specific function. Commonly used horizons include (1) a layer of rocks or gravel at or near the top of the cap to armor it against erosion, (2) a sandy layer which serves as a relatively stable substrate that resists burrowing and bioturbation, (3) a fine-grained lower layer which promotes chemical isolation through sorption processes, and (4) geotextiles that limit burrowing and material mixing (Fig. 10.5) (Cushing 1999). Emplacement of the materials can be accomplished using either mechanical methods (e.g., clamshells or their release from a barge), or in the case of granular materials, their discharge in the form of a slurry from the end of a pipe. The latter hydraulic methods may allow for the more precise placement of the cap materials, but have the potential to resuspend the contaminated sediment along the water-sediment interface (USEPA 2005).

An important consideration in cap design is the ability of the upper most cap materials to be used as habitat by desirable organisms. This may require slight modifications in the cap design from what would be used for solely isolating and immobilizing the contaminated sediment. For example, gravel is often used to protect the cap from erosion, even where natural channel bed sediment is composed of silt and clay. In this situation, however, the gravel is unlikely to serve as an

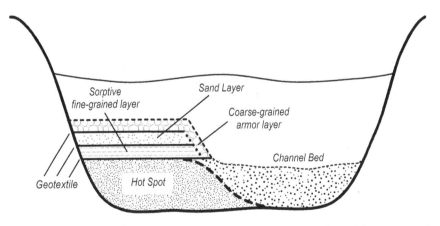

Figure 10.5. Hypothetical example of multi-layer in situ cap. Each type of material serves a specific function. Gravel is typically used for armor, sand for increasing thickness and stability, and silts/clays for metal sorption. Geotextiles reduce mixing and burrowing by bottom dwelling organisms. Relative thickness of layers is exaggerated

adequate substrate for the existing bottom-dwelling organisms. It may therefore be necessary to place a layer of finer sediment over the gravel to improve the cap's overall function as an aquatic habitat. In doing so, care must be taken to insure that bottom-dwelling organisms can not burrow through the cap and bring contaminated sediment to the surface.

As is the case for any engineered method of containment, in situ caps require constant and continued monitoring to insure that the structures are performing as intended. The costs of contaminant monitoring and maintenance can be high, and need to be considered in the overall project costs.

10.3.2.   Soil and Sediment Capping

Soil and sediment capping is similar to the in situ (subaqueous) capping methods described above, with the exception that it is performed in a subareal environment. It is more frequently used than in situ capping, particularly in residential environments. The capping materials may be applied directly over the contaminated soils, or alternatively, following excavation, either because some residual contamination remains, or because it was decided that it was not necessary to excavate deeper materials. The direct capping over contaminated sediment (without sediment removal) is most cost effective as expenditures for excavation, treatment, and disposal are avoided.

A perfectly designed cap is intended to (1) physically isolate the contaminated materials from direct contact with biota, (2) stabilize the contaminants in place to inhibit their remobilization by wind or water, and (3) limit the potential for contaminant remobilization by geochemical processes by reducing the infiltration of water to the subsurface. A variety of materials have been used for soil and sediment capping (Vangronsveld and Cunningham 1998). They include all of those cited for in situ capping, plus such materials as asphalt and concrete.

A significant difference between the selection of materials for soil and in situ (subaqueous) capping is the need for the former to limit infiltration of precipitation through the cap to the underlying contaminated materials. In other words, the cap should reduce the potential for infiltrating waters to leach sediment-borne trace metals or other contaminants into the groundwater system. Thus, capping materials are often selected for their hydraulic characteristics. In addition, multi-layer caps may be required to limit the downward flow of water while meet the intended land-use requirements of the site following cleanup. Asphalt and concrete, for example, limit infiltration while serving as the perfect surface layer in areas which will ultimately be utilized as a parting lot. Both materials, however, require significant maintenance as settling or frost heave often lead to the development of fractures. In other instances, the surface layer may need to support vegetation and protect the cap from erosion, while subsequent layers must reduce infiltration and increase chemical sequestration.

As was the case for in situ capping, cap failure is primarily driven by geomorphic processes associated with cap erosion, or through the chemical migration of contaminants through the cap or into the groundwater system. Where caps are situated in a flood-prone region, the cap must be able to withstand the erosive forces associated with extreme events predicted for the site. Caps may be inappropriate (1) along dynamic rivers characterized by high rates of lateral migration as the cap may be undercut and breach relatively rapidly, (2) where the cap possess a high potential to be buried by contaminated overbank flood deposits, and (3) in zones of groundwater discharge where advective flow may increase the transport of dissolved constituents through the cap to the ground surface or an adjacent water body.

Economically, soil and sediment capping is most applicable over relatively small areas. Thus, it is often utilized where there is a high potential for humans or other biota to come into contact with the contaminated particles. Its application becomes less attractive over large areas, such as contaminated floodplains and terraces, where it is likely to be hindered by a lack of available capping materials near the site.

10.3.3.   In Situ Solidification and Stabilization

In situ solidification and stabilization is intended to decrease contaminant mobility and solubility by mixing soils and sediment with a binding agent that (1) encases the material in a stabilized mass (solidification), and or (2) reacts with the material and any remaining water to reduce the solubility and mobility of the trace metals through changes in material chemistry (stabilization) (USEPA 1994; Iskandar and Adriano 1997). In most cases, the process reduces permeability of the contaminated sediment and, therefore, their contact with groundwater. The binding agents that are used in the process are similar to those described in Chapter 9 for ex situ solidification/stabilization. They are typically applied either by means of an injector system, or through the use of an auger which mixes the reagents in place.

In most cases, the utilized augers are equipped with injection ports through which the reagents are shot into the sediment as the materials are mixed. The actual process is founded on methods that have been utilized for some time in the construction field for installing cement footers and grout curtains. The augers are surprisingly large, ranging from 0.75 to 3.65 m in diameter (Evans 1997). Smaller sized augers are usually required for materials dominated by silts and clays, or when augering to greater depths.

In situ solidification and stabilization is less frequently used than the ex situ approach, in part because the mixing of materials in place is likely to be more expensive for depths up to about 3 m (Evans 1997), which is where contaminants are most often concentrated. The decrease in cost at greater depths primarily results from the savings associated with treating larger volumes of materials. In the U.S., the FRTR (2006) estimated that in situ solidification and stabilization ranged from \$50 to \$80 per m^3 for shallow operations, and \$190–\$330 per m^3 for greater depths, depending on the reagents which were used, their availability, the project size, and nature of the contaminant (Table 10.1). The local availability of the reagents is especially important in controlling overall costs.

With respect to contaminated rivers, the approach is most applicable to floodplain and terrace deposits. It has not, to our knowledge, been applied to sediment within the channel bed. Its application to fine-grained materials is questionable as it is often difficult to adequately mix and combine the binding agents with silt- and clay-rich materials.

Some investigators considered in situ vitrification a thermal form of solidification and stabilization (Evans 1997; FRTR 2006). As described in Chapter 9, *vitrification* is the melting of a soil mass at extremely high temperatures (between 1,300 and 2,000° C). Upon cooling, the contaminants are contained within a highly durable, glass-like material which is chemically inert, of low permeability and from which metals can only be leached with extreme difficulty. In the case of in situ vitrification, the method usually involves the emplacement of two graphite electrodes in the ground, and the heating of the geological materials located between them (Fig. 10.6). Although it can be applied to a broad range of metals, it can be expensive, particularly in materials possessing a high moisture content.

10.3.4. Phytostabilization and Immobilization

Phytostabilization is a form of phytoremediation, which has also been referred to as *in situ inactivation* and *phytorestoration*. The approach is founded on the ability of plants to reduce metal mobility and bioavailability through three primary mechanisms (Fig. 10.7). First, vegetation is used to cover the landscape and protect it from the erosional affects of wind and water. Reductions in mobility are brought about by limiting the transport of sediment-borne trace metals to the surrounding environment. In addition, a thick cover of vegetation limits contact with the underlying soil, particularly by small children, thereby reducing potential health and

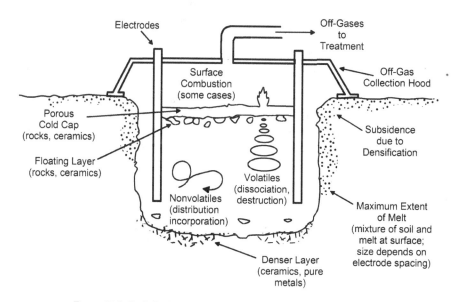

Figure 10.6. Basic in situ vitrification scheme (From Houthoofd et al. 1991)

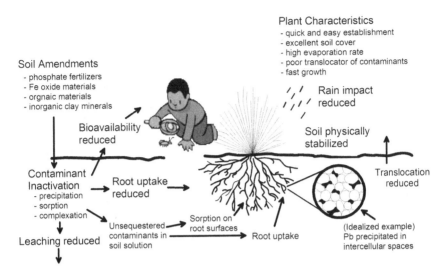

Figure 10.7. Basic functions of plants and soil amendments in the phytostabilization of metal
contaminated soils and sediment (From Berti and Cunningham 2000)

ecological risks. Second, plant roots possess an enormous surface area. They may
therefore help to immobilize metals in the subsurface by adsorbing metals onto
their root surfaces, or by accumulating and precipitating metals within the roots
themselves. Moreover, root exudates may change the chemistry of the surrounding
soils and pore waters in such a way as to reduce the transport of metals through
the geological materials. The reductions in transport result primarily from changes
in pH, the oxidation and reduction potential (Eh), or microbial activity in root the
zone (Evans 1997).

The third mechanism commonly cited as a means of reducing mobility is through
a decrease in water percolation and leaching as vegetation becomes established.
A common effect of vegetation, however, is to increase the infiltration capacity
of the soil. Thus, the development of a thick vegetal cover may either increase
or decrease leaching depending on whether the effects on infiltration are offset
by enhanced evapotranspiration. It is important, then, to select plants that possess
relatively high transpiration rates which effectively dewater the underlying soils
and sediment (Berti and Cunningham 2000).

With regards to plant selection, the types chosen for phytostabilization are very
different from those used in phytoextraction. Perhaps the most significant difference
is that for phytostabilization it is important to use plants that do not accumulate
metals in their above ground tissues (Berti and Cunningham 2000). Limiting root
to shoot metal translocation is necessary in order to minimize the possible transfer
of metals to biota that may consume the above ground shoots. Other important
characteristics noted by Berti and Cunningham (2000) is that the plants should (1)
be tolerant of high metal concentrations in the soils or sediment, (2) grow quickly
to establish a thick cover over the landscape, (3) have a dense rooting system, and

(4) possess high transpiration rates. In addition, the plant communities should be easy to establish and be self-propagating in order to limit future maintenance and maintenance costs at the site. It turns out that many of the plants used for soil conservation and management purposes are well suited for phytostabilization, including various species of grass (Salt et al. 1995). An added benefit of these plant types is that they can be established using existing agricultural equipment and methods, thereby reducing implementation costs (Vangronsveld and Cunningham 1998).

The development of phytostabilization as a remediation alternative was founded on the reclamation of mining and milling wastes, and landscapes surrounding historic smelters. The primary problem in these areas was the establishment of vegetation on barren soils as the physiochemical conditions of the soils, including high metal concentrations, inhibited plant growth. It was determined, however, that the addition of certain soil amendments transformed soluble and/or easily exchangeable metal species into less bioavailable forms, decreasing the uptake and toxicity of the metals in plants (Vangronsveld and Cunningham 1998). In other words, the amendments not only allowed for plant establishment, but reduced the mobility of the metals in the environment. As a result, phytostabilization often includes the addition of various amendments to the soil and sediment.

Significant effort in recent years has gone into defining the amendments that are most effective for specific trace metals (Ma et al. 1993; Mench et al. 1998). Currently, the most commonly used amendments include phosphates (Mench et al. 1998), lime (Albesel and Cotteni 1985), Fe and Mn oxyhydroxides (Berti and Cunningham 1997), various organic materials such as sewage sludge, manure or compost (Berti and Cunningham 1994), and natural or synthetic aluminosilicates (e.g., clay minerals and zeolites) (Gworek 1992; Mench et al. 1994). Exactly which amendment is most suited to the site depends on the targeted metal(s), the geochemical nature of the soils or sediment, and the plant species which will be used in phytostabilization. Berti and Cunningham (2000) argue that in general the most effective amendments are (1) inexpensive, (2) easy to safely handle, (3) nontoxic to the plants selected for phytostabilization, (4) easily obtainable, preferably from a local source, and (5) unlikely to possess any additional threat to the environment. Moreover, it is important for the amendment to rapidly immobilize the metal contaminants following their application without requiring additional treatments. The application of lime, for example, may raise the pH of the soil and pore waters, and thereby reduce the mobility of many cations. However, lime is readily leached from the soil by percolating groundwater and must be reapplied. In addition, while lime may reduce bioavailability in the soil, it may not be effective at reducing its availability under strongly acidic conditions. Thus, its potential to reduce the availability of Pb in the human stomach if the soil particles are ingested is limited (Brown and Chaney 1996). In contrast, Pb may be transformed by the addition of various phosphate amendments into highly insoluble Pb phosphate minerals which remain stable even in very acid environments (Ma et al. 1993). It may therefore remain unavailable in the human stomach (Berti and Cunningham 2000), although the toxicity of certain Pb phosphate minerals may also be of concern.

Although phytostabilization has been applied to the reclamation of mine and smelter impacted areas, it has not been extensively used as a remedial alternative along contaminated rivers or at other types of metal contaminated sites. Given that it is a relatively inexpensive and easily applied method, it appears to be underutilized at the present time, particularly along rivers in rural areas where the alluvial deposits possess relatively low levels of contamination. It has also been pointed out that phytostabilization can serve as an excellent interim measure (Evans 1997), reducing metal mobility until site assessments have been completed, or until more aggressive measures are undertaken.

10.4. MONITORED NATURAL RECOVERY

Monitored natural recovery (MNR) represents a significant departure from other remedial methods in that cleanup of the site rests completely on naturally occurring processes that contain, destroy, or reduce the bioavailability or toxicity of contaminated sediment (USEPA 2005). In many ways, it is similar to the no-action alternative as contaminated sediment is neither removed nor treated in any way. Most recent proponents of MNR argue, however, that is not equivalent to the no-action approach (see, for example, Brown 1999; Magar 2001). A decision of no-action, for example, may be based on the fact that the reduction in risk associated with sediment removal or treatment are outweighed by the potential consequences of remediation. Although this may also be the case for MNR, natural recovery additionally assumes on the basis of historical data and a variety of predictive methods that (1) site recovery is currently underway as a result of natural processes, and (2) the rate of recovery will result in the desired outcomes in an appropriate period of time. In addition, MNR may be accompanied by an adaptive management approach where if continued monitoring of key indicators determines that natural recovery is not occurring, or not occurring at an acceptable rate, then other remedial technologies can be applied. Moreover, it is likely to be combined with a number of institutional controls, such as advisories on fish consumption or restricted use of the site.

The application of MNR is a relatively recent remedial option that from a practical perspective emphasizes to local stakeholders that site conditions are expected to improve. In the U.S. it has been cited as a potential remedy at approximately 12 Superfund sites across the country, and is often used in combination with other technologies which are applied to more heavily contaminated areas of the fluvial environment (USEPA 2005). Its use will undoubtedly increase in the coming years as a result of its cost-effective nature. The primary expenditure for the technique revolves around the implementation of a detailed monitoring program and a detailed analysis of current and future site conditions. Other significant advantages of MNR include its non-invasive nature and its limited impact on the local community. While more aggressive remedial strategies such as sediment removal significantly disrupt stream and riparian ecosystems, and require areas for sediment staging or treatment, MNR leaves the existing terrains intact. Short-term impacts associated with sediment remobilization are eliminated and potential dispersal during the treatment or transport of the contaminated materials is avoided. The primary limitation of MNR is that contaminated materials remain at the site and continue to pose a threat. Moreover, MNR is typically a slow process in comparison to other remedial alternatives, possibly requiring decades to reach the remediation objectives.

The natural processes which promote recovery of the aquatic environment can broadly be placed into three categories: physical, chemical, and biological. All three of these process types are interconnected. The outcome of these interactions may actually increase the environmental risks, rather than reduce it (as would be the case if the process were operating in isolation) (USEPA 2005). Thus, both the independent and combined effects of these processes must be evaluated with regards to improving water and sediment quality.

Physical processes which are commonly cited as being of interest to MNR include sedimentation, erosion, dispersion, bioturbation, advection, and volatilization. Sedimentation, in this case, is envisioned to bury contaminated materials with clean or cleaner materials, thereby containing the contaminants in place, and reducing the potential for biota to come into contact with the more heavily contaminated sediment. The long-term effectiveness of this process depends on the likelihood of remobilizing the newly deposited sediment, and the potential for contaminants to be moved through the deposited materials by means of chemical processes (i.e., in solution). It is therefore most effective in stable, depositional environments (USEPA 2005) such as floodplains or terraces. The other processes cited above tend to dilute contaminant concentrations by distributing them over a larger area, mixing them with additional sediment, or transferring them to other media (e.g., water or air). In the case of trace metals, the processes involved in particle transport are likely to be of most importance. Erosion of contaminated bank materials and their subsequent downstream transport, for example, is a process that while adding contaminants to the aquatic environment, may ultimately reduce contaminant concentrations in the alluvial deposits by mixing the contaminated bank materials with clean sediment from tributaries or upstream areas. Contaminant concentrations can also be diluted by the spreading of the contaminated sediment, eroded from the bank, over a larger area.

As described in Chapter 2, there are number of chemical processes that may either increase or decrease trace metal mobility. Monitored natural recovery primarily focuses on redox related reactions that can decrease the solutibility and bioavailability of metal and organometallic compounds (USEPA 2005). A commonly cited example is the formation of sulfide minerals under reducing conditions; the sulfide compounds bind the metals in a non-available form provided that the reducing environment is maintained. Similarly, biological processes driven by sulfate-reducing bacteria may not only influence the redox potential of the environment, but directly influence metal mobility by forming stable metal sulfide complexes. Other biological processes that may be of significance to metal sequestration include natural phytoremediation, biological stabilization, and in the case of organic contaminants, biodegradation (USEPA 2005).

In most situations, chemical and biological processes are relatively slow and, thus, recovery based solely on these process types alone would require decades or centuries to complete (Magar 2001). As a result, MNR often focuses on contaminant burial and dilution by fluvial geomorphic processes, although the other processes must also be examined.

An inherent requirement of MNR is that the riverine environment is currently recovering from the impacts from a contaminant source. Data must therefore be collected which adequately demonstrates a decrease in contaminants levels within the alluvial sediment, the river water, and/or its biota. In many instances, this can be shown empirically by plotting changes in metal concentrations within a specific media (e.g., fish tissues) through time; where decreases occur, rates of natural recovery, usually reported in half-times, can be determined (Fig. 10.8). In other cases, it may be possible to piece together water or sediment quality records obtained over a period of years to determine if statistical decreases in contaminants levels exist. Unfortunately, monitoring data for many sites is insufficient to demonstrate system recovery. In these situations, it may be possible to use proxy records of changing contaminant loads. Two important examples include the development and interpretation of pollution histories from overbank floodplain deposits, and

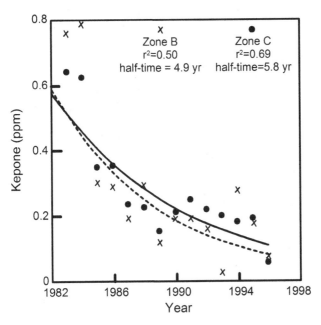

Figure 10.8. Change in Kepone concentrations within the channel bed sediment of the James River. Decreasing concentrations reflect the influence of natural recovery processes (From Brown 1999)

changes in trace metal concentrations in progressively younger alluvial deposits (as described in Chapters 6 and 7).

A more difficult task that must be undertaken to implement MNR is to determine that recovery will continue in the future and at an acceptable rate. There is no set approach through which these predictions can be undertaken. Perhaps one of the most simplistic methods is to extrapolate previously documented (historical) trends into the future. A shortcoming of this approach is that it assumes the physiochemical environment will remain similar to that of the past, which may or may not be the case. Thus, it may be necessary to use a modeling approach that allows for the prediction of system behavior. The utilized models range from relatively simple mathematical or conceptual expressions to highly sophisticated, multidimensional computer-based algorithms. The latter is commonly applied to complicated, high-risk sites where it is essential to gain a more complete understanding of the processes of contaminant transport and fate than may be possible from field or empirical data alone. Actually, these routines may consist of several different kinds of models which are linked together to describe changes in flow, sediment transport, and chemical partitioning and load over a range of rainfall-runoff conditions.

There are literally hundreds of chemical transport and fate models that have been published during the past several decades. One of the most frequently used models is the Water Quality Analysis Simulation Program (WASP), developed by the USEPA (Ambrose at al. 1993). In general, the output from these mathematical models is used for two primary purposes. First, the application may yield important insights into how the system functions by considering the simultaneous variation in multiple parameters in response to changes in certain driving forces, such as rainfall. Second, they may be used to make a wide range of predictions, such as the contaminant load during a 100-year event, or the change in contaminant loads or bioavailability in response to some form of site remediation.

The use of models for predictive purposes is an important tool in the assessment of human and ecosystem risk and river recovery. It is extremely important to recognize, however, that all models generate output possessing a degree of uncertainty and, in general, the more sophisticated the model, the more uncertain the output. This uncertainty results from multiple factors (USEPA 2005), including: (1) an incomplete understanding of the modeled processes; (2) the inability to develop mathematical expressions that describe the modeled processes without using simplifying or approximating expressions, (3) uncertainties or errors in the input data, particularly with regards to how well they represent the site conditions, and, when used for prediction, (4) the inability to accurately determine the nature of the site's future conditions (e.g., rainfall, land-use, or contaminant sources).

The uncertainty inherent in models means that their results must be used with a degree of caution, in many cases a high degree of caution. Moreover, decisions based on the modeled output must be consistent with the overall understanding of how the system functions, and should be supported by other lines of evidence.

10.5.   SUMMARY

The application of in situ remediation technologies has increased in recent years and is likely to grow as a result of their relatively low cost and their less disruptive nature in comparison to the more commonly used sediment removal and disposal approach. There are only a limited number of in situ technologies which can be applied to metal contaminated river systems. Of these, capping of contaminated sediment and soils (either in the subaqueous or subareal environment) using natural or artificial materials is the most frequently used method. Other techniques that hold promise include soil flushing, electrokinetics, solidification/stabilization, and phytoremediation. Two forms of phytoremediation (phytoextraction and phytostabilization) are receiving considerable attention at the present time, and may be particularly well-suited to fluvial environments where low-levels of contaminants are spread over larges areas of the floodplain.

Monitored natural recovery represents a significant departure from the other in situ methods as it relies on natural processes to destroy, contain, or reduce the bioavailability or toxicity of contaminated sediment. The most effective natural processes in most riverine environments include deposition and burial of sediment-borne contaminants. Burial isolates the contaminated materials in place, and decreases the potential for them to be remobilized in the dissolved and particulate form. Erosion and mixing of contaminated sediment with clean or cleaner sediment is also an important process. The implementation of MNR requires that the riverine environment has already begun to recover from the impacts of contamination, and that recovery will continue in the future at a rate that will achieve the remediation objectives in an appropriate timeframe. A determination of both requires a detailed analysis of fluvial processes at a wide-range of spatial and temporal scales.

10.6.   SUGGESTED READINGS

Berti WR, Cunningham SD (2000) Phytostabilization of metals. In: Raskin I, Ensley BD (eds) Phytore-mediation of toxic metals: using plants to clean up the environment, Wiley, pp 71–88.

Blaylock MJ, Huang JW (2000) Phytoextraction of metals. In: Raskin I, Ensley BD (eds) Phytoremedi-ation of toxic metals: using plants to clean up the environment, Wiley, pp 53–70.

USEPA (2005) Contaminated sediment remediation guidance for hazardous waste sites. EPA-540-R-05-012

Iskandar IK, Adriano DC (1997) Remediation of soils contaminated with metals – a review of current practices in the USA. In: Iskandar IK, Adriano DC (eds) Remediation of soils contaminated with metals. Science Reviews, Northwood, U.K. pp 1–26.

Vangronsveld J, Cunningham SD (1998) Introduction to the concepts. In: Vangronsveld J, Cunningham SD (eds) Metal-contaminated soils, in situ inactivation and phytorestoration. Springer-Verlag and R.G. Landes Company, Berlin pp 1–15.

APPENDIX A

DATING TECHNIQUES

A.1.   INTRODUCTION

River deposits contain a long, albeit discontinuous, record of fluvial landscape evolution (Stokes and Walling 2003). As such, they hold the key to understanding how a fluvial system has responded to past disturbances, and how it may respond in the future. Deciphering this record, then, is an important component of the geochemical-geomorphological approach in that it provides insights into the precise mechanisms through which contaminated particles were distributed and where they are currently stored within the watershed. In order to fully exploit this record, it is essential to date the alluvial sequence and determine which of the deposits correspond to periods of sediment-borne trace metal influx. The importance of these relations stem, of course, from the fact that trace metal concentrations are likely to vary closely with the history of contaminant releases in the watershed, as was demonstrated for floodplains and terraces in Chapters 6 and 7, respectively. For most sites, a detailed geomorphological history of the watershed will not be available prior to site assessment, and will therefore need to be constructed as part of the characterization and assessment program. Reconstructing a history of geomorphological events is a difficult process that requires a considerable amount of experience in examining and characterizing alluvial deposits. It is therefore best performed by one who is trained in the analysis of alluvial stratigraphy and geomorphology. In general, the analysis involves four basic steps: (1) the delineation, characterization, and mapping of fluvial landforms, including those that may serve as contaminant sinks, (2) determining the relative and absolute age of the landforms and their deposits, (3) correlating sediment-borne trace metal concentrations to landforms, deposit age, and deposit sedimentology, and (4) combining the temporal and spatial data to construct a geomorphic history of the system that depicts the timing and nature of contaminant movement through the river valley. One of the more difficult aspects of this process is determining the age of specific stratigraphic units that comprise floodplains and terraces.

Historically, two types of dating were generally recognized: relative and absolute (see Stokes and Walling 2003 for a different classification of dating methods). Relative dating compares the ages of two or more stratigraphic units to determine which is younger and which is older. The dating methods are founded on a set of principles involving basic logic, and typically rely on the spatial relationships of the units to one another. Two of the more important concepts include the principle of superposition and the principle of cross-cutting relationships. The former states that sediment is deposited on top of earlier, older deposits. Thus, the younger deposits within any alluvial sequence will be found at the top, whereas the older deposits will be found at the bottom. The principle of cross-cutting relationships states that a disrupted or eroded deposit is older than the cause of the disruption. For example, the cross-cutting relationship indicates that terrace deposits truncated during a period of channel entrenchment must be older than the period of downcutting. Moreover, the truncated deposits must be older than the deposits found abutting the eroded truncation surface and which are found at a lower elevation. It follows, then, that younger terraces and their deposits are positioned at lower elevations than older terraces (although a few minor exceptions have been reported in the literature) (e.g., Kochel et al. 1987). When working with alluvial sequences, other approaches have been used to assess the relative age of the deposits, including the analysis of soil profile development and the degree of mineral or rock weathering. These types of techniques compare changes in the character of landforms and/or their sediments to develop a relative sequence of ages (see, for example, Birkeland 1999).

Absolute dating methods generally rely on a physiochemical parameter that changes at a constant rate through time; they can therefore be used to estimate the age of the material with which they are associated on an absolute (i.e., numerical) time-scale. A wide range of absolute dating methods have been developed since the 1960s (Table A.1), each of which vary in their time of applicability, the precision and accuracy of the resulting age estimates, and the nature of the event being dated (Stokes and Walling 2003). Unfortunately, only a handful of these can be used to date recent alluvial deposits. In the sections below, we will briefly examine the theory behind these methods, and the age range for which the techniques may be applied, to provide insights into the methods that may be used at a contaminated site. Additional information can be obtained from a number of discussions on the subject, including Mahaney (1984), Carroll and Lerche (2003), and Stokes and Walling (2003).

A.2.   SHORT-LIVED RADIONUCLIDES (Pb-210 and Cs-137)

In Chapter 3, it was shown how various short-lived radionuclides, such as ^{210}Pb and ^{137}Cs, could be used to estimate the rates of upland erosion and deposition. The primary method with which we were concerned was the measurement of the total nuclide inventory for an undisturbed or reference site, and its comparison to other localities for which sedimentation or erosion had occurred. The magnitude of erosion or deposition could then be estimated as a function of the negative or positive change in the inventory, respectively. Although minor variations in the method are required, the procedure can also be applied to determine the rates and spatial patterns in sedimentation on floodplains and terraces (Stokes and Walling

Table A.1. Summary of selected dating methods applicable to Quaternary alluvial deposits

Method	Approximate age range (years)	Basis of method
Dendrochronology	10^0 to 10^4	Counting of annually produced tree rings and correlation to sedimentary deposit
Varve chronology	10^1 to 8×10^3	Counting of seasonally deposited layers back from present
Radiocarbon (C-14)	10^2 to 5×10^4	
Cosmogenic nuclides		
^{10}Be	10^4 to 10^7	Formation and decay of nuclides
^{38}Cl	5×10^3 to 5×10^5	in rocks exposed to cosmic rays
^{3}He	5×10^2 to 10^6	
^{14}C	5×10^3 to 3×10^4	
Potassium-argon; argon-argon	10^4 to 5×10^7	Radioactive decay of K in K-bearing minerals
Lead-210	0 to 150	Radioactive decay of Pb-210
Cesium-137	0 to 1950	Accumulation and decay of Cs-137
Uranium series	0 to 5×10^5	Measurement of radioactive decay of U and U-daughter produces in sedimentary minerals
Uranium-lead; thorium-lead	10^4 to 4×10^7	Measurement of Pb enrichment due to decay of U and Th
Fission Track	3×10^3 to 4×10^7	Measurement of damage trails due to U fission decay
Luminescence (TL, OSL, IRSL)	0 to 10^6	Measurement of electron accumulation in crystal defects
Obsidian hydration	0 to 10^6	Increase in thickness of hydration rind on obsidian particles
Lichenometry	0 to 10^4	Growth of lichens on bare rock surfaces
Paleomagnetism, secular variations	0 to 7.5×10^3	Secular variations in Earth's magnetic field
Paleomagnetism, reversal stratigraphy	5×10^5 to 2.5×10^6	Reversal of Earth's magnetic field
Tephrochronology	$0 \; 2.5 \times 10^6$	Correlation and dating of tephra layers

Adapted and modified from Stokes and Walling (2005)

2003). An alternative to the total inventory approach, which is of interest here, is to date surficial sediment by determining the distribution of the nuclides with depth. This typically involves the collection of a sediment core, its subsectioning into 15–20 units, and their analysis, generally for ^{210}Pb and ^{137}Cs.

The vertical distribution of ^{210}Pb and ^{137}Cs differ significantly from one another as a result of the differing sources. In the case of ^{137}Cs, the profile typically observed in upland areas is characterized by a peak in concentration which occurs just below the ground surface, and which then decreases to non-detectable levels within a few tens of centimeters in depth (Fig. 3.17 of Chapter 3). The ^{137}Cs profiles constructed for floodplain deposits are also characterized by a peak in concentration. However, the peak typically occurs at a greater depth than is the case for the uplands. Moreover, the total inventory and the depth over which it is distributed tend to be significantly greater than that found at an upland reference site (Fig. A.1). The difference in the observed profiles occurs because the upland soils receive ^{137}Cs solely by means of atmospheric deposition. In contrast, ^{137}Cs within the floodplain deposits is derived both from the atmosphere and from the deposition of ^{137}Cs bound to particles which are deposited on the floodplain during overbank flooding. In the latter instance, the ^{137}Cs represents radiocesium which was atmospherically deposited over the watershed and was subsequently eroded and transported downstream as part of the river's load (Stokes and Walling 2003). It follows, then, that the observed peak in ^{137}Cs represents the floodplain surface at the time of peak fallout (i.e., 1963 in the northern hemisphere and 1964/1965 in the southern hemisphere). The maximum depth to which ^{137}Cs occurs represents its first occurrence in the environment in the 1950s (Popp et al. 1988; Ely et al. 1992). The depth to which these horizons occur represents the thickness to which sediment has accumulated since the horizons were

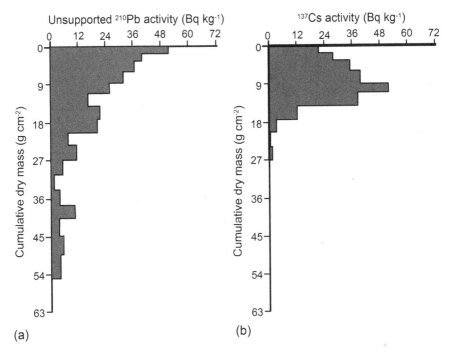

Figure A.1. Excess ^{210}Pb (A) and ^{137}Cs (B) profiles for two sediment cores extracted from the floodplain of the river Culm (Modified from Stokes and Walling 2005)

formed. Because the thickness of sediment deposition is known for a specific time interval, it is possible to calculate an average rate of floodplain aggradation. Such calculations, however, assume that sedimentation is essentially a continuous process (Stokes and Walling 2003). Moreover, the estimates must consider the possibility of the post-depositional redistribution of ^{137}Cs in the profile. As a result of this latter concern, most investigations use the peak in ^{137}Cs to calculate rates of sedimentation rather than its deepest occurrence. This follows because the position of the peak is less likely to be affected by post-depositional processes than the maximum depth to which ^{137}Cs first appears in the sediment record.

Lead-210 occurs naturally as one of the many radionuclides found within the ^{238}U decay series. It is most closely linked to the decay of ^{226}Ra in the soil to ^{222}Rn, the latter of which escapes to the atmosphere. ^{222}Rn subsequently decays through a series of short-lived radionuclides to ^{210}Pb before it is redeposited across the landscape by precipitation or dry deposition (Appleby 2001). This atmospheric ^{210}Pb is commonly referred to as the unsupported or excess ^{210}Pb in the soil and sediment because it represents the ^{210}Pb that occurs in excess of that which is found in equilibrium of ^{226}Ra. It is this excess ^{210}Pb this is used to date sedimentary deposits, and is typically determined by subtracting the supported activity, estimated from the analysis of ^{226}Ra, from the total ^{210}Pb activity.

The potential advantages of using ^{210}Pb over ^{137}Cs are (1) its potential to cover longer-time frames, and (2) its ability to be used in parts of the world, particularly the southern hemisphere, where bomb-derived fallout was limited (Stokes and Walling 2003). Unlike ^{137}Cs, the atmospheric flux of ^{210}Pb to the ground surface is assumed to be constant when considered over a period of a year or more. The near constant flux eliminates the occurrence of a peak in ^{210}Pb activity at depth within the sediment cores. Rather, ^{210}Pb activity is highest at the surface where sediment has been most recently exposed to the atmosphere, and decreases downward in materials that have been buried and removed from the atmospheric inputs (Fig. A.1). The downward decline in activity is associated with the progressively older age of the deeper sediment and radioactive decay such that the total ^{210}Pb activity can be defined by:

(1) $$C_{tot} = C_{tot}(0)e^{-\lambda t} + C_{sup}(1 - e^{-\lambda t})$$

where C_{sup} is the supported ^{210}Pb activity, λ is the ^{210}Pb radioactive decay constant and $C_{tot}(0)$ is the total ^{210}Pb activity of the sediment at the time that it was buried (Appleby 2001). In most cases, the total ^{210}Pb activity and the ^{226}Ra activity occurs after six to seven half-lives, allowing the dating of sediment over a period of 130–150 years (Appleby 2001). Where the initial ^{210}Pb concentrations are low, or the ^{226}Ra concentrations are high, a distinction between the supported and unsupported ^{210}Pb activity cannot be determined in sediment older than 60 to 90 years (three to four half-lives).

Within lake and floodplain environments, the surficial sediment may exhibit unsupported ^{210}Pb concentrations that are significantly different from that expected from the local atmospheric flux. These differences are predominately related to inputs associated with the erosion and transport of ^{210}Pb enriched sediment from the catchment, and losses from the water column, particularly in lakes, via outflow and depositional focusing of sediment accumulation. Where these processes are essentially constant, it can be assumed that each layer of sediment will have the same initial unsupported ^{210}Pb concentration. It is therefore a relatively simple matter to estimate sediment age and the rates of sediment accumulation on the basis of the measured variations in the unsupported ^{210}Pb content of the sediment with depth. For example, the mean sedimentation rate (r) can be determined from the slope of the line created by plotting the unsupported ^{210}Pb activity versus sediment depth (usually measured as cumulative dry mass, g/cm^2) (Fig. A.2.) The age of each layer can then be determined by:

(2) $t = m/r$

where t is the age of the layer and m is the depth of the layer below the ground surface. In most cases, however, sediment accumulation rates vary through time, making the estimation of deposit age more difficult. For these types of sites, a number of models have been developed, the two most common of which are the constant rate of ^{210}Pb

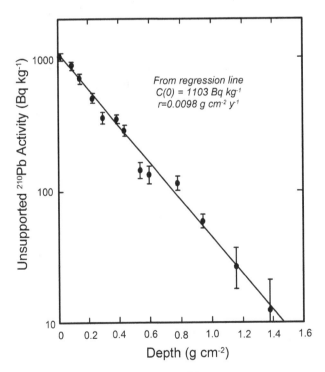

From regression line
$C(0) = 1103$ Bq kg^{-1}
$r = 0.0098$ g cm^{-2} y^{-1}

Figure A.2. Regression of unsupported ^{210}Pb activity versus sediment depth in a lake core from Øvre Neadalsvatn, Norway (From Appleby 2001)

supply (CRS) (Appleby and Oldfield 1978; Robbins 1978; Appleby et al. 1979) and the constant initial concentration (CIC) (Appleby and Oldfield 1992).

Lead-210 dating of sediment has been most extensively and effectively applied to reservoir, lake, and wetland environments. In the case of floodplains, He and Walling (1996) suggested that any attempt to date specific horizons in a sediment core was unlikely to yield reliable results. The difficulties arise from (1) the deposition of sediment during infrequent runoff events, (2) variations in depositional rates during overbank flooding as a result of variations in flood magnitude and frequency, and (3) the post-depositional remobilization of sediment which may complicate the vertical distribution of ^{210}Pb in the sediment profile. He and Walling (1996) argued, however, that average long-term, annual rates of sedimentation can be estimated by ^{210}Pb using a constant initial concentration, constant sedimentation rate (CICCS) model. Although the method will not provide continuous ages through the core, it possesses the important advantage of only requiring the measurement of the total ^{210}Pb inventory from the deposits, rather than the measurement of ^{210}Pb in multiple subsections (Stokes and Walling 2003).

A.3.   RADIOCARBON (C-14)

The most widely used method for dating deposits greater than approximately 100 to 150 years, but less than about 35 to 50 thousand years of age, is radiocarbon or ^{14}C. ^{14}C is the product of the continual bombardment of ^{14}N in the atmosphere by high-energy cosmic rays. The ^{14}C produced is unstable and decays back to ^{14}N at a constant rate described by its half-live, estimated to be $5,730 +/- 40$ years. In other words, half of the ^{14}C within any given reservoir or medium will be converted back to ^{14}N in about 5,730 years.

A certain amount of ^{14}C, along with ^{12}C, is incorporated in CO_2, which allows both isotopes to be accumulated in plants until equilibrium is reached with the atmosphere (Smol 2002); animals also accumulate ^{14}C as they consume organic matter in which it resides. Upon death of the plant or animal, ^{14}C replenishment ends and its levels within the organic materials decline as it decays to N. It follows, then, that the time since the plant or animal died can be determined provided that the initial ^{14}C :^{12}C ratio and the decay rate are known.

Oddly enough, the true half-life of ^{14}C is not used to determine the age of an organic carbon bearing substance. Rather, the half-life of ^{14}C originally determined by William Libby (sometimes referred to as the conventional half-life) of 5,568 years is used to keep all of the radiocarbon dates which have been determined since the 1950s consistent.

Originally, it was assumed that the production of ^{14}C was a constant, but the radiocarbon dating of tree rings (the age of which is precisely known) showed that it was not. Variations in ^{14}C production can be caused by a range of factors including changes in solar activity, the geomagnetic flux, fluctuations in CO_2 uptake by the oceans and human activity (Brown 1997). As a result, the determination of age in terms of calendar dates requires a calibration process which is conducted on a regional basis. The correction factors are based on a number of parameters including tree rings, ^{14}C dated marine sediments, and ^{14}C/U-Th dated corals (Stuiver et al. 1998; Smol 2002). For a larger number of studies, uncalibrated ages can be used, in which case the dates are reported in years before present (YBP) where the present is set at 1950 AD.

As with any technique there are a number of problems that can affect the results. One of the most important issues is that of contamination. Contamination can occur through a host of processes including rootlet penetration, fungal growth, and the infiltration of humic acids containing younger organic carbon from overlying horizons. Older organic carbon may also contaminate the material, particularly in areas containing calcareous soils or bedrock. In these areas, plants may not only accumulate ^{14}C from the atmosphere, but carbon in the water that was derived from rocks millions of years old. This so-called "hard-water" effect will lead to dates that are significantly older than the material's actual age (MacDonald et al. 1991).

It is important to recognize that the date derived for a particular alluvial strati-graphic unit using ^{14}C is based on the date derived from some organic carbon bearing particle or substance within the deposit. In other words, the method is dating substances contained within the deposit, not the actual time of deposition. An important assumption, then, is that the death of the plant or animal from which the date is derived corresponds very closely with the time of deposition of the sediment in which it resides. This is not always the case, however. For example, charcoal reworked from older deposits can significantly pre-date the timing of deposition, thereby leading to erroneous age estimates. This reworking problem tends to be less significant with the dating of non-woody macrofossils of local origin and which exhibit limited degrees of abrasion (Brown 1997). One of the means which is now being used to overcome the problem of old in-washed carbon from water, contamination by rootlets, and reworked materials is the dating of carbon-bearing materials by accelerator mass spectrometry (AMS). AMS uses very small quantities of material, such as individual macrofossils, which are known to be of local origin. In addition, the ^{14}C content of the material is directly measured, rather than its radioactivity, allowing for a rather precise estimation of the material's age for samples containing as little as 1 to 3 mg of carbon (Hedges 1991).

A.4.   LUMINESCENCE TECHNIQUES (TL, IRSL, AND OSL)

Grains of sediment buried within an alluvial deposit are continuously bombarded by very low levels of ionizing radiation as a result of radioactive substances which exist within the sediment mixture. This bombardment leads to the trapping of electrons within individual minerals that accumulate over time in a predictable manner. Upon erosion and exposure to sunlight, the trapped electrons are released within a short-period of time (seconds to minutes) in a process referred to as bleaching. Thus, it is theoretically possible to determine the time since a well-bleach grain was buried provided that the amount of built up electrons can be measured and the ionizing dose rate is known. Mathematically, the age of burial (B_{age}) is expressed as:

(3) $B_{age} = De/Dr$

where De is the measured equivalent dose for a grain or group of grains, and Dr is the dose rate provided by the ionizing radiation. The dose rate (Dr) can be estimated by determining the concentration of radioactive elements in surrounding materials, usually by mass spectrometry or high-resolution gamma spectrometry (e.g., Olley et al. 2004). The equivalent dose can be determined by measuring the amount of excess energy released in the form of photons, called *luminescence*, as the trapped electrons return to their ground state. The various methods depend on the mechanism used to release the trapped electrons. When luminescence is initiated by thermal stimulation it is called *thermoluminescence* (TL). Luminescence produced by optical stimulation is called *optically stimulated luminescence* (OSL), and infrared stimulation is referred to as *infrared stimulated luminescence* (IRSL). It is probably fair to say that TL and OSL are most widely used methods. TL relies on the application of heat to release the electrons. As the electrons are emitted, an extra amount of light is given off which reflects the quantity of stored electrons. The amount of extra light that is emitted can be determined by re-heating the mineral grains. Since all of the electrons were released the first time the grains were heated, there will be no extra light emitted during repeated heating of the particles. In the case of OSL, visible light is used to stimulate luminescence in quartz, while infra-red radiation is used for feldspars. Again the amount of "extra" light is measured.

As might be expected, a large number of factors can affect the accuracy of the obtained dates, including grain mineralogy, the nature of the ionizing radiation (which is affected by water content of the sediment), the occurrence of surface coatings on grains, and the degree of bleaching before burial. The latter is particularly important in river systems as early studies questioned whether short periods of exposure of sediment to light during transport (particularly in turbid waters) were sufficient to completely bleach the particles, a requirement of obtaining accurate dates (Berger 1984; Bailiff 1992). However, TL has generated ages in some locations which are consistent with other dating methods (Nanson and Young 1987; Nanson et al. 1991). Perhaps more importantly, recent advances in OSL dating, including the measurement of the equivalent dose for individual grains, indicate that it may be a particularly powerful tool for dating deposits in fluvial environments (Murray and Olley 2002; Olley et al. 1999, 2004), particularly in larger catchments, provided an appropriate methodology is used (Stokes and Walling 1993).

A.5.   MARKER BEDS (TEPHRA, TRACE METALS)

In many instances it is possible to use one or more distinct horizons that have been deposited nearly instantaneously (at least on the geological time scale) as an indicator of sediment age. Such layers, commonly referred to as *marker beds*, can be particularly useful because they provide a time-line that may cross-cut multiple depositional environments, thereby allowing for a determination of landscape morphology and character at a specific instant in the past. In addition, their occurrence in a sedimentary package can be used to correlate terraces and other alluvial deposits along a river channel, while their absence indicates that the deposits are either younger or older than the identified horizon. Perhaps the most commonly used and easily recognizable marker beds are composed of tephra (Fig. A.3). Tephra consists of a wide range of pyroclastic material ejected during a volcanic eruption, including pieces of pumice and scoria, glass fragments, and both felsic and mafic minerals. In general, tephra of basaltic composition is black, those of andesitic origin are grey, and rhyolitic materials are white (Turney and Lowe 2001). The particle size of the tephra also varies considerably, ranging from blocks and bombs (>64 mm), lapilli (2–64 mm), and ash (<2 mm). Grain size of the tephra, and its thickness, normally decrease with distance from the source of the volcanic eruption (Turney and Lowe 2001).

Interest in the use of tephra as an age dating tool, referred to as *tephrochronology*, has grown significantly in the past three decades because (1) tephra is deposited nearly instantaneous from the atmosphere over large areas, and therefore provides a precise timeline that cross-cuts multiple depositional environments, (2) multiple tephra units may occur in a region of volcanic activity allowing for the relative or

Figure A.3. Tephra layer (light-colored sediment in center of photo) positioned in alluvial terrace along the Rio Pilcomayo, Bolivia

absolute dating of deposits spanning a wide range of ages, and (3) individual tephra units can readily be identified in stratigraphic sequences. Historically, fingerprinting of tephra relied on the morphology of glass shards, but subsequent studies have demonstrated that shard shape and color are rarely adequate to identify individual eruptive events because their general morphology may be common to other tephras of similar composition (Turney and Lowe 2001). As a result, tephra identification has utilized other parameters including its mineralogy and chemical composition. By themselves, tephra beds can only be used to correlate alluvial deposits; they therefore represent a relative age dating tool. However, tephra beds can be dated using a variety of methods including $^{40}K/^{40}Ar$, $^{40}Ar/^{39}Ar$, thermoluminescence, and fission track analysis (Berger 1992). The age of the tephra units can also be determined indirectly by dating the deposits in which they are contained. Once the age of a tephra unit is adequately constrained, it becomes a powerful absolute dating tool because tephra are time-parallel units (i.e., exhibit a spatially similar age across the landscape), and, thus, a determination of their age at one location is sufficient to determine the tephra's age at all other locations.

Clearly, the fingerprinting and dating of tephra beds is a significant effort. Fortunately, regional chronological sequences have been established for a large number of areas where tephra is commonly found. In addition, there are now a number of labs to which samples can be sent for analysis by tephrachronologists, dramatically simplifying their use as a correlation and dating tool.

In addition to tephra, zones of trace metal enrichment have been used as marker beds. In most instances, horizons distinguished on the basis of elemental concentration are associated with rivers impacted by mining operations. From this perspective, the contaminant character of the deposits provides information on the geomorphic processes within the river. These types of marker beds (i.e., those defined on geochemistry) are more difficult to use than tephra because they are usually not visibly identifiable in the field, and the dispersal of contaminated sediment may require considerable periods of time. Their age, then, may vary with distance from their source. Nonetheless, a number of studies have been able to correlate one or more horizons of elevated trace metal content to specific periods of mining activity, thereby providing an absolute age for the material (Lewin and Macklin 1987; Popp et al. 1988; Macklin and Dowsett 1989). Such geochemical marks have been shown to be particularly useful for floodplain and terrace sediment deposited by vertical accretion processes.

A.6. DENDROCHRONOLOGY

Dendrochronology refers to the science of assigning calendar dates to the annual growth rings in trees (Stokes and Smiley 1968). Its basic principles were developed by an astronomer, Andrew Ellicott Douglass, during the first three decades of the 1900s. Since that time, it has been applied to a growing list of fields including archaeology, climatology, forest ecology, hydrology, seismology, and geomorphology. It has also been linked to the geochemical analysis of the wood to provide a wealth of information regarding environmental change. One of the areas of growing interest is in the development of *dendrogeomorphic analysis* which refers to the use of dendrochronology to estimate the timing of important geomorphic events (e.g., floods and mass movements) and the rates at which erosional and depositional processes function. The latter involves the dating of recent unconsolidated deposits, the methods of which fall into two broad categories. The first involves the determination of the time at which a geomorphic surface became stabilized by dating the period of germination of the oldest woody species growing on the surface. The technique is based on the realization that zones of deposition or scour are generally devoid of vegetation because the reworking of the surface materials either inhibits germination or rapidly destroys any woody species that germinates before they can be firmly established. At some point, however, the frequency of surface disruption by erosional or depositional processes will slow to the point where they can become rapidly colonized by woody plants (Everitt 1968; Johnson 1994; Scott et al. 1997). The minimum age of surface stabilization can then be determined by dating the oldest trees and shrubs on the surface using dendrochronologic methods (Hupp and Bornette 2003). A common feature to which this method has been applied along meandering rivers is point bars. In these areas, point bars are characterized by bands of woody plants which increase in age with distance from the channel. The maximum age of each band can be determined and used to both date the point bar deposits and estimate the rates of lateral channel migration. This method may be particularly useful in relating point bar and floodplain deposits to periods of contaminant influx along meandering channels.

The second method which is commonly used to date alluvial deposits applies to trees that have been buried by sediment, particularly those growing on floodplains which have been subjected to the periodic deposition of suspended materials during overbank flooding. At the time of germination, tree roots grow just below the ground surface where they are close to sources of water and nutrients. With time, a network of roots will develop which radiates out from the point of germination, creating a basal root flare. This basal root flare, or collar, then, represents a marker of the original ground surface at the time of germination (Fig. A.4). Subsequent deposition, however, may bury the basal root flare, imparting a straight or telephone looking appearance to the tree-trunk at the ground surface (Hupp and Bornette 2003). Burial of the basal roots allow average rates of deposition to be determined by (1) measuring the depth to which the basal root flare is buried, (2) determining the time of germination by dendrochronologically dating an increment core or slab taken near the ground surface, and (3) dividing the depth of burial by the age of the tree. In most instances, numerous trees will need to be analyzed to filter out local site variations. This general approach has been modified to determine temporal trends in sedimentation rates by organizing the collected data into specific age categories ranging from the youngest to the oldest trees present. Average depositional rates can then be determined for each age class, such as a decade, and used to determine changes in depositional rates through time (Hupp and Morris 1990; Hupp and Bazemore 1993).

Figure A.4. Adventitious roots on a Sycamore tree in Southern Illinois buried in response to massive aggradation following forest clearing near the turn of the 20th century. Tree could be dated using dendrochronology to determine the onset of aggradation and the rates of sedimentation

In some cases, basal root flares are associated with buried soils (Fig. A.4). Given that soil formation requires a stable geomorphic surface, the determined ages are likely to indicate the onset of a period of channel instability (i.e., the timing of a threshold crossing event). Thus, dating of the trees not only provides an average rate of deposition for the overlying sediment, but the time at which increased rates of deposition on the surface began. Unfortunately, in many cases, buried paleosurfaces are devoid of living trees. However, a procedure referred to as cross-dating can be used to determine the approximate timing of germination for trees associated with the paleosurface and which are no longer living. Cross-dating essentially matches the pattern of tree-ring widths in the dead and exhumed tree-trunks to that of an existing tree-ring chronology, thereby allowing the age of the trees in question to be determined. Orbock Miller et al. (1993), for example, found a large number of tree stumps buried beneath post-settlement alluvial in the Drury Creek watershed of southern Illinois (Fig. A.5). The basal root flares of these stumps were associated with a buried paleosol. Using cross-dating techniques, they were able to determine the approximate time at which a significant period of channel aggradation began in response to land-uses changes in the watershed at the end of the 1800s and the beginning of the 1900s.

Figure A.5. Buried tree stump in Southern Illinois. Cross-dating allowed the stumps to be dated, thereby allowing a determination of the onset of valley aggradation

The use of dendrochronology and dendrogeomorphology has thus far been highly underutilized in contaminant studies. Nonetheless, one would suspect that these and other biogeomorphic methods will be applied much more frequently to contaminated sites in the future because they are relatively inexpensive, and have been shown to provide accurate results in comparison to data from repeated cross-sections (Hupp and Bazemore 1993).

APPENDIX B

UNIT CONVERSIONS AND ELEMENTAL DATA

Table B.1. List of unit abbreviations

English unit	Abbreviation	SI unit	Abbreviation
Acre	ac	Hectare	Ha
Fahrenheit	F	Celsius	C
Inches	in.	Centimeters	cm
Yards	yd	Meters	m
Ounces	oz	Grams	gm
Pounds	lb	Kilograms	kg
Gallons	gal	Liter	l
		Milliliter	ml

Table B.2. Common conversions

Distance/Length	Area
$1\,nm = 10^{-9}\,m$	$1\,ft^2 = 9.280 \times 10^{-2}\,m^2$
$1\,\mu m = 10^{-6}\,m$	$1\,acre = 0.4046\,hectares$
$1\,mm = 10^{-3}\,m$	$1\,mi^2 = 2.5899\,km^2$
$1\,cm = 10^{-2}\,m$	
$1\,km = 10^3\,m$	
$1\,mile = 1.609\,km$ or $1,609.34\,m$	
$1\,in. = 25.4\,mm$ or $2.54\,cm$	
Flow/Volume	Mass/Weight
$1\,ml = 10^{-3}\,L$	$1\,pg = 10^{-12}\,g$
$1\,L = 1000\,cm^3$	$1\,ng = 10^{-9}\,g$
$1\,gal = 3.785\,L$	$1\,\mu g = 10^{-6}\,g$
$1\,quart\,(US) = 0.9463\,L$	$1\,mg = 10^{-3}\,g$
$1\,ft^3 = 0.0283\,m^3$	$1\,kg = 10^3\,g$
$1\,cfs = 28.32\,L/s$ or $0.0283\,m^3/s$	$1\,metric\,ton = 10^3\,kg$
	$1\,short\,ton = 2,000\,lbs$ or $907.184\,kg$
	$1\,lb = 453.592\,g$
	$1\,troy\,oz = 31.1035\,g$

(Continued)

365

Table B.2. (Continued)

Radioactivity	Pressure
1rad (absorbed dose) $= 10^{-2}$ J/kg	1 bar $= 10^5$ Pa or 0.9869 atm
1pCi $= 10^{-12}$ Ci or 0.037 Bq	1 Pa $= 10^{-5}$ bar
1curie $= 3.7 \times 10^{10}$ dps	1 atm $= 760$ mm Hg or 1.01325×10^5 Pa
1becquerel (Bq) $= 1.0$ dps	

Table B.3. Units of concentration

Concentrations (volume)		Concentrations (weight)	
Unit	*Symbols*	*Unit*	*Symbols*
moles per liter	mol/L or M	moles per kg	mol/kg
millimoles per liter	mmol/L or mM	milliequvalents per kilogram	meq/kg
micromoles per liter	μmol/L or μM	micrograms per kg or ppb	μg/kg or ppb
micrograms per liter	μg/L	milligrams per kg or parts per million	g/kg $=$ ppm

Table B.4. Load calculation constants

Units of concentration	Units of flow	Constant
μg/L	m^3/s	0.000864
mg/L	m^3/s	0.0864
g/L	ft^3/s	86.4
μg/L	ft^3/s	0.000002447
mg/L	ft^3/s	0.002447
g/L	ft^3/s	2.447

Adapted from Richards (2001)
Load $=$ constant $\times$ concentration $\times$ flow

Table B.5. Elemental data

Element	Symbol	Atomic number	Atomic weight
Hydrogen	H	1	1.0079
Helium	He	2	4.0026
Lithium	Li	3	6.941
Beryllium	Be	4	9.0122
Boron	B	5	10.811
Carbon	C	6	12.011
Nitrogen	N	7	14.007
Oxygen	O	8	15.999
Fluorine	F	9	18.998
Neon	Ne	10	20.180
Sodium	Na	11	22.980
Magnesium	Mg	12	24.305
Aluminum	Al	13	26.982

Silicon	Si	14	28.086
Phosphorus	P	15	30.974
Sulfur	S	16	32.065
Chlorine	Cl	17	35.453
Argon	Ar	18	39.948
Potassium	K	19	39.098
Calcium	Ca	20	40.078
Scandium	Sc	21	44.956
Titanium	Ti	22	47.867
Vanadium	V	23	50.942
Chromium	Cr	24	51.996
Manganese	Mn	25	54.938
Iron	Fe	26	55.845
Cobalt	Co	27	58.693
Nickel	Ni	28	58.693
Copper	Cu	29	63.546
Zinc	Zn	30	65.39
Gallium	Ga	31	69.723
Germanium	Ge	32	72.84
Arsenic	As	33	74.922
Selenium	Se	34	78.96
Bromine	Br	35	79.904
Krypton	Kr	36	83.80
Rubidium	Rb	37	85.468
Strontium	Sr	38	87.62
Yttrium	Y	39	88.906
Zirconium	Zr	40	91.224
Niobium	Nb	41	92.906
Molybdenum	Mo	42	95.94
Technetium	Te	43	(98)
Ruthenium	Ru	44	101.07
Rhodium	Rh	45	102.91
Palladium	Pd	46	106.42
Silver	Ag	47	107.87
Cadmium	Cd	48	112.41
Indium	In	49	114.82
Tin	Sn	50	118.71
Antimony	Sb	51	121.60
Tellurium	Te	52	127.60
Iodine	I	53	126.90
Xenon	Xe	54	131.29
Cesium	Cs	55	132.91
Barium	Ba	56	137.33
Lanthanum	La	57	137.91
Cerium	Ce	58	140.12
Praseodymium	Pr	59	140.91
Neodymium	Nd	60	144.24
Promethium	Pm	61	(145)
Samarium	Sm	62	150.36
Europium	Eu	63	150.96
Gadolinium	Gd	64	157.25
Terbium	Tb	65	158.93

(Continued)

Table B.5. (Continued)

Element	Symbol	Atomic number	Atomic weight
Dysprosium	Dy	66	162.50
Holmium	Ho	67	164.93
Erbium	Er	68	167.26
Thulium	Tm	69	168.93
Ytterbium	Yb	70	173.04
Lutetium	Lu	71	174.97
Hafnium	Hf	72	178.49
Tantalum	Ta	73	180.95
Tungsten	W	74	183.84
Rhenium	Re	75	186.21
Osmium	Os	76	190.23
Iridium	Ir	77	192.22
Platinum	Pt	78	195.08
Gold	Au	79	196.97
Mercury	Hg	80	200.59
Thallium	Tl	81	204.38
Lead	Pb	82	207.2
Bismuth	Bi	83	208.98
Polonium	Po	84	(209)
Astatine	At	85	(210)
Radon	Rn	86	(222)
Francium	Fr	87	(223)
Radium	Ra	88	(226)
Actinium	Ac	89	(227)
Thorium	Th	90	232.94
Protactinium	Pa	91	232.04
Uranium	U	92	238.03
Neptunium	Np	93	(237)
Plutonium	Pu	94	(244)
Americium	Am	95	(243)
Curium	Cm	96	(247)
Berkelium	Bk	97	247
Californium	Cf	98	(251)
Einsteinium	Es	99	(252)
Fermium	Fm	100	(257)
Mendelevium	Md	101	(258)
Nobelium	No	102	(259)
Lawrencium	Lr	103	(262)
Rutherfordium	Rf	104	(261)
Dubnium	Db	105	(262)
Seaborgium	Sg	106	(266)
Bohrium	Bh	107	(264)
Hassium	Hs	108	(265)
Meitnerium	Mt	109	(267)
Ununnilium	Uun	110	285
Ununbium	Uub	112	285
Ununquadium	Uuq	114	289

(xxx) – The mass number of the longest-lived isotope of the element

REFERENCES

Abumaizar RJ, Khan LI (1996) Laboratory investigation of heavy metal removal by soil washing. Journal of the Air Waste Management Association 46:765–768

Ackers P, Charlton FG (1971) The slope and resistance of small meandering channels. Institute of Civil Engineers, Supp 15, paper 73625

Adams DD, Darby DA (1980) A dilution-mining model for dredged sediments in freshwater systems. In: Baker RA (ed) Contaminated Sediments, vol 1, Science Publishers, Ann Arbor

Albasel N, Cottenie A (1985) Heavy metal uptake from contaminated soils as affected by peat, lime, and chelates. Soil Science Society of America Journal 49:386–390

Alexander CS, Prior JC (1971) Holocene sedimentation rates in overbank deposits in the black bottom of the lower Ohio River, southern Illinois. American Journal of Science 270:361–372

Allégre CJ, Dupré B, Négrel P, Gaillardet J (1996) Sr-Nd-Pb isotope systematics in Amazon and Congo river systems: constraints about erosion processes. Chemical Geology 13:93–112

Allen HE, Chen PH (1993) Remediation of metal contaminated soil by EDTA incorporating electrochemical recovery of metal and EDTA. Environmental Progress 12:284–293

Allen HE, Huang CP, Bailey GW, Bowers AR (1995) Metal speciation and contamination of soil. Lewis Publishers, Boca Raton

Allen JRL (1965) A review of the origin and characteristics of recent alluvial sediments. Sedimentology 3:163–198

Allen JRL (1985) Principles of physical sedimentology. George Allen and Unwin, London

Ambrose RB Jr, Wool TA, Martin JL, (1993) The water quality analysis simulation program WASP5, version 5.00, model documentation. USEPA Environmental Research Laboratory, Athens, Georgia

Anderson MG, Burt TP (1978) The role of topography in controlling throughflow generation. Earth Surface Processes 3:331–344

Anderson MG, Burt TP (1982) The role of throughflow in storm runoff generation: an evaluation of a chemical mixing model. Earth Surface Processes and Landforms 7:565–574

Anderson MG, Burt TP (1990) Subsurface runoff. In: Anderson MG, Burt TP (eds) Process studies in hillslope hydrology. John Wiley and Sons, Chichester, pp 365–400

Appelo CAJ, Postma D (1994) Geochemistry, groundwater and pollution. AA Balkema, Rotterdam

Appleby PG (2001) Chronostratigraphic techniques in recent sediment. In: last WM, Smol JP (eds) Tracking Environmental Change Using Lake Sediments, vol 1, Basin Analysis, Coring, and Chronological Techniques. Kluwer, London, pp. 171–203

Appleby PG, Oldfield F (1978) The calculations of ^{210}Pb dates assuming a constant rate of supply of unsupported ^{210}Pb to the sediment. Catena 5:1–8

Appleby PG, Oldfield F (1992) Applications of 210Pb to sedimentation studies. In Ivanovich M, Haron RS (eds) Uranium-Series Disequilibrium: Applications to Earth, Marine, and Environmental Studies. Oxford University Press, Oxford, pp 731–778

Appleby PG, Oldfield F, Thompson R, Huttenen P, Tolonen K (1979) ^{210}Pb dating of annually laminated lake sediments from Finland. Nature 280:53–55

Andrews ED (1979) Scour and fill in a stream channel, East Fork River, western Wyoming. US Geological Survey Professional Paper 1117

Andrews ED (1980) Effective and bankfull discharges of streams in the Yampa river basin, Colorado and Wyoming. Journal of Hydrology 46:311–330

Andrews ED (1983) Entrainment of gravel from naturally sorted river bed material. Geological Society of America Bulletin 94:1125–1231

Andrews ED, Erman DC (1986) Persistence in the size distribution of surficial bed material during an extreme snowmelt flood. Water Resources Research 22:191–197

Appelgren BG, Barchi S (1993) Technical, policy and legal aspects of chemical time bombs with emphasis on the institutional action required in eastern Europe. Land Degradation and Rehabilitation 4:437–440

ASCE (1998) River width adjustment, 1: processes and mechanisms by American Society of Civil Engineers Task Committee on Hydraulics, Bank Mechanics, and Modeling of River Width Adjustment. Journal of Hydraulic Engineering 124:881–902

Ashmore PE (1991) How do gravel-bed rivers braid? Canadian Journal of Earth Science 28:326341

Ashmore PE, Day TJ (1988) Effective discharge for suspended sediment transport in streams of the Saskatchewan River Basin. Water Resources Research 24:864–870

Ault WA, Senechal RG, Erleback WE (1970) Isotopic composition as a natural tracer of lead in the environment. Environmental Science and Technology 4:305–313

Averett DE, Perry BD, Torrey EJ, Miller JA (1990) Review of removal, containment and treatment technologies for remediation of contaminated sediment in the Great Lakes. US Army Engineering Waterways Experiment Station, Miscellaneous Paper EL-90-25, Vicksburg, Mississippi

Axe L, Anderson PR (1998) Intraparticle diffusion of metal contaminants in amorphous oxide minerals. In: Jenne EA (ed) Adsorption of metals by geomedia: variables, mechanisms, and model applications. Academic Press, San Diego, pp 193–208

Babcock WH (1986) Tenmile Creek: a study of stream relocation. Water Resources Bulletin 22:405–415

Bagnold RA (1980) An empirical correlation of bedload transport rates in flumes and natural rivers. Proceedings of the Royal Society 372A:453–473

Bailiff I (1992) Luminescence dating of alluvial deposits. In: Needham S, Macklin MM (eds) Archaeology under alluvium. Oxbow Books, Oxford, pp 27–36

Baker VR, Ritter DF (1975) Competence of rivers to transport coarse bedload material. Geological Society of America Bulletin 86:975–978

Baker AJM, McGrath SP, Reeves RD (1994) The possibility of in situ heavy metal uptake by tolerant plants. In: Shaw AJ (ed) Heavy Metal Tolerance in Plants: Evolutionary Aspects, Resource Conservation and Recycling 11:41–49

Balling RC Jr, Wells SG (1990) Historic rainfall patterns and arroyo activity within the Zuni River drainage basin, New Mexico. Annals of the Association of American Geographers 80:603–617

Barnes HA (1968) Roughness characteristics of natural channels. US Geological Survey Water Supply Paper 1849

Bartley GE (1989) Trace element speciation: analytical methods and problems. CRC Press, Boca Raton, Florida

Bartley R, Rutherfurd I (2005) Re-evaluation of the wave model as a tool for quantifying the geomorphic recovery potential of streams disturbed by sediment slugs. Geomorphology 64:221–242

Basham EL, Lechler PJ, Miller JR (1996) Database of historic mercury amalgamation mining of gold and silver ores in the United States. In: Proceedings of the Mercury as a Global Pollutant Conference, August 4–8, Hamburg, pp 141

Bateman AM (1950) Economic mineral deposits, 2nd edition. New York, Wiley and Sons

Bathurst JC (1993) Flow resistance through the channel network. In: Beven K, Kirkby MJ (eds) Channel Network Hydrology, John Wiley and Sons, New York, pp 69–98

Bathurst RGC (1975) Carbonate sediments and their diagenesis, 2nd edn. Elsevier Science Publication, Amsterdam

Bazemore DE, Eshleman KN, Hollenbeck KJ (1994) The role of soil water in stormflow generation in forested headwater catchment: synthesis of natural tracer and hydrometric evidence. Journal of Hydrology 162:47–75

Beasley DB, Huggins LF (1982) ANSWERS user's manual, EPA-905/9-82-001. US Environmental Protection Agency, Region V, Chicago

Behrendt H, Opitz D (1999) Retention of nutrients in rivers systems: dependence on specific runoff and hydraulic load. Hydrobiologia 410:111–122

Berger G (1984) Thermoluminescence dating studies of glacial silts from Ontario. Canadian Journal of Earth Science 21:1393–1399

Berger G (1992) Dating volcanic ash by use of thermoluminescence. Geology 20:11–14

Berger JJ (1990) Evaluating ecological protection and restoration projects: a holistic approach to the assessment of complex, multi-attribute resource management problems. Doctoral dissertation, University of California, Davis

Bernhardt ES et al (2005) Synthesizing US restoration efforts. Science 308:636–637

Berti WR, Cunningham SD (1994) Remediating soil with green plants. In: Cothern DR (ed) Trace Substances, Environment and Health, Science Reviews, Northwood, UK, pp 43–51

Berti WR, Cunningham SD (1997) In-place inactivation of Pb in Pb-contaminated soils. Environmental Science and Technology 31:1359–1364

Berti WR, Cunningham SD (2000) Phytostabilization of metals. In: Raskin I, Ensley BD (eds) Phytoremediation of toxic metals: using plants to clean up the environment. John Wiley and Sons, Inc, pp 71–88

Bestcha RL, Platts WS, Kauffman JB, Hill MT (1993) Artificial stream restoration – money well spent or an expensive failure? In: Proceedings on environmental restoration, Universities Council on Water Resources 1994 Annual Meeting, August 2–5, Big Sky, Montana, pp 76–89

Beven K (2004a) Infiltration excess at the Horton Hydrology Laboratory (or not?). Journal of Hydrology, 293:219–234

Beven K (2004b) Erratum to "Infiltration excess at the Horton Hydrology Laboratory (or not?)". Journal of Hydrology, 294:294–295

Beven K (2004c). Robert E. Horton and abrupt rises of ground water. Hydrological Processes, 18:3687–3696

Beven K (2004d) Robert E. Horton's perceptual model of infiltration processes. Hydrological Processes, 18:3447–3460

Birkeland PW (1999) Soils and geomorphology. Oxford University Press, New York

Bisson PA, Nielson JL, Palmason RA, Grove LE (1982) A system of naming habitat types in small streams, with examples of habitat utilization by salmonids during low stream-aquatic habitat inventory information. Proceedings American Fisheries Society, Portland, Oregon, pp 62–73

Bhangu I, Whitfield PH (1997) Seasonal and long-term variations in water quality of the Skeena River at USK, British Columbia. Water Resources 31:2187–2194

Biedenharn DS, Throne CR (1994) Magnitude-frequency analysis of sediment transport in the Lower Mississippi River. Regulated Rivers: Research and Management 9:237–251

Birch GF, Taylor SE, Matthai C (2001) Small-scale spatial and temporal variance in the concentration of heavy metals in aquatic sediments: a review and some new concepts. Environmental Pollution 113:357–372

Blaylock MJ, Salt DE, Dushenkov S, Zakharova O, Gussman C, Kapulnik Y, Ensley B, Raskin I (1997) Enhanced accumulation of Pb in Indian mustard by soil-appllied chelating agents. Environmental Science and Technology 31:860–865

Blaylock MJ, Huang JW (2000) Phytoextraction of metals. In: Raskin I, Ensley BD (eds) Phytoremediation of toxic metals: using plants to clean up the environment. John Wiley and Sons, New York Inc, pp 53–70

Boggs S Jr (2001) Principles of sedimentology and stratigraphy, 3rd edn. Prentice Hall, Upper Saddle River, New Jersey

Bohlen WF (1978) Factors governing the distribution of dredge resuspended sediment. In: Proceedings of the 16th coastal engineering research conference, August 17–September 3 1978, Hamburg, Germany. American Society of Civil Engineers, New York, pp 2001–2019

Bolton SH, Breteler RJ, Vigon BW, Scanlon JA, Clark SL (1985) National perspective on sediment quality. Report Prepared for the US Environmental Protection Agency, Washington, DC

Bølviken B, Bogen J, Demetriades A, DeVos W, Ebbing J, Hindel R, Langedal M, Locutra J, O'Connor P, Ottesen RT, Pulkkinen E, Salminen R, Schermann O, Swenne R, Van der Sluys J, Volden J (1996) Regional geochemical mapping of Western Europe towards the year 2000. Journal of Geochemical Exploration 56:141–166

Bottrill LJ, Walling DE, Leeks GJL (2000) Using recent overbank deposits to investigate contemporary sediment sources in larger river basins. In: Foster IDL (ed) Tracers in geomorphology. John Wiley and Sons, Chichester, pp 369–387

Bourg ACM, Loch JPG (1995) Mobilization of heavy metals as affected by pH and redox conditions. In: Salomons W, Stigliani WM (eds) Biogeodynamics of pollutants in soil and sediments: risk assessment of delayed and non-linear responses. Springer, Berlin, pp 87–102

Boyd KF, Skidmore PB (1999) Development of channel restoration strategies in contaminated floodplain environments. In: Olsen DS, Potyondy JP (eds) Wildland hydrology. American Water Resources Association, Herdon, Virginia, TPS-99-3, pp 301–308

Bradley SB, Cox JJ (1987) Heavy metals in the Hamps and Manifold valleys, North Straffordshire, UK: partitioning of metals in floodplain soils. The Science of the Total Environment 65:135–153

Bradley SB, Cox JJ (1990) The significance of the floodplain to the cycling of metals in river Derwent catchment, UK. The Science of the Total Environment 97/98:441–453

Bray DJ (1982) Regime equations for gravel bed rivers. In: Hey RD, Bathurst JC, Thorne CR (eds) Gravel bed rivers. John Wiley and Sons, Chichester, pp 517–580

Breckenridge RP, Crockett AB (1998) Determination of background concentrations of inorganics in soils and sediments at hazardous waste sites. Environmental Monitoring and Assessment 51:621–656

Brewer PA, Taylor MP (1997) The spatial distribution of heavy metal contaminated sediment across terraced floodplains. Catena 30:229–249

Brick CM, Moore JN (1996) Diel variation of trace metals in the upper Clark Fork River, Montana. Environmental Science Technology 30:1953–1960

Brierley GJ (1991) Floodplain sedimentology of the Squamish River, British Columbia: relevance of element analysis. Sedimentology 38:735–750

Brook EJ, Moore JN (1988) Particle-size and chemical control of As, Cd, Cu, Fe, Mn, Ni, Pb, and Zn in bed sediment from the Clark Fork River, Montana (USA). Science of the Total Environment 76:247–266

Brooks PD, McKnight DM, Bencala KE (2001) Annual maxima in Zn concentrations during spring snowmelt in streams impacted by mine drainage. Environmental Geology 40:1447–1454

Brookstrom AA, Box SE, Campbell JK, Foster KI, Jackson BL (2001) Lead-rich sediments, Coeur d'Alene River valley, Idaho: area, volume, tonnage, and lead content. USGS Open-file report 01-140

Brown AG (1997) Alluvial geoarchaeology: floodplain archaeology and environmental change. Cambridge Manuals in Archaeology, Cambridge University Press, Cambridge

Brown G, Quine T (1999) Fluvial processes and environmental change. John Wiley and Sons, Chichester

Brown MP (1999) The role of natural attenuation/recovery processes in managing contaminated sediments, *http://www.smwg.org/*

Brown SL, Chaney RL, Baker AJM (1994) Phytoremediation potential of Thlaspi caerulescens and bladder campion for zinc- and cadmium-contaminated soils. Journal of Environmental Quality 22:768–792

Brunsden D, Kesel RH (1973) Slope development on a Mississippi River bluff in historic time. Journal of Geology 81:576–597

Bucher B, Wolff CG, Cawlfield L (2000) Channel remediation and restoration design for Silver Bow Creek, Butte, Montana. Maximum Technologies, Helena, Montana, pp 3–11

Buckman HO, Brady NC (1960) The nature and properties of soils, 6th edn. Macmillan Publishing Company, Houndmills, UK

Bull WB (1991) Geomorphic responses to climate change. Oxford University Press, Oxford

Burnside CD (1985) Mapping from aerial photographs, 2nd edn. Wiley, New York

Burt TP, Pinay G (2005) Linking hydrology and biogeochemistry in complex landscapes. Progress in Physical Geography 29:297–316

Buttle JM (1994) Isotope hydrograph separations and rapid delivery of pre-event water from drainage basins. Progress in Physical Geography 18:16–41

Carling PA (1991) An appraisal of the velocity-reversal hypothesis for stable pool-riffle sequences in the River Severn, England. Earth Surface Processes and Landforms 16:19–31

Carling PA, Wood N (1994) Simulation of flow over pool-riffle topography: a consideration of the velocity reversal hypothesis. Earth Surface Processes and Landforms 19:319–332

Carling PA, Williams JJ, Kelsey A, Glaister MS, Orr HG (1998) Coarse bedload transport in a mountain river. Earth Surface Processes and Landforms 23:141–157

Carroll J, Lerche I (2003) Sediment processes quantification using radionuclides. Radioactivity in the environment vol 5. Elsevier, Amsterdam

Castelle AJ, Johnson AW, Conolly C (1994) Wetland and stream buffer size requirements: a review. Journal of Environmental Quality 23:878–882

Chang HH (1986) River channel changes: adjustments of equilibrium. Journal of Hydraulic Engineering 122:43–55

Chang HH (1988) Fluvial processes in river engineering. John Wiley and Sons, New York

Chao TT, Theobald PK Jr (1976) The significance of secondary iron and manganese oxides in geochemical exploration. Economic Geology 71:1560–1569

Chorley RJ, Kennedy BA (1971) Physical geography. Prentice-Hall International, London

Chow TJ, Earl JL (1972) Lead isotopes in North American coals. Science 176:510–511

Chow TJ, Johnstone MS (1965) Lead isotopes in gasoline and aerosols of Los Angeles basin, California. Science 147:502–503

Church M, Jones D (1982) Channel bars in gravel-bed rivers. In: Hey R, Bathurst J, Thorne C (eds) Gravel-bed rivers. John Wiley and Sons, New York, pp 291–338

Ciszewski D (2001) Flood-related changes in heavy metal concentrations within sediments of the Biala Przemsza River. Geomorphology 40:205–218

Cleland J (2000) Results of contaminated sediment cleanups relevant to the Hudson River. Report to Scenic Hudson, Poughkeepsie, NY

Cochran WG (1977) Sampling techniques, 3rd edn. John Wiley and Sons, New York

Collins AL (1995) The use of composite fingerprints for tracing the source of suspended sediment in river basins. Unpublished PhD Thesis, University of Exceter

Collins AL, Walling DE (2002) Selecting fingerprinting properties for discriminating potential suspended sediment sources in river basins. Journal of Hydrology 261:218–244

Collins AL, Walling DE, Leeks GJL (1997a) Source type ascription for fluvial suspended sediment based on a quantitative composite fingerprinting technique. Catena 29:1–27

Collins AL, Walling DE, Leeks GJL (1997b) Fingerprinting the origin of fluvial suspended sediment in larger river basins: combining assessment of spatial provenance and source type. Geografiska Annaler 79A:239–254

Collins AL, Walling DE, Leeks GJL (1997c) Use of the geochemical record preserved in floodplain deposits to reconstruct recent changes in river basin sediment sources. Geomorphology 19:151–167

Collins AL, Walling DE, Leeks GJL (1997d) Sediment sources in the Upper Severn catchment: a finger-printing approach. Hydrology and Earth Systems Sciences 1:509–521

Collins AL, Walling DE, Leeks GJL (1998) Use of composite fingerprints to determine the provenance of the contemporary suspended sediment load transported by rivers. Earth Surface Processes and Landforms 23:31–52

Collins MA (1995) Dredging-induced near-field resuspended sediment concentrations and source strengths. US Army Engineering Waterways Experiment Station, Miscellaneous Paper D-95-2, Vicksburg, Mississippi

Copeland RR (1994) Applications of channel stability methods–case studies. Technical Report HL-94-11, US Army Corps of Engineers, Waterways Experimental Station, Vicksburg, Mississippi

Corbett JO (1988) Uncertainty in risk analysis: an alternative approach through pessimisation. Journal of Radiological Protection 8:107–17

Costa JE (1984) The physical geomorphology of debris flows. In: Costa JE, Fleisher PJ (eds) Developments and applications of geomorphology, Springer-Verlag Berlin, pp. 268–317

Costa JE, Jarrett RD (1981) Debris flows in small mountain stream channels of Colorado, and their hydrologic implications. Bulletin of the Association of Engineering Geology 18:309–322

Costerton JW, Cheng KJ, Geesey GG, Ladd TI, Nickel JC, Dasgupta M, Marrie TJ (1987) Bacterial biofilms in nature and disease. Annual Reviews in Microbiology 41:435–464

Coulthard TJ, Macklin MG (2003) Modeling long-term contamination in river systems from historical metal mining. Geological Society of America Bulletin 31:451–454

Covelli S, Fontolan G (1997) Application of a normalization procedure in determining regional geochemical baselines. Environmental Geology 30:34–45

Cowan EJ (1991) The large-scale architecture of the fluvial Westwater Canyon Member, Morrison Formation (Jurassic), San Juan Basin, New Mexico. In: Miall AD, Tyler N (eds) The three-dimensional facies architecture of terrigenous clastic sediments and its implications for hydrocarbon discovery and recovery. Society of Economic Paleontologists and Mineral, Concepts in Sedimentology and Paleontology 3:80–93

Crockett TR (1993) Modeling near field sediment resuspension in cutterhead suction dredging operations. MS Thesis, University of Nebraska, Lincoln

Cundy DF, Bohlen WF (1982) A numerical simulation of the dispersion of sediments suspended by estuarine dredging operations. In: Hamilton P, MacDonald KB (eds) Estuarine and wetlands processes. Plenum, New York, pp 339–352

Cushing BS (1999) State of current contaminated sediment management practices. Sediment Management Working Group Technical Paper, *http://www.smwg.org*, 32pp

Davies BE, Lewin J (1974) Chronosequences in alluvial soils with special reference to historical pollution in Cardiganshire, Wales. Environmental Pollution 6:49–57

Davis JA, Kent DB (1990) Surface complexation models in aqueous geochemistry. In: Hochella MF, White AF (eds) Mineral-water interface geochemistry. Reviews in Mineralogy 23:177–260

Davis JA, Leckie JO (1978) Surface ionization and complexation at the oxide/water interface, III: adsorption of anions. Journal of Colloid and Interface Science 74:32–42

Day SJ, Fletcher WK (1989) Effects of valley and local channel morphology on the distribution of gold in stream sediments from Harris Creek, British Columbia, Canada. Journal of Geochemical Exploration 32:1–16

Degens BP, Donohue RD (2002) Sampling mass loads in river: a review of approaches for identifying, evaluating, and minimizing estimation error. Water Resources Technical Series, Water Rivers Commission, Report No WRT 24

de Groot A, Zshuppe K, Salomons W (1982) Standardization of methods of analysis for heavy metals in sediments. Hydrobiologia 92:689–695

de Jong E, Kachanoski RG (1988) Estimates of soil erosion and depositon for some Saskatchewan soils. Canadian Journal of Soil Science 63:607–617

Dennis IA, Macklin MG, Coulthard TJ, Brewer PA (2003) The impact of the October–November 2000 floods on contaminant metal dispersal in the River Swale catchment, North Yorkshire, UK. Hydrological Processes 17:1641–1657

Deutsch WJ (1997) Groundwater geochemistry: fundamentals and applications to contamination. Lewis Publishers, New York

de Vries JJ (1995) Seasonal expansion and contraction of stream networks in shallow groundwater systems. Journal of Hydrology 170:15–26

DeWalle DR, Pionke HB (1994) Streamflow generation on a small agricultural catchment during autumn recharge, II Stormflow periods. Journal of Hydrology 163:23–42

Dickin AP (1997) Radiogenic isotope geology. Cambridge University Press, Cambridge

Dietrich WE (1987) Mechanics of flow and sediment transport in river bends. In: Richards K (ed) River channels: environment and process. Basil Blackwell, Ltd, Oxford, pp 179–227

Dietrich WE, Smith JD (1983) Influence of the point bar on flow through curved channels. Water Resource Research 19:1173–1192

Domenico PA, Schwartz FW (1998) Physical and chemical hydrogeology, 2nd edn. John Wiley and Sons, New York

Donigian AS Jr, Imhoff JC, Bicknell BR, Kittle JL (1984) Application guide for hydrological simulations program – Fortran (HSPF). Prepared for USEPA, EPA-600/3-84-065, Environmental Research Laboratory, Athens, GA

Douglas G, Palmer M, Caitcheon G (2003a) The provenance of sediments in Moreton Bay, Australia: a synthesis of major, trace element and Sr-Nd-Pb isotopic geochemistry, modeling and landscape analysis. Hydrobiologia 494:145–152

Downs PW, Gregory KJ (1993) The sensitivity of river channels in the landscape system. In: Thomas DSG, Allison RJ (eds) Landscape sensitivity. John Wiley and Sons, Chichester, pp 15–30

Downs PW, Gregory KJ (2004) River channel management: towards sustainable catchment hydrosystems. Arnold, New York

Droppo IG, Ongley ED (1994) Flocculation of suspended sediment in rivers of southeastern Canada. Water Research 28:1799–1809

Duffus JH (2002) Heavy metals-a meaningless term? Pure and Applied Chemistry 74:793–807

Dunne T, Leopold LB (1978) Water in environmental planning. WH Freeman and Company, New York

Dunnette DA (1992) Assessing global river water quality: overview and data collection. In: Dunnette DA, O'Brien RJ (eds) The science of global change: the impact of human activities on the environment. American Chemical Society, Washington, DC, pp 240–259

Dushenkov V, Nanda Kumar PBA, Motto H, Raskin I (1995) Rhizofiltration: the use of plants to remove heavy metals from aqueous streams. Environmental Science and Technology 29:1239–1245

Dushenkov S, Kapulnik Y (2000) Phytofiltration of metals. In: Raskin I, Ensley BD (eds) Phytoremediation of toxic metals: using plants to clean up the environment. John Wiley and Sons, Chichester

EC, MENVIQ (1992) Interim criteria for quality assessment of St Lawrence River sediment. Environment Canada and Ministere de l'Environement du Quebec, Ottawa

Ehlers R (1956) An evaluation of stream improvement devices constructed eighteen years ago. California Fish and Game 42:203–217

Einstein HA (1950) The bedload function for sediment transportation in open channels flows. US Department of Agriculture Technical Bulletin 1026

Elliot GL, Campbell BL, Loughran RJ (1990) Correlation of erosion measurements and soil caesium-137 content. Journal of Applied Radiation and Isotopes 41:714–717

Elsenbeer H, Lack A, Cassel K (1995) Chemical fingerprints of hydrological compartments and flow paths at La Cuenca, western Amazonia. Water Resources Research 31:3051–3058

Ely LL, Webb RH, Enzel Y (1992) Accuracy of post-bomb ^{137}Cs and ^{14}C in dating fluvial deposits. Quaternary Research 38:196–204

Engler RM, Brannon JM, Rose J, Bigham G (1977) A practical selective extraction procedure for sediment characterization. In: Yen TF (ed) Chemistry of marine sediments. Ann Arbor Science Publisher Inc, Ann Arbor

Environmental News Service (2006) $50 Million cleanup of Ashtabula River sediment begins. June 7, 2006

Erickson PM (1992) Waste treatability in solidification/stabilization technology. In: Proceedings of the 1992 USEPA/A&WMA International Symposium, Air & Waste Management Association, Pittsburg, PA, pp 336–337

Erskine W, Melville M, Page KJ, Mowbray PD (1982) Cutoff and oxbow lake. Australian Geographer 15:174–180

Eubanks CE (2006) A soil bioengineering guide for streambank and lakeshore stabilization. USDA Forest Service, Technology and Development Program Report

Evans M (1997) Recent developments for in situ treatment of metal contaminated soils. Report for USEPA, Office of Solid Waste and Emergency Response, Technology Innovation Office, Washington DC

Evans C, Davies TD (1998) Causes of concentration/discharge hysteresis and its potential as a tool for analysis of episode hydrochemistry. Water Resources Research 34:129–137

Evans D, Davies BE (1994) The influence of channel morphology on the chemical partitioning of Pb and Zn in contaminated river sediments. Applied Geochemistry 9:45–52

Everitt BL (1968) Use of cottonwood in an investigation of recent history of a flood plain. American Journal of Science 266:417–439

Eynon G, Walker RG (1974) Facies relationships in Pleistocene outwash gravels, southern Ontario: a model for bar growth in braided rivers. Sedimentology 21:43–70

Facetti J, Dekov VM, Van Grieken R (1998) Heavy metals in sediments from the Paraguay River: a preliminary study. The Science of the Total Environment 209:79–86

Fahnestock RK (1963) Morphology and hydrology of a glacial stream – White River, Mount Rainer, Washington. US Geological Survey Professional Paper 422-A

Farrell KM (1987) Sedimentology and facies architecture of overbank deposits of the Mississippi River, False River region, Louisiana. In: Ethridge FG, Flores RM, Harvey MD (eds) Recent developments in fluvial sedimentology. Society of Economic Paleontologists and Mineralogists Special Publication 39:111–120

Fein JB, Doughney J, Yee N, Davis TA (1997) Chemical equilibrium model for metal adsorption onto bacterial surfaces. Geochimica et Cosmochimica Acta 61:3319–3328

Ferguson R (1987) Hydraulic and sedimentary controls of channel pattern. In: Richards K (ed) River channels: environmental and processes. Basil Blackwell, London, pp 129–158

Ferguson RI (1993) Understanding braiding processes in gravel-bed rivers: progress and unsolved problems. In: Best JL, Bristow CS (eds) Braided rivers. Geological Society of America Special Publication No 75, pp 73–87

Ferring CR (1986) Rates of fluvial sedimentation: implications for archaeological variability. Geoarchaeology 1:259–274

Ferring CR, Peter DE (1982) The Late Holocene prehistory of the Dyer Site (34PS96), Ti Valley, Oklahoma. Report to the National Park Service, Denver, Institute of Applied Science, N Texas State University

Fielding CR (1993) A review of recent research in fluvial sedimentology. Sedimentary Geology 85:3–14

FISRWB (The Federal Interagency Stream Restoration Working Group) (1998) Stream corridor restoration: principles, processes, and practices. National Engineering Handbook, NEH 653

Fitzgerald WF, Mason RP, Vandal GM (1991) Atmospheric cycling and air-water exchange of mercury over mid-continental lacustrine regions. Water, Air and Soil Pollution 56:745–767

Flemming HC, Schmitt J, Marshall KC (1996) Sorption properties of biofilms. In: Calmano W, Förstner U (eds), Sediments and toxic substances: environmental effects and ecotoxicty, Springer, Berlin pp 115–157

Foth and Van Dyke (2000) Summary Report: Fox River Deposit N. Prepared for Wisconsin Department of Administration, Wisconsin Department of Natural Resources, April 2000

Förstner U, Wittmann GTW (1979) Metal pollution in the aquatic environment, 1st edn, Springer-Verlag, New York

Förstner U, Wittmann GTW (1981) Metal pollution in the aquatic environment, 2nd edn. Springer-Verlag, New York

Förstner U (1995) Non-linear release of metals from aquatic sediments. In: Salomons W, Stigliani WM (eds) Biogeodynamics of pollutants in soil and sediments: risk assessment of delayed and non-linear responses. Springer, Berlin, pp 247–307

Foster IDL, Walling DE (1994) Using reservoir deposits to reconstruct changing sediment yields and sources in the catchment of the Old Mill Reservoir. South Devon, U.K. over the past 50 years. Hydrological Sciences Journal 39 347–368

Fredericks DJ, Perrens SJ (1988) Estimating erosion using caesium-137: II, estimating rates of soil loss. IAHS Publication International Association of Hydrological Sciences Press, Wallingford 174:233–240

Freeze RA, Cherry JA (1979) Groundwater. Prentice-Hall, Englewood Cliffs

Frissell'CA, Nawa RK (1992) Incidence and causes of physical failure of artificial habitat structures in streams of western Oregon and Washington. North American Journal of Fisheries Management 12:182–197

FRG (1999) Effectiveness of sediment removal: an evaluation of EPA Region 5 claims regarding twelve contaminated sediment removal projects. Fox River Group and JP Doody, Blasland, Bouck & Lee, Inc

FRRAT (2000) Evaluation of the effectiveness of remediation dredging: the Fox River Deposit N demonstration project, November 1998–January 1999. Fox River Remediation Advisory Team, Water Resources Institute, University of Wisconsin-Madison

FRTR (2006) Remediation technologies screening matrix and reference guide, version 4. Federal Remediation Technologies Roundtable, http://www.frtr.gov/matrix2/top_page.html

Gaiero DM, Ross GR, Depetris PJ, Kempe S (1997) Spatial and temporal variability of total non-residual heavy metals content in stream sediments from the Suquia River system, Cordoba, Argentina. Water, Air, and Soil Pollution 93:303–319

Gallart F, Benito G, Martín-Vide JP, Benito A, Prió JP, Regüés D (1999) Fluvial geomorphology and hydrology in the dispersal and fate of pyrite mud particles released by the Aznalcllar mine tailings spill. The Science of the Total Environment 242:13–26

Galm JR, Flynn P (1978) The cultural sequences at the Scott (34Lf-11) and Wann (34Lf-27) sites and prehistory of the Wister Valley. Archaeology Research and Management Center, Norman, Oklahoma University, OK, Research Series 3

Garnett L (2002) Cuyahoga River in Ohio no longer a fire hazard but a tourist attraction. National Public Radio Transcript, June 24, 1999

GTC (1990) The global ecology handbook. The global tomorrow coalition. Beacon Press, Boston

Germanoski D (1990) Comparison of bar-forming processes and differences in morphology in sand- and gravel-bed braided rivers. Ecological Society of America Abstracts with Programs, Annual Meeting 27:A110

Germanoski D (2000) Bar forming processes in gravel-bed braided rivers, with implications for small-scale gravel mining. In: Anthony DJ, Harvey M, Laronne J, Mosley P (eds) Applying geomorphology to environmental management. Water Resources Publications, LLC, Highlands Ranch, co, pp 1–26

Gerrard J (1993) Soil geomorphology, present dilemmas and future challenges. Geomorphology 7:61–84

Gibbs RJ (1977) Transport phases of transition metals in the Amazon and Yukon Rivers. Geological Society of America Bulletin 88:829–843

Giger W, Roberts PV (1977) Characterization of refractory organic carbon. In: Mitchell R (ed) Water pollution Microbiology, vol 2. Wiley-Interscience, New York

Gilbert GK (1877) Geology of the Henry Mountains (Utah). US geographical and gological survey of the Rock Mountain region. US Government Printing Office, Washington, DC

Gilbert GK (1917) Hydraulic-mining debris in the Sierra Nevada. US Geological Survey Professional Paper 105

Gilbert RO (1987) Statistical methods for environmental pollution monitoring. Van Nostrand Reinhold, New York

Gilvear D, Bryant R (2003) Analysis of aerial photography and other remotely sensed data. In: Kondolf GM, Piégay H (eds) Tools in fluvial geomorphology. John Wiley and Sons, Chichester

Gomez B, Church M (1989) An assessment of bed load sediment transport formule for gravel bed rivers. Water Resources Research 25:1161–1186

Gomez B, Eden DN, Hicks DM, Trutrum NA, Peacock DH, Wilmshurst J (1999) Contribution of floodplain sequestration to the sediment budget of the Waipaoa River, New Zealand. In: Marriott SB, Alexander J (eds) Floodplains: interdisciplinary approaches, Geological Society Special Publication 163, Geological Society of London, London, pp. 69–88

Graf WL (1988) Definition of flood plain along arid-region rivers. In: Baker V, Kochel RC, Patton PC (eds) Flood geomorphology. John Wiley and Sons, New York, pp 231–242

Graf WL (1990) Fluvial dynamics of Thorium-230 in the Church Rock event, Puerco River, New Mexico. Annals of the Association of American Geographers 80:327–342

Graf WL, Clark SL, Kammerer MT, Lehman T, Randall K, Tempe R, Schroeder A (1991) Geomorphology of heavy metals in the sediments of Queen Creek, Arizona, USA. Catena 18:567–582

Graney JR, Halliday AN, Keeler GJ, Nriagu JO, Robbins JA, Norton SA (1995) Isotopic record of lead pollution in lake sediments from the northeastern United States. Geochimica Cosmochimica Acta 59:1715–1728

Grasso D (1993) Hazardous waste site remediation – source control. Lewis Publishers, Boca Raton, Florida

Graves W, Eliab P (1977) Sediment study: alternative delta water facilitites-peripheral canal plan, Sacramento, CA. California Department of Water Resources, Central District

Gregory SV, Swanson FJ, McKee WA, Cummins KW (1991) An ecosystem perspective of riparian zones: focus on links between land and water. Bioscience 41:540–551

Grim RE (1968) Clay mineralogy. McGraw Hill, New York

Gross DW, Ross AR, Allen PB, Naney JW (1972) Geomorphology of the Central Washita River Basin. Proceedings of the Oklahoma Academia of Science 52:145–149

Guilbert JM, Park CF Jr (1986) The Geology of ore deposits. WH Freeman and Company, New York

Gulson BL, Tiller KG, Mizon KJ, Merry RH (1981) Use of lead isotope ratios in soils to identify the source of lead contamination near Adelaide, South Australia. Environmental Science and Technology 15:691–696

Gundersen P, Steinnes E (2001) Influence of temporal variations in river discharge, pH, alkalinity, and Ca on the speciation and concentration of heavy metals in some mining polluted rivers. Aquatic Geochemistry 7:173–193

Gunn AM, Winnard DA, Hunt DTE (1988) Trace metal speciation in sediments and soils: an overview from a water industry perspective. In: Kramer JR, Allen EH (eds) Metal speciation: theory, analysis, and application. Lewis Publishers, Chelsea, Michigan

Gworek B (1992) Lead inactivation by zeolites. Plant and Soil 143:71–74

Hagerty DJ (1980) Multifactor analysis of bank caving along a navigable stream. In: National Watersways Roundtable Proceedings, US Army Engineering Water Resources Support Center, Institute for Water Resources, IWR-80-1:463–492

Hagerty DJ, Hamel JV (1989) Geotechnical aspects of river bank erosion. In: Proceedings of the national conference on hydraulic engineering, New Orleans, August 14–18

Hall SA, Lintz C (1984) Buried trees, water table fluctuations and 3000 years of changing climate in west central Oklahoma. Quaternary Research 22:129–133

Hamilton JB (1989) Response of juvenile Steelhead to instream deflectors in a high gradient stream. In: Gresswell RE, Barton, BA, Kershner JL (eds) Practical approaches to riparian resource management: an educational workshop. US Bureau of Land Management, Billings, pp 149–158

Hansmann W, Köppel V (2000) Lead-isotopes as tracers of pollutants in soils. Chemical Geology 171:123–144

Harman WA, Jennings GD, Patterson JM, Clinton DR, Slate O, Jessup AO, Everhart JR, Smith RE (1999) Bankfull hydraulic geometry relationships for North Carolina streams. In: Olsen DS, Potyondy JP (eds) Wildland hydrology. American Water Resources Association, Herdon, Virginia, TPS-99-3, pp 401–408

Harper D, Smith C, Barham P, Howell R (1995) The ecological basis for the management of the natural river environment. In: Harper DM, Ferguson AJD (eds) The ecological basis for river management. John Wiley and Sons, Chichester

Harrison J, Heijnis H, Caprarelli G (2003) Historical pollution variability form abandoned mine sites, Greater Blue Mountains World Heritage Area, New South Wales, Australia. Environmental Geology 43:680–687

Hassan MA, Church M (1992) The Movement of Individual Grains on the Streambed, In: Billi P Hey RD, Thorne CR, Tacconi P, Dynamics of Gravel-bed Rivers, John Wiley and Sons, Ltd, New York, pp. 159–175

Hawkes HE, Webb JS (1962) Geochemistry in mineral exploration. Harper, New York

He Q, Walling DE (1996) Use of fallout Pb-210 measurements to investigate longer-term rates and patterns of overbank sediment deposition on floodplains of lowland rivers. Earth Surface Processes and Landforms 21:141–154

Headley JV, Gandrass J, Kuballa J, Peru KM, Gong Y (1998) Rates of sorption and partitioning of contaminants in river biofilm. Environmental Science and Technology 32:3968–3973

Healy TW (1974) Principles of adsorption organics at solid-solution interfaces. Journal of Macromolecular Science and Applied Chemistry A8:603–619

Heathwaite AL (1997) Sources and pathways of phosphorus loss from agriculture. In: Tunney H, Caton OT, Brookes PC, Johnston AE, eds, Phosphorus loss to water from agriculture. CAB International, UK, pp 205–224

Heathwaite AL, Dils RM (2000) Characterizing phosphorus loss in surface and subsurface hydrological pathways. The Science of the Total Environment 251–252:523–538

Hedges REM (1991) AMS dating: present status and potential applications. Quaternary Proceedings 1:5–11

Helgen SO, Moore JN (1996) Natural background determination and impact quantification in trace metal-contaminated river sediments. Environmental Science and Technology 30:129–135

Heller D, Roper B, Konnoff D, Weiman K (2000) Durability of Pacific northwest instream structures following floods. International Conference on Wood in Rivers, Oregon State University, October 2000

Heppell, CM, Burt TP, Williams RJ (2000) The use of tracers to aid the understanding of the herbicide leaching in clay soils. In: Foster IDL, ed, Tracers in geomorphology, Wiley, Chichester, pp. 309–332

Herbich JB, DeVries J (1986) An evaluation of the effects of operational parameters on sediment resuspension during cutterhead dredging using a laboratory model dredge system. Report No CDS 286, Texas A&M University, College Station

Hewlett JD (1961) Watershed management. In: Report for 1961 Southwestern Forest Experimental Station, US Forest Service, Asheville, North Carolina, pp. 62–66

Hewlett JD, Hibbert AR (1967) Factors affecting the response of small watersheds to precipitation in humid areas. In: Sopper WE, Lull HW (eds) Forest hydrology Pergamon Press, New York, pp 275–290

Hey RD (2006) Fluvial geomorphological methodology for natural stable channel design. Journal of the American Water Resources Association 42:357–374

Hey RD, Thorne CR (1975) Secondary flow in river channels. Area 7:191–195

Hey RD, Thorne CR (1986) Stable channels with mobile gravel beds. Journal of Hydraulic Engineering 112:671–689

Hjulström F (1939) Transportation of dentritus by moving water. In: Trask P (ed) Recent marine sediments. American Association of Petroleum Geologists, Tulsa, OK

Hobbs RJ, Norton DA (1996) Towards a conceptual framework for restoration ecology. Restoration Ecology 4:93–110

Hodges RA, Whittle JH, Avalle DL (1997) Site investigation and ground remediation – an integrated approach. In: Yong RN, Thomas HR (eds) Geoenvironmental engineering. Thomas Telford, London, pp 26–31

Hodson ME (2004) Heavy metals: geochemical bogey men? Environmental Pollution 129:341–343

Hoey T (1992) Temporal variation in bedload transport rates and sediment storage in gravel river beds. Progress in Physical Geography 16:319–338

Hoffman JS, Fletcher WK (1979) Selective sequential extraction of Cu, Zn, Fe, Mn, and Mo from soils and sediments. In: Watterson JR, Theobald PK (eds) Geochemical exploration 1978. Association of Exploration Geochemistry, Rexdale, Ontario

Holden J, Burt TP, Cox NJ (2001) Macroporosity and infiltration in blanket peat: the implications of tension disc infiltrometer measurements. Hydrological Processes 15:289–303

Hollings CS (1978) Adaptive management assessment and management. John Wiley and Sons, Chichester

Hooper RP, Christophersen N, Peters NE (1990) Modeling stream-water chemistry as a mixture of soilwater end-members-an application to the Panola Mountain catchment, Georgia, USA. Journal of Hydrology 116:321–343

Horowitz AJ (1991) A primer on sediment-trace element chemistry, 2nd edn. Lewis, Chelsea, Michigan

Horowitz AJ (1995) The use of suspended sediment and associated trace elements in water quality studies, IAHS special publication 4. IAHS Press, Wallingford, UK

Horowitz AJ, Elrick KA (1988) Interpretation of bed sediment trace metal data: methods of dealing with the grain size effect. In: Lichtenberg JJ, Winter JA, Weber CC, Fradkin L (eds) Chemical and biological characterization of sludges, sediments, dredge spoils, and drilling muds. American Society for Testing and Materials, pp 114–128

Horowitz AJ, Rinella FA, LaMothe P, Miller TL, Edwards TK, Roche RL, Rickert DA (1990) Variations in suspended sediment and associated trace element concentrations in selected riverine cross sections. Environmental Science and Technology 24:1313–1320

Horowitz AJ, Elrick KA, Smith JJ (2001) Annual suspended sediment and trace element fluxes in Mississippi, Columbia, Colorado, and Rio Grande drainage basins. Hydrological Processes 15:1169–1207

Horton RE (1933) The role of infiltration in the hydrological cycle. American Geophysical Union Transactions 14:446–460

Horton RE (1945) Erosional development of streams and their drainage basins. Hydrophysical approach to quantitative morphology. Geological Society of America Bulletin 56:275–370

House WA, Warwick MS (1998) Hysteresis of the solute concentration/discharge relationship in rivers during storms. Water Research 32:2279–2290

Houthoofd JM, McCready JH, Roulier MH (1991) Remedial Action, Treatment, Disposal of Hazardous Waste: Proceedings of the 7th Annual RREL Hazardous Waste Research Symposium

Howard AD (1996) Modelling channel evolution and floodplain morphology. In: Anderson MG, Walling DE, Bates PD (eds) Floodplain processes. John Wiley and Sons, Chichester, pp 15–62

Howe WH, Knopf FL (1991) On the imminent decline of Rio Grande cottonwoods in central New Mexico. The Southwestern Naturalist 36:218–224

Huang JW, Chen J, Berti, WR, Cunningham SD (1997) Phytoremediation of lead-contaminated soils: role of synthetic chelates in lead phytoextraction. Environmental Science and Technology 31:800–805

Huber WC, Dickinson RE (1988) Storm water management model user's manual, version 4. EPA/600/3-88/001a (NITS PB88-236641/AS), Environmental Protection Agency, Athens, Georgia

Hudson-Edwards KA, Macklin MG, Curtis CD, Vaughan DJ (1998) Chemical remobilization of contaminated metals within floodplain sediments in an incising river system: implications for dating and chemostratigraphy. Earth Surface Processes and Landforms 23:671–684

Hudson-Edwards KA, Macklin MG, Finlayson R, Passmore DG (1999) Mediaeval lead pollution in the River Ouse at York, England. Journal of Archaeological Science 26:809–819

Hudson-Edwards KA, Macklin MG, Taylor MP (1999b) 2000 years of sediment-borne heavy metal storage in the Yorkshire Ouse basin, NE England, UK. Hydrological Processes 13:1087–1102

Hudson-Edwards KA, Macklin MG, Miller JR, Lechler PJ (2001) Sources, distribution and storage of heavy metals in the Rio Pilcomayo, Bolivia. Journal of Geochemical Exploration 72:229–250

Hudson-Edwards KA, Macklin MG, Jamieson HE, Brewer PA, Coulthard TJ, Howard AJ, Turner JN (2003) The impact of tailings dam spills and clean-up operations on sediment and water quality in river systems: the Ros Agrio-Guadiamar, Aznalcóllar, Spain. Applied Geochemistry 18:221–239

Huggett RJ (2003) Fundamentals of geomorphology. Routledge, London

Hudson-Edwards KA, Macklin MG, Curtis CD, Vaughan DJ (1996) Processes of formation and distribution of Pb, Zn, Cd, and Cu bearing mineral species in the Tyne basin, NE England: implications for metal-contaminated river systems. Environmental Science and Technology 30:72–80

Hupp CR, Bazemore DE (1993) Spatial and temporal aspets of sediment deposition in West Tennessee forested wetlands. Journal of Hydrology 141:179–196

Hupp CR, Bornette G (2003) Vegetation as a tool in the interpretation of fluvial geomorphic processes and landforms in humid temperate areas. In: Kondolf MG, Piegay H, (eds) Tools in fluvial geomorphology, John Wiley and Sons, Chichester, pp. 269–288

Hupp CR, Morris EE (1990) A dendrogeomorphic approach to measurement of sedimentation in a forested wetland, Black Swamp, Arkansas. Wetlands 10:107–124

Hupp CR, Osterkamp WR (1996) Riparian vegetation and fluvial geomorphic processes. Geomorphology 14:277–295

IJC (1997) Overcoming obstacles to sediment remediation in the Great Lakes Basin. White Paper by the International Joint Commission, Sediment Priority Action Committee, Great Lakes Water Quality Board

Imeson AC (1974) The origin of sediment in a moorland catchment with special reference to the role of vegetation. In: Fluvial processes in instrumented watersheds. Institute of British Geographers Special Publications No 6, pp 69–72

Iskandar IK, Adriano DC (1997) Remediation of soils contaminated with metals – a review of current practices in the USA. In: Iskandar IK, Adriano DC (eds) Remediation of soils contaminated with metals. Science Reviews, Northwood, UK, pp 1–26

Iversen TM, Kronvang MB, Madsen BI, Markmann P, Nielsen MB (1993) Re-establishment of Danish streams: restoration and maintenance measures. Aquatic Conservation: Marine and Freshwater Ecosystems 3:73–92

James LA (1989) Sustained storage and transport of hydraulic gold mining sediment in the Bear River, California. Annals of the Association of American Geographers 79:570–592

James RO, Healy TW (1972) Adsorption of hydrolysable metal ions at the oxide-water interface. Journal of Colloid and Interface Science 40:65–81

Jenne E (1976) Trace element sorption by sediments and soils – sites and processes. In: Chappell W, Peterson K (eds) Symposium on molybdenum, vol 2. Marcel-Dekker, New York, pp 425–553

Jernelov A, Ramel C (1994) Mercury in the environment, Synopsis of the Scientific Committee on Problems of the Environment (SCOPE) meeting on mercury held at the Royal Academy of Sciences, Stockholm, October 28–30, 1993, Ambio 23:166

Johansson K (1977) The fundamental chemical and physical characteristics of Swedish lakes; heavy metal content in lake sediment from some lakes on the Swedish west coast and its connection with the atmospheric supply. Abstract, SIL-Congress, Copenhagen, p 133

Johnson WC (1994) Woodland expansion in the Platte River, Nebraska: patterns and causes. Ecological Monograph 64:45–84

Juang RS, Wang SW (2000) Metal recovery and EDTA recycling from simulated washing effluents of metal-contaminated soils. Water Research 34:3795–3803

Kabata-Pendias A, Pendias H (1984) Trace elements in soils and plants. CRC Press, Boca Raton, Florida

Kachanoski RG (1987) Comparison of measured soil 137-caesium losses and erosion rates. Canadian Journal of Soil Science 67:199–203

Kamnev AA (2003) Phytoremedation of heavy metals: an overview. Recent Advances in Marine Biotechnology 8:269–319

Keith L (1988) Principles of environmental sampling. American Chemical Society, Washington, DC

Keith L, Crummett W, Deegan J, Libby R, Taylor J, Wentler G (1983) Principles of environmental analysis. Analytical Chemistry 55:2210–2218

Kelley DW, Nater EA (2000) Source apportionment of Lake Bed sediments to watersheds in an Upper Mississippi Basin using a chemical mass balance method. Catena 41:277–292

Kendall C, Bullen T (2003) Applications of environmental isotopes for tracing anthropogenic contaminants in ground waters and surface waters. GSA Short-Course, Geological Society of America, 2003 Annual Meeting, Seattle

Kendall C, Doctor DH (2004) Stable isotope applications in hydrological studies. In: Drever JJ (ed) Treatise on Geochemistry, vol 5, Surface and Ground Water, Weathering, Erosion, and Soils, pp 5–46

Kendall C, McDonnel JJ (1998) Isotope tracers in catchment hydrology. Elsevier, Amsterdam

Kennedy RG (1894) The prevention of silting in irrigation canals in India. Proceedings of the Institute of Civil Engineering 119:281–295

Kesel RH, Dunne KC, McDonald RC, Allison KR, Spincer BE (1974) Lateral erosion and overbank deposition on the Mississippi River in Louisiana caused by 1973 flooding. Geology 2:461–464

Kesel RH, Yodis EG, McCrow DJ (1992) An approximation of the sediment budget of the lower Mississippi River prior to human modification. Earth Surface Processes and Landforms 17:711–722

Kersten M, Förstner U (1989) Speciation of trace elements in sediments. In: Batley GE (ed) Trace element speciation: analytical methods and problems, CRC Press, Inc, Boca Raton, pp 245–317

Kersten M, Garbe-Schonberg CD, Thomsen S, Anagnostou C, Sioulas A (1997) Source apportionment of Pb pollution in the coastal waters of Elefsis Bay, Greece. Environmental Science and Technology 31:1295–1301

Kimball BA (1997) Use of tracer injections and synoptic sampling to measure metal loading from acid mine drainage. US Geological Survey Fact Sheet FS-245-296

Kimball BA, Runkel RL, Walton-Day K, Bencala KE (2002) Assessment of metal loads in watersheds affected by acid mine drainage by using tracer injection and synoptic sampling: Cement Creek, Colorado, USA. Applied Geochemistry 17:1183–1207

Kinnell PIA (2000) AGNPS-UM: applying the USLE-M within the agricultural non-point source pollution model. Environmental Modelling and Software 15:331–341

Klimek K, Zawilinska L (1985) Trace elements in alluvia of the Upper Vistula as indicators of palaeo-hyology. Earth Surface Processes and Landforms 10:273–280

Knighton D (1991) Channel bed adjustment along mine-affected rivers of northeast Tasmania. Geomorphology 4:205–219

Knighton D (1998) Fluvial forms and processes: a new prospective. Arnold, London

Knighton AD, Nanson G (1993) Anastomosis and the continuum of channel pattern. Earth Surface Processes and Landforms 18:613–625

Knisel WG (1980) CREAMS: a field scale model for chemicals, runoff and erosion from agricultural management systems. USDA, Washington, DC

Knox JC (1987) Historical valley floor sedimentation in the Upper Mississippi valley. Annals of the Association of American Geographers 77:224–244

Kochel RC (1987) Holocene debris flows in central Virginia. In: Costa JE, Wieczorek GF (eds) Debris flows/avalanches: process, recognition, and mitigation, Geological Society of America Reviews in Engineering 7, pp. 139–155

Kochel RC, Ritter DF, Miller JR (1987) Role of tree dams in the construction of pseudo-terraces and variable geomorphic response to floods in Little River Valley, Virginia. Geology 15:718–721

Kochel RC, Miller JR, Lord M, Martin T (2005) Geomorphic problems with in-stream structures using Natural Channel Design strategy for stream restoration projects in North Carolina. Geological Society of America Abstracts with Programs 37:329

Komar PD (1989) Flow-competence evaluations of the hydraulic parameters of floods: an assessment of the technique. In: Beven KJ, Carling PA (eds) Floods: hydrological, sedimentological, and geomorphological implications. Wiley, Chichester, pp 107–134

Komar PD, Shih SM (1992) Equal mobility versus changing bedload grain sizes in gravel-bed streams: In: Billie P, Hey RD, Thorne CR, Tacconi P (eds) Dynamics of gravel-bed rivers. Wiley and Sons, Ltd, New York, pp 73–106

Kondolf GM, Micheli ER (1995) Evaluating stream restoration projects. Environmental Management 19:1–15

Kondolf GM, Smeltzer MW, Railsback SF (2001) Design and performance of a channel reconstruction project in coastal California gravel-bed stream. Environmental Management 28:761–776

Kramer JR, Allen HE (1988) Metal speciation: theory, analysis, and application. Lewis Publishers, Chelsa, Michigan

Krauskopf KB, Bird DK (1995) Introduction to geochemistry, 3rd edn. McGraw-Hill, Inc, New York

Krishman R, Parker HW, Tock RW (1996) Electrode assisted soil washing. Journal of Hazardous Materials 48:111–119

Kubasek NK, Silverman GS (1997) Environmental law, 2nd edn. Prentice Hall, Inc, Upper Saddle River, NJ

Lacerda LD, Salomons W (1992) Mercurio na Amazonia: uma bomba relogio quimica? (mercury in the Amazon: a chemical time bomb?). Serio Tecnologia Ambiental, vol 3, CETEM/CNPq, Rio de Janeiro

Ladd SC, Marcus WA, Cherry S (1998) Differences in trace metal concentrations among fluvial concentrations among fluvial morphologic units and implications for sampling. Environmental Geology 36:259–270

Lajczak A (1995) The impact of river regulation, 1850–1990, on the channel and floodplain of the upper Vistula River, southern Poland. In: Hickin E (ed) River geomorphology. John Wiley and Sons, Chichester, pp 209–233

Lamber CP, Walling DE (1987) Floodplain sedimentation: a preliminary investigation of contemporary deposition with the lower reaches of the River Culm, Devon, UK. Geografiska Annaler 69A:47–59

Lambert BA, Leven BA, Green RM (2000) New methods of cleaning up heavy metal in soils and water. Environmental Science and Technology Briefs for Citizens, Hazardous Substance Research Centers, Kansas State University

Lane EW (1957) A study of the shape of channels formed by natural streams in erodible materials. MRD Sediments Series 9, US Army Corps of Engineers, Engineering Division, Missouri River, Omaha, Nebraska

Lane SN, Chandler JH, Richards KS (1998) Landform monitoring, modeling and analysis: land form in geomorphological research. In: Lane SN, Richards KS, Chandler KS (eds) Landform monitoring, modelling, and analysis. John Wiley and Sons, Chichester, pp 1–17

Lane SN, Richards KS, Chandler JH (1993) Developments in photogrammetry; the geomorphological potential. Progress in Physical Geography 17:306–328

Langbein WB et al (1949) Annual runoff in the United States. US Geological Circular 52

Langbein WB, Schumm SA (1958) Yield of sediment in relation to mean annual precipitation. Transactions of the American Geophysical Union 39:1076–1084

Langmuir D (1997) Aqueous environmental geochemistry. Prentice Hall, Upper Saddle River, New Jersey

Lawler DM (1991) A new technique for the automatic monitoring of erosion and deposition rates. Water Resources Research 27:2125–2128

Lawler DM (2005) The importance of high-resolution monitoring in erosion and deposition dynamics studies: examples from estuarine and fluvial systems. Geomorphology 64:1–23

Lee KN (1993) Compass and gyroscope: integrating science and politics for the environment. Journal of the North American Benthological Society 11:111–137

Leece SA (1997) Spatial patterns of historical sedimentation and floodplain evolution, Blue River, Wisconsin. Geomorphology 18:265–277

Leenaers H, Schouten CJ, Rang MC (1988) Variability of the metal content of flood deposits. Environmental Geology and Water Science 11:95–108

Lehman D (1963) Some principles of chelation chemistry. Proceedings of the Soil Science of America 27:167–170

Leopold LB, Emmett WW (1977) Bedload measurements, East Fork River, Wyoming. National Academic of Science Proceedings 73:1000–1004

Leopold LB, Maddock T Jr (1953) The hydraulic geometry of stream channels and some physiographic implications. US Geological Survey Professional Paper 252

Leopold LB, Wolman MG (1957) River channel patterns; braided, meandering, straight. US Geological Survey Professional Paper 282-B

Leopold LB, Wolman MG, Miller JP (1964) Fluvial processes in geomorphology. WH Freeman, San Francisco

Lewin J, Macklin MG (1987) Metal mining and floodplain sedimentation in Britain. In: Gardiner V (ed) International geomorphology. John Wiley and Sons, pp 1009–1027

Lewin J, Macklin MG, Johstone E (2005) Interpreting alluvial archives: sedimentological factors in the British Holocene fluvial record. Quaternary Science Reviews 24:1873–1889

Letcher RA, Jakeman AJ, Merritt WS, McKee LD, Eyre BD, Baginska B (1999) Review of techniques to estimate catchment exports. New South Wales Environmental Protection Authority Report, EPA 99/73

Libbey LK, McQuarrie ME, Pilkey OH, Rice TM, Samspon DW, Stutz ML, Trembanis AC (1998) Another view of the maturity of our science. Shore and Beach 66:2–4

Lischeid G, Kolb A, Alewell C (2002) Apparent translatory flow in groundwater recharge and runoff generation. Journal of Hydrology 265:195–211

Litchfield C (2005) Thirty years and counting: bioremediation in its prime? BioScience 55:273–279

Lomborg B (2001) The skeptical environmentalist: measuring the real state of the world. Cambridge University Press, Cambridge

Long ER, Morgan LG (1991) The potential for biological effects of sediment-sorbed contaminants tested in the National Status and Trends Program. NOAA Technical Memorandum NOS OMA 52, National Oceanic and Atmospheric Administration, Seattle, WA, 175pp

Longmore ME (1982) The caesium-137 dating technique and associated applications in Australia-a review. In: Ambrose W, Duerden P (eds) Archaeometry: an Australian perspective. Australian National University Press, Canberra, pp 310–321

Loring DH (1990) Lithium-a new approach for the granulometric normalization of trace metal data. Marine Chemistry 29:155–168

Loring DH, Rantala RTT (1992) Manual for the geochemical analysis of marine sediments and suspended particular matter. Earth-Science Reviews 32:235–283

Luck JM, Othman D (1998) Geochemistry and water dynamics II. Trace metals and Pb-Sr isotopes as tracers of water movements and erosion processes. Chemical Geology 150:263–282

Lueder DR (1959) Aerial photographic interpretation: principles and applications. McGraw-Hill, New York

Ma QY, Traina SJ, Logan TJ, Ryan JA (1993) In situ lead immobilization by apatite. Environmental Science and Technology 271:1803–1810

MacDonald GM, Beukens RP, Kieser WE, Vitt DH (1991) Comparative radiocarbon dating of terrestrial plant macrofossils and aquatic moss from the 'ice-free corridor' of western Canada. Geology 15:837–840

MacDonald DD, Ingersoll CG, Berger TA (2000) Development and evaluation of consensus-based sediment quality guidelines for freshwater ecosystems. Archives of Environmental Contamination and Toxicology 39:20–31

Mackin JH (1948) Concept of the graded river. Geological Society of America Bulletin 59:463–512

Macklin MG (1996) Fluxes and storage of sediment-associated heavy metals in floodplain systems: assessment and river basin management issues at a time of rapid environmental change. In: Anderson MG, Walling DE, Bates PD (eds) Floodplain processes. John Wiley and Sons, Ltd, pp 441–460

Macklin MG, Klimek K (1992) Dispersal, storage and transformation of metal-contaminated alluvium in the upper Vistula basin, southwest Poland. Applied Geography 12:7–30

Macklin MG, Lewin J (1989) Sediment transfer and transformation of an alluvial valley floor; the river South Tyne, Northumbria, UK. Earth Surface Processes and Landforms 14:233–246

Macklin MG, Ridgway J, Passmore DG, Rumsby BT (1994) The use of overbank sediment geochemical mapping and contamination assessment: results from selected English and Welsh floodplains. Applied Geochemistry 9:689–700

Macklin MG, Brewer PA, Hudson-Edwards KA, Bird G, Coulthard TJ, Dennis IA, Lechler PJ, Miller JR, Turner JN (2006) A geomorphological-geochemical approach to river basin management in mining-affected rivers. In James LA, Marcos WA (eds) The Human Role in Changing Fluvial Systems, Elsevier, Amsterdam, pp 423–447

Macklin MG, Dowsett RB (1989) The chemical and physical speciation of trace metals in fine grained overbank flood sediments in the Tyne Basin, north-east England. Catena 16:135–151

Macklin MG, Hudson-Edwards KA, Jamieson HE, Brewer P, Coulthard TJ, Howard AJ, Renenda VH (1999) Physical stability and rehabilitation of sustainable aquatic and riparian ecosystems in the Rio Guadiamar, Spain, following the Aznalcóllar mine tailings dam failure. In: Mine, water and environment. International Mine Water Association Congress, Seville, Spain, pp 271–278

Magar VS (2001) Natural recovery of contaminated sediments. Journal of Environmental Engineering 6:473–474

Mahaney WC (1984) Quaternary dating methods. Elsevier, Amsterdam

Malm O (1998) Gold mining as a source of mercury exposure in the Brazilian Amazon. Environmental Research, Section A 77:73–78

Manahan SE (2000) Environmental chemistry, 7th edn. Lewis Publishers, CRC Press, Boca Raton

Marcantonio F, Flowers G, Thien L, Ellgaard E (1998) Lead isotopes in tree rings: chronology of pollution in Bayou Trepangnier, Louisiana. Environmental Science and Technology 32:2371–2376

Marcus WA (1987) Copper dispersion in ephemeral stream sediments. Earth Surface Processes and Landforms 12:217–228

Markham AJ, Thorne CR (1992) Geomorphology of gravel-bed river bends. In: Billi P, Hey RD, Thorne CR, Tacconi P (eds) Dynamics of gravel-bed rivers. John Wiley and Sons, Ltd, New York, pp 433–456

Marmiroli N, McCutcheon SC (2003) Making phytoremediation a successful technology. In: McCutcheon SC, Schnoor JL (eds) Phytoremediation: transformation and control of contaminants. John Wiley and Sons, New York, pp 3–58

Marriott S (1996) Analysis and modelling of overbank deposits. In: Anderson MG, Walling DE, Bates PD (eds) Floodplain processes. John Wiley and Sons, Chichester, pp 63–93

Marron DC (1987) Floodplain storage of metal-contaminated sediments downstream of a gold mine at lead, South Dakota. In: Averett RC, McKnight DM (eds) Chemical quality of water and the hydrological cycle, Lewin Publishers, Chelsea, pp. 193–209

Marron DC (1989) Physical and chemical characteristics of a metal-contaminated overbank deposit, west-central Dakota, USA. Earth Surface Processes and Landforms 14:419–432

Marron DC (1992) Floodplain storage of mine tailings in the Belle Fourche River system: a sediment budget approach. Earth Surface Processes and Landforms 17:675–685

Matschullat J, Ottenstein R, Reimann C (2000) Geochemical background – can we calculate it? Environmental Geology 39:990–1000

McCuen RH (1998) Hydrologic analysis and design, 2nd edn. Prentice Hall, Upper Saddle River

McDiffett WF, Beidler AW, Dominick TF, McCrea KD (1989) Nutrient concentration-stream discharge relationships during storm events in a first order stream. Hydrobiologia 179:97–102

McGovern GJ (1991) Sedimentology, evolution, and sedimentation rates of a delta in a Mississippi River Oxbow Lake, southern Illinois. Master's Thesis, Department of Geology, Southern Illinois University, Carbondale

McGowan JH, Garner LE (1970) Physiographic features and stratification types of coarse-grained point bars: modern and ancient examples. Sedimentology 14:77–111

McKergow LA, Prosser IP, Hughes AO, Brodie J (2005) Sources of sediment to the Great Barrier Reef World Heritage Area. Marine Pollution Bulletin 51:200–211

Meade RH (1994) Suspended sediments of the modern Amazon and Orinoco rivers. Quaternary International 21:29–39

Mench M, Vangronsveld J, Didier V, Clijsters H (1994) Evaluation of metal mobility, plant availability and immobilization by chemical agents in limed-silty soil. Environmental Pollution 86:279–286

Mench M, Vangronsveld J, Lepp NW, Edwards R (1998) physico-chemical aspects and efficiency of trace element immobilization by soil amendments. In: Vangronsveld J, Cunningham SD (eds) Metal-Contaminated Soils: In Situ Inactivation and Phytorestoration, Springer-Verlag, Berlin, pp. 151–181

Menzel RG (1960) Transport of strontium-90 in runoff. Science 131:451–500

Mertes LAK (1990) Hydrology, hydraulics, sediment transport and geomorphology of the central Amazon floodplain. PhD Dissertation, University of Washington, Seattle

Meybeck M, Helmer R (1989) The quality of rivers: from pristine stage to global pollution. Paleogrography, Paleoclimatology, and Paleoecology, Global and Planetary Change Section 75:283–309

Meyer-Peter E, Muller R (1948) Formulas for bed-load transport. International Association for Hydraulic Structures, Research Proceedings, 2nd Meeting, Stockholm, pp 39–65

Meyers CR, Nealson KH (1988) Bacterial manganese reduction and growth with manganese oxide as the sole electron acceptor. Science 240:1319–1321

Miall AD (1978) Fluvial sedimentology. Canadian Society of Petroleum Geology Memoir 5

Miall AD (1985) Architectural-element analysis: a new method of facies analysis applied to fluvial deposits. Earth Science Reviews 22:261–308

Miall AD (1992) Alluvial deposits. In: Walker RG, James NP (eds) Facies models. Geological Association of Canada, Newfoundland, pp 119–142

Miall AD (1996) Geology of fluvial deposits. Springer-Verlag, Berlin

Miall AD (2000) Principles of sedimentary basin analysis, 3rd edn. Springer-Verlag, Berlin

Middelkoop H, Asselmann NEM (1994) Spatial and temporal variability of floodplain sedimentation in the Netherlands. Report GEOPRO 1994.05, Faculty of Geographical Sciences, Rijkuniversiteit Utrecht

Miller DE (1999) Deformable stream banks: can we call it restoration without them? In: Olsen DS, Potyondy JP (eds) Wildland Hydrology. American Water Resources Association, Herdon, Virginia, TPS-99-3, pp 293–300

Miller DE, Skidmore PB, White DJ (2001) Channel design. White Paper submitted to Washington Department of Fish and Wildlife, Washington Department of Ecology, and Washington Department of Transportation

Miller JR (1991) Development of anastomosing channels in south-central Indiana. Geomorphology 4:221–229

Miller JR (1997) The role of fluvial geomorphic processes in the transport and storage of heavy metals from mine sites. Journal of Geochemical Exploration, Special Issue 58:101–118

Miller JR, Lechler JR (1998) Mercury partitioning within alluvial sediments of the Carson River Valley, Nevada: implications for sampling strategies in tropical environments. In: Wasserman J, Silva-Filho EV, Abrao JJ (eds) Environmental Geochemistry in the Tropics. Lecture Notes in Earth Science, No 72, Springer-Verlag, pp 211–233

Miller JR, Lechler PJ, Desilets M (1998) The role of geomorphic processes in the transport and fate of mercury within the Carson River Basin, West-Central Nevada. Environmental Geology 33:249–262

Miller JR, Lechler PJ, Rowland J, Desilets M, Hsu LC (1996) Dispersal of mercurycontaminated sediments by fluvial geomorphic processes: A case study from Six Mile Canyon, Nevada, USA. Water, Air, and Soil Pollution, 82:1–16

Miller JR, Barr R, Grow D, Lechler P, Richardson D, Waltman K, Warwick J (1999) Effects of the 1997 flood on the transport and storage of sediment and mercury within the Carson River Valley, west-central Nevada. The Journal of Geology 107:313–327

Miller JR, Lechler PJ, Hudson-Edwards KA, Macklin MG (2002) Lead isotopic fingerprinting of heavy metal contamination, Río Pilcomayo basin, Bolivia. Geochemistry: Exploration, Environment, Analysis 2:225–233

Miller JR, Lechler PJ, Bridge G (2003) Mercury contamination of alluvial sediments within the Essequibo and Mazaruni River Basins, Guyana. Water, Air, and Soil Pollution 148:139–166

Miller J, Lord M, Yurkovich S, Mackin G, Kolenbrander L (2005) Historical trends in sedimentation rates and sediment provenance, Fairfield Lake, western North Carolina. Journal of American Water Resources Association 41:1053–1075

Millroy SJ (1995) Science-based risk assessment: a piece of the superfund puzzle. National Environmental Policy Institute, Washington, DC

Montgomery DR (1997) The influence of geological processes on ecological systems. In: Rain forests of home: profile of a North American bioregion, Island Press, Washington, DC, pp 43–68

Moody J, Pizzuto J, Meade R (1999) Ontongeny of a floodplain. Geological Society of American Bulletin 111:291–303

Moor HC, Schaller T, Sturm M (1996) Recent changes in stable lead isotope ratios in sediments of Lake Zug, Switzerland. Environmental Science and Technology 30:2928–2933

Moore JN, Brook EJ, Johns C (1989) Grain size partitioning of metals in contaminated, coarse-grained river floodplain sediments: Clark Fork River, Montana, USA. Environmental Geology and Water Science 14:107–115

Mount JF (1995) California rivers and streams: the conflict between fluvial processes and land use. University of California Press, Berkley

Mudroch A, MacKnight SD (1991) CRC handbook of techniques for aquatic sediment sampling. CRC Press, Ann Arbor

Müller K, Deurer M, Hartmann H, Back M, Spiteller M, Frede HG (2003) Hydrological characterization of pesticide loads using hydrograph separation at different scales in a German catchment. Journal of Hydrology 273:1–17

Murray AS, Olley JM (2002) Percision and accuracy in the optically stimulated luminescence dating of sedimentary quartz: a status review. Geochronometria 21:1–16

Nagano T, Yanase N, Tsuduki K, Nagao S (2003) Particulate and dissolved elemental loads in the Kuji River related to discharge rate. Environment International 28:649–658

Nagorski SA, Moore JN, McKinnon TE (2003) Geochemical response to variable streamflow conditions in contaminated and uncontaminated streams. Water Resources Research 39:1–13

Naiman RJ, Décamps H (1997) The ecology of interfaces: riparian zones. Annual Review of Ecology and Systematics 28:621–658

Nanson GC (1986) Episodes of vertical accretion and catastrophic stripping: a model of disequilibrium flood-plain development. Bulletin of the American Geological Society 97:1467–1475

Nanson GC, Young RW (1981) Overbank deposition and floodplain formation on small coastal streams of New South Wales. Zeitschrift fur Geomorphologie 25:332–345

Nanson GC, Young RW (1987) Comparison of thermoluminescence and radiocarbon age-determinations from late-Pleistocene alluvial deposits near Sydney, Australia. Quaternary Research 27:263–269

Nanson GC, Croke J (1992) A genetic classification of floodplains. Geomorphology 4:459–486

Nanson GC, Knighton AD (1996) Anabranching rivers: their cause, character and classification. Earth Surface Processes and Landforms 21:217–239

Nanson GC, Price, DM, Short SA (1991) Comparative uranium-thorium and thermoluminescence dating of weathered Quaternary alluvium in the tropics of Northern Australia. Quaternary Research 35:347–366

Nash DB (1994) Effective sediment-transporting discharge from magnitude-frequency analysis. Journal of Geology 102:79–95

Neale CN, Bricka RM, Chao AC (1997) Evaluating acids and chelating agents for removing heavy metals form contaminated soils. Environmental Progress 16:274–280

Ng HYF, Clegg SB (1997) Atrazine and metolachlor losses in runoff events from an agricultural watershed: the importance of runoff components. The Science of the Total Environment 193:215–228

Nicholas AP, Ashworth PJ, Kirkby MJ, Macklin MG, Murray T (1995) Sediment slugs: large-scale fluctuations in fluvial sediment transport rates and storage volumes. Progress in Physical Geography 19:500–519

Norvell WA (1991) Reactions of metal chelates in soils and nutrient soils. In: Mortvedt JJ, Cox FR, Shuman LM, Welch RM (ed) Micronutrients in Agriculture, 2nd edn, Soil Science Society of America, Madison

NRC (1992) Restoration of aquatic ecosystems. National Research Council, The National Academy Press, Washington, DC

NRC (1997) Contaminated sediments in ports and waterways: cleanup strategies and technologies. National Research Council, Committee on Contaminated Marine Sediment, The National Academy Press, Washington, DC

NRC (1997b) Innovations in ground water and soil cleanup: from concept to commercialization. National Research Council, National Academy Press, Washington, DC

NRC (2003) Environmental cleanup at navy facilities: adaptive site management. National Research Council, The National Academies Press, Washington, DC

NRC (2005) Superfund and mining megasites: lessons from the Coeur d'Alene River Basin. National Research Council, The National Academies Press, Washington, DC

Nriagu JO (1994) Mercury pollution from the past mining of gold and silver in the Americas. Science of the Total Environment 149:167–181

Odgaard AJ, Jain SC, Luzbetak DJ (1989) Hydraulic mechanisms of riverbank erosion. In: Proceedings of the national conference on hydraulic engineering, New Orleans, August 14–18

Office of Water Data Coordination (1982) National handbook of recommended methods for water data acquisition, Chapter 5. US Geological Survey, Reston VA 5-10-5-11

Olías M, Nieto JM, Sarmiento AM, Cerón JC, Cánovas CR (2004) Seasonal water quality variations in a river affected by acid mine drainage: the Odiel River (South West Spain). Science of the Total Environment 333:267–281

Oliver B (1973) Heavy metal levels of Ottawa and Rideau River sediment. Environmental Science and Technology 7:135–137

Olley J, Caitcheon GG, Roberts RG (1999) The origin of dose distribution in fluvial sediments and the prospects of dating single grains from fluvial deposits using optically stimulated luminescence. Radiation Measurements 30:207–217

Olley JM, Pietsch T, Roberts RG (2004) Optical dating of Holocene sediments from a variety of geomorphic settings using single grains of quartz. Geomorphology 60:337–358

Orbock Miller S, Ritter DF, Kochel R, Miller JR (1993) Fluvial response to land-use changes and climatic variations within the Drury Creek watershed, southern Illinois. Geomorphology 6:309–329

Ottesen RT, Bogen J, Bølviken B, Volden T (1989) Overbank sediment: a representative sample medium for regional geochemical mapping. Journal of Geochemical Exploration 32:257–277

Otto DME, Buttner JK, Arquette DM, Moon TW (1996) Impaired inducibility of xenobiotic and antioxidant responses in rainbow trout to polychlorinated biphenyl contaminated sediments in the St Lawrence River. Chemospher 33:2021–2032

Owens PN, Walling DE, Leeks GJL (1999) The use of floodplain sediment cores to investigate recent historical changes in overbank sedimentation rates and sediment sources in the catchment of the River Ouse, Yorkshire, UK. Catena 36:21–47

Owens PN, Walling DE, Leeks GJL (1999b) Deposition and storage of fine-grained sediment within the main channel system of the river Tweed, Scotland. Earth Surface Processes and Landforms 24:1061–1076

Owens PN, Walling DE, Carton J, Meharg AA, Wright J, Leeks GJL (2001) Downstream changes in the transport and storage of sediment-associated contaminants (P, Cr and PCBs) in agricultural and industrialized drainage basins. The Science of the Total Environment 206:177–186

Paine DP (1981) Aerial photography and image interpretation for resource management. Wiley, New York

Palermo MR (1991) Design requirements for capping. US Army Corp of Engineers, Dredging Research Technical Notes, DRP-05-03, Vicksburg, MS

Palermo MR, Fracingues NR, Averett DE (2004) Operational characteristics and equipment selection factors for environmental dredging. Journal of Dredging Engineering, Western Dredging Association 15(4)

Palmer MA, Bernhardt ES, Allan JD, Lake PS, Alexander G, Brooks S, Carr J, Clayton S, Dahm CN, Follstad Shan J, Galat DL, Loss SG, Kondolf GM, Lave R, Meyer JL, O'Donnell TK, Pagano L, Sudduth E (2005) Standards for ecologically successful river restoration. Journal of Applied Ecology 42:208–217

Paola C, Seal R (1995) Grain size patchiness as a cause of selective deposition and downstream fining. Water Resources Research 31:1395–1407

Parker G, Klingeman PC, McLean DG (1982) Bedload and size distribution in paved gravel-bed streams. In: Proceedings of the American society of civil engineers, Journal Hydraulics Division, HY4 108:544–571

Parker RS (1976) Experimental study of drainage system evolution: unpublished report, Colorado State University

Parr AD, Richardson C, Lane DD, Baughman D (1987) Pore water uptake by agricultural runoff. Journal of Environmental Engineering Division, American Society of Civil Engineering 113:49–63

Patton PC, Schumm SA (1975) Gully erosion, northwestern Colorado: a threshold phenomenon. Geology 3:88–90

Peart MR, Walling DE (1986) Fingerprinting sediment source: the example of a drainage basin in Devon, UK. In: Hadley RF (ed) Drainage basin sediment delivery. IAHS-AISH Publication 159

Perkins RW, Thomas CW (1980) Worldwide fallout. In: Hanson WC (ed) Transuranic elements in the environment. USDOE/TIC-22800, Washington, DC, US Department of Energy, pp 53–82

Persaud D, Jaagumagi R, Hayton A (1993) Guidelines for the protection and management of aquatic sediment quality in Ontario. Water Resources Branch, Ontario Ministry of the Environment, Toronto, 27pp

Peters RW (1999) Chelant extraction of heavy metals from contaminated soils. Journal of Hazardous Materials 66:151–210

Peterson G, Wolfe J, DePinto J (1999) Decision tree for sediment management. Sediment Management Working Group Technical Paper, *http://www.smwg.org*, 37pp

Petts J, Cairney T, Smith M (1997) Risk-based contaminated land investigation and assessment. John Wiley and Sons, Chichester

Petelet EB, Ben Othman D, Luck JM (1995) Trace element, Pb isotope constraints on the natural vs anthropogenic sources of matter transfer in the dissolved and suspended matter of a Mediterranean river (Véne, S France). EUG 7:247

Petelet E, Luck JM, Négrel P, Ben Othman D (1996) Global geochemical approach to an hydrological problem: major, trace isotopic (H_2, O_2, Sr, Pb) constraints on surficial/groundwater interaction in the Hérault watershed (Southern France). Journal Conference Abstracts 1:460

Petit D, Mennesier JP, Lamberts L (1984) Stable Pb isotopes in pond sediments as tracer of past and present atmospheric Pb pollution in Belgium. Atmospheric Environment 18:1189–1193

Phillips J, Renwick W (1992) Geomorphic systems. In: Proceedings of the 23 Binghamton Symposium, Elsevier, New York

Pilkey OH, Thieler ER (1996) Mathematical modeling in coastal geology. Geotimes 41:5

Pitlick J (1993) Response and recovery of a subalpine stream following a catastrophic flood. Geological Society of America Bulletin 105:657–670

Pizzuto JE (1984) Bank erodibility of shallow sandbed streams. Earth Surface Processes and Landforms 9:113–124

Pizzuto JE (1987) Sediment diffusion during overbank flows. Sedimentology 34:301–317

Popp CL, Hawley JW, Love DW, Dehn M (1988) Use of radiometric (Cs-137, Pb-210), geomorphic and stratigraphic echniques to date recent oxbow sediments in the Rio Puerco drainage, Grants Uranium region, New Mexico. Environmental Geology and Water Science 11:253–269

Preston SD, Bierman VJ Jr, Silliman SE (1989) An evaluation of methods for the estimation of tributary mass loads. Water Resources Research 25:1379–1389

Probstein RF, Hicks RE (1993) Removal of contaminants from soils by electric fields. Science 260:498

Prosser IP, Rustomji P, Young WJ, Moran CJ, Hughes A (2001) Constructing river bain sediment budgets for the national land and water resources audit. CSIRO Land and Water Technical Report 15/01, CSIRO, Canberra

Provost LP (1984) Statistical methods in environmental sampling. In: Schweitzer EG, Santolucito JA (eds) Environmental sampling for hazardous wastes. ACS Symposium Series 267, America Chemical Society, Washington, DC

Raghavan R, Coles E, Dietz D (1991) Cleaning excavated soil using extraction agents: a state-of-the-art review. Journal of Hazardous Materials 26:81–87

Rang MC, Schouten CJ (1989) Evidence for historical heavy metal pollution in floodplain soils: the Meuse. In: Petts GE (ed) Historical change of large alluvial rivers: Western Europe. John Wiley and Sons, pp 127–142

Rantz SE et al (1982) Measurement and computation of streamflow: vol 2, Computation of discharge. US Geological Survey Water-Supply Paper 2175, Washington, DC

Reading HG (1978) Sedimentary environments and facies. Blackwell

Reimann C, Garrett RG (2005) Geochemical background-concept and reality. Science of the Total Environment 350:12–27

Reimann C, Filzmoser P, Garrett RG (2005) Background and threshold: critical comparison of methods of determination. Science of the Total Environment 346:1–16

Reineck HE, Singh IB (1980) Depositional sedimentary environments with reference to terrigenous clastics, 2 edn. Springer-Vergla, New York

Rem P, Marc P, van Hoek C, van Kooy L (2002) A decision supported system for wet soil cleaning. Environmental Pollution 120:717–723

Renard KG, Foster GA, Weesies DK, McCool DK, Yoder DC (1997) Predicting soil erosion by water: a guide to conservation planning with the revised universal soil loss equation. Agriculture Handbook 703, USDA, Washington, DC

RETEC (2002) Final feasibility study: Lower Fox River and Green Bay, Wisconsin remedial investigation and feasibility study. The RETEC Group, Inc, Appendix B

Reuter JH, Perdue EM (1977) Importance of heavy metal-organic matter interactions in natural waters. Geochimica, Cosmochimica, Acta 41:325–334

Rice CP, White DS (1987) PCB availability assessment of river dredging using caged clams and fish. Environmental Toxicology and Chemistry 6:259–274

Rice KC (1999) Trace-element concentrations in streambed sediment across the conterminous United States. Environmental Science and Technology 33:2499–2504

Richards RP (2001) Estimation of pollutant loads in rivers and streams: a guidance document for NPS programs. Report to US Environmental Protection Agency, Grant X998397-01-0, Region VIII

Ritchie JC, McHenry JR (1990) Application of radioactive fallout caesium-137 for measuring soil erosion and sediment accumulation rates and patterns: a review. Journal of Environmental Quality 19:215–233

Ritchie JC, Ritchie CA (2005) Bibliography of publications of 137-caesium studies related to erosion and sediment deposition. USDA-ARS Occasional Paper HRSL-2005-01

Ritter DF (1978) Process geomorphology. Wm C Brown, Dubuque

Ritter DF (1986) Process geomorphology, 2nd edn. Wm C Brown Publishers, Dubuque, Iowa

Ritter DF (1988) Floodplain erosion and deposition during the December 1982 floods in southeast Missouri. In: Baker V, Kochel RC, Patton PC (eds) Flood geomorphology. John Wiley and Sons, New York, pp 243–259

Ritter DF, Kinsey WF, Kauffman ME (1973) Overbank sedimentation in the Delaware River valley during the last 6,000 years. Science 179:374–375

Ritter DF, Kochel RC, Miller JR (1999) The disruption of Grassy Creek: implications concerning catastrophic events and thresholds. Geomorphology 29:323–338

Ritter DF, Kochel RC, Miller JR (2002) Process geomorphology, 4nd edn. McGraw-Hill, New York

Robbins JA (1978) Geochemical and geophysical applications of radioactive lead. In: Nriagu JO (ed) Biogeochemistry of Lead in the Environment, Elsevier Scientific, Amsterdam, pp 285–393

Roberge J, Plamondon AP (1987) Snowmelt runoff pathway in a boreal forest hillslope, the role of pipe throughflow. Journal of Hydrology 95:39–54

Rogowski AS, Tamura T (1965) Movement of ^{137}Cs by runoff, erosion and infiltration on the alluvial Captina silt loam. Health Physics 11:1333–1340

Romagnoli R, Doody JP, VanDewalker HM, Hill SA (2002) Environmental dredging effectiveness: lessons learned. In: Porta A, Hinchee RE, Pellei M (eds) Management of contaminated sediment. Battelle Press, Columbus, pp 181–188

Roni P (2005) Monitoring stream and watershed restoration. American Fisheries Society, Bethesda

Rose AW, Suhr NH (1971) Major element content as a means of allowing for background variation in stream-sediment geochemical exploration. In: Boyle RW (ed) Geochemical exploration. Canadian Institute of Mining and Minerals, Special Volume 11

Rosgen DL (1996) Applied river morphology. Wildland Hydrology, Colorado

Rossman KJR, Chisholm W, Boutron CF, Candelone JP, Gorlach U (1993) Isotopic evidence for the source of lead in Greenland snow since the late 1960's. Nature 362:333–335

Rowan JS, Goodwill P, Franks SW (1999) Floodplain evolution and sediment provenance reconstructed from channel fill sequences: the Upper Clyde basin, Scotland. In: Brown AG, Quine TA (eds) Fluvial processes and environmental change. John Wiley and Sons, Chichester

Rumsby B (2000) Vertical accretion rates in fluvial systems: a comparison of volumetric and depth-based estimates. Earth Surface Processes and Landforms 25:617–631

Russell M, Colglazier EW, English MR (1991) Hazardous waste remediation: the task ahead. Waste Management Research Institute, The University of Tennessee, Knoxville, 41pp

Russell MA, Walling DE, Hodgkinson RA (2001) Suspended sediment sources in two small lowland agricultural catchments in the UK. Journal of Hydrology 252:1–24

Salomons W, Eysink Q (1981) Pathways of mud and particulate trace metals from rivers to the southern North Sea Basin. Special Publication of the International association of Sedimentology 5:429–450

Salomons W, Förstner U (1984) Metals in the hydrocycle. Springer-Verlag, Berlin

Salomons W, Stigliani WM (1995) Biogeodynamics of pollutants in soil and sediments: risk assessment of delayed and non-linear responses, Springer, Berlin

Salt DE, Prince RC, Pickering IJ, Raskin I (1995) Mechanisms of cadmium mobility and accumulation of Indian mustard. Plant Physiology 109:1427–1433

Salt DE, Pickering IJ, Prince RC, Gleba D, Dushenkov S, Smith RD, Raskin I (1997) Metal accumulation by aquacultured seedlings of Indian mustard. Environmental Science and Technology 31: 1636–1644

Sanders T, Ward R, Loftis J, Steele T, Adrian D, Yevjevich V (1983) Testing of networks for monitoring water quality. Water Resources Publications, Littleton, CO

Schulin R, Geiger G, Furrer G (1995) Heavy metal retention by soil organic matter under changing environmental conditions. In: Salomons W, Stigliani WM (eds) Biogeodynamics of pollutants in soil and sediments: risk assessment of delayed and non-linear responses. Springer, Berlin, pp 53–85

Schumm SA (1965) Quaternary paleohydrology. In: Wright HE, Frey DG (eds) The Quaternary of the United States. Princeton University Press, Princeton, pp 783–794

Schumm SA (1968) River adjustments to altered hydrologic regime, Murrumbidgee River and paleochannels, Australia. US Geological Survey Professional Paper 598

Schumm SA (1969) River metamorphosis. Journal of the Hydraulics Division, American Society of Civil Engineers, vol 95, HY1, pp 255–273

Schumm SA (1973) Geomorphic thresholds and complex response of drainage systems. In: Morisawa M (ed) Fluvial geomorphology. New York State University Publications in Geomorphology, Binghamton, New York, pp 299–309

Schumm SA (1977) The fluvial system. John Wiley and Sons, New York

Schumm SA (1981) Evolution and response of the fluvial system; sedimentologic implications. In: Ethridge F, Flores R (eds) Recent and ancient nonmarine depositional environments: models for exploration. Society of Economic Paleontologist and Mineralogists Special Publication 31:19–39

Schumm SA (1985) Patterns of alluvial rivers. Annual Review of Earth and Planetary Sciences 13:5–27

Schumm SA, Brakenridge GR (1987) River responses. In: Ruddiman WF, Wright HE Jr (eds) North America and adjacent oceans during the last deglaciation. Geological Society of America, vol K-3, Boulder, Colorado

Schumm SA, Khan HR (1972) Experimental study of channel patterns. Geological Society of America Bulletin 83:1755–1770

Schumm SA, Lichty RW (1965) Time, space, and causality in geomorphology. American Journal of Science 263:110–111

Schumm SA, Parker RS (1973) Implications of complex response of drainage systems for Quaternary alluvial stratigraphy. Natural Physical Science 243:99–100

Scott MI, Auble GT, Friedman JM (1997) Flood dependency of cottonwood establishment along the Missouri River, Montana, USA. Ecological Applications 7:677–690

Sear DA (1994) River restoration and geomorphology. Aquatic Conservation: Marine and Freshwater Ecosystems 4:169–177

Sear D, Carver S (1996) The release and dispersal of Pb and Zn contaminated sediments within an Arctic braided river system. Applied Geochemistry 11:187–195

Seidel H, Loser C, Zehnsdorf A, Hoffman P, Schmerold R, (2004) Bioremediation processes for sediments contaminated by heavy metals: feasibility study on a plot scale. Environmental Science and Technology 38:1582–1588

Sharpley, Williams JR (1990) EPIC-Erosion/Productivity Impact Calculator: 1. Model Documentation, USDA Tech Bull # 1768.12.2

Sherrell RM, Ross JM (1999) Temporal variability of trace metals in New Jersey Pinelands streams: relationships to discharge and pH. Geochimica Cosmochimca Acta 63:3321–3336

Shields FD Jr, Copeland RR, Klingeman PC, Doyle MW, Simon A (2003) Design for stream restoration. Journal of Hydraulic Engineering 129:575–584

Shiller AM (1997) Dissolved trace elements n the Mississippi River: seasonal, interannual, and decadal variability. Geochemica et Cosmochimica Acta 61:4321–4330

Shiono K, Knight DW (1991) Turbulent open channel flows with variable depth across the channel. Journal of Fluid Mechanics 22:617–646

Shirahata H, Elias RW, Patterson C (1980) Chronological variations in concentrations and isotopic compositions and anthropogenic atmospheric Pb in sediments of a remote subalpine pond. Geochimica Cosmochimca Acta 44:149–167

Shotyk W, Weiss D, Appleby PG, Cheburkin AK, Frei R, Gloor M, Kramers JD, Reese S, Knaap WOVD (1998) History of atmospheric lead deposition since 12,370 14 C yr BP from a peat bog, Jura Mountains, Switzerland. Science 281:1635–1640

Simon A, Curini A (1998) Pore pressure and bank stability: the influence of matric suction. In: Abt SR (ed) Water Resources in Engineering, vol 1, American Society of Civil Engineers, New York, pp 358–363

Simon A, Curini A, Darby S, Langendoen EJ (1999) Streambank mechanics and the role of bank and near-bank processes in incised channels. In: Darby SE, Simon A (eds) Incised river channels, John Wiley and Sons, Chichester, pp 123–151

Skidmore PB, Shields FD, Doyle MW, Miller DE (2001) A categorization of approaches to natural channel design. In: Hayes DF (ed) Proceedings of the 2001 Wetlands Engineering and River Restoration Conference, CD-ROM, American Society of Civil Engineers, Reston, VA

Slaney P, Millar R, Koning W, D'Aoust D (2000) Effectiveness of large wood restoration at a large steram in the southern interior of British Columbia. International Conference on Wood in Rivers, Oregan State University, October, 2000

Slattery MC, Burt TP, Walden J (1995) The application of mineral magnetic measurements to quantify within-storm variations in suspended sediment sources. In: Leibundgut CH (ed) Tracer technologies for hydrological systems. IAHS Publication No 229, IAHS Press, Wallingford, pp 143–151

Slingerland R, Smith ND (1986) Occurrence and formation of water-laid placers. Annual Review of Earth and Planet Science 14:113–147

Smal H (1993) Enviromental conditions and chemical time bomb hazards in Poland. Land Degradation and Rehabilitation 4:269–274

Smith DG (1986) Anastomosing river deposits, sedimentation rates and basin subsidence, Magdalena River, northwestern Colombia, South America. Sedimentary Geology 46:177–196

Smith DG, Smith ND (1976) Sedimentation in anastamosed river systems: examples from alluvial valleys near Banff, Alberta. Journal of Sedimentary Petrology 50:157–164

Smith ND (1970) The braided stream depositional environment: comparison of the Platte River with some Silurian clastic rocks, north-central Appalachians. Geological Society of America Bulletin 81:2993–3014

Smith ND (1974) Sedimentology and bar formation in the upper Kicking Horse River, a braided outwash stream. Journal of Geology 82:205–223

Smith ND, Beukes N (1983) Bar to bank flow convergence zones: a contribution to the Origin of alluvial placers. Economic Geology 78:1342–1349

Smith ND, Cross TA, Dufficy JP, Clough SR (1989) Anatomy of an avulsion. Sedimentology 36:1–23

Smith SL, MacDonald DD, Keenleyside KA, Ingersoll CG, Field J (1996) A preliminary evaluation of sediment quality assessment values for freshwater ecosystems. Journal of Great Lakes Research 22:624–638

Smol JP (2002) Pollution of lake and rivers: a paleoenvironmental perspective. Arnold, London

Soesilo JA, Wilson SR (1997) Site remediation: planning and management. Lewis Publishers, CRC Press, Inc, Boca Raton

Sorensen JA, Glass GE, Schmidt KW, Huber JK, Rapp GR Jr (1990) Airborne mercury deposition and watershed characteristics in relation to mercury concentrations in water, sediment plankton, and fish of eighty northern Minnesota lakes. Environmental Science and Technology 24:1716–1727

Squillance PJ, Thurman EM (1992) Herbicide transport in rivers: importance of hydrology and geochemistry in nonpoint-source contamination. Environmental Science and Technology 26:538–545

Stanley DJ, Krinitzsky EI, Compton JR (1966) Mississippi River bank failure, Fort Jackson, Louisiana. Geological Society of America Bulletin 77:850–866

Statzner B, Gore JA, Resh VH (1988) Hydraulic stream ecology: observed patterns and potential application. Journal of the North American Benthological Society 4:307–360

Steding DJ, Dunlap CE, Flegal AR (2000) New isotopic evidence for chronic lead contamination in the San Francisco Bay estuary system: implications for the persistence of past industrial lead emissions in the biosphere. Proceedings of the National Academia of Science 97:11181–11186

Steele MC, Pichtel J (1998) Ex-situ remediation of a metal-contaminated superfund soil using selective extractants. Journal of Environmental Engineering ASCE 124:639–645

Steiger J, Gurnell AM (2002) Spatial hydrogeomorphological influences on sediment and nutrient deposition in riparian zones: observations from the Garonne River, France. Geomorphology 49:1–23

Steiger J, Gurnell AM, Petts GE (2001) Sediment deposition along the channel margins of a reach of the middle River Stevern, UK. Regulated Rivers: Research Management 17:441–458

Steinmann M, Stille P (1997) Rare earth element behavior and Pb, Sr, Nd isotope systematics in a heavy metal contaminated soil. Applied Geochemistry 12:607–632

Steuer JJ (2000) A mass-balance approach for assessing PCB movement during remediation of a PCB-contaminated deposit on the Fox River, Wisconsin. USGS Water-Resources Investigations Report 00-4245

Stigliani WM (1991) Chemical time bombs: definition, concepts, and examples. Executive Report 16, International Institute of Applied Systems Analysis, Laxenburg, Austria

Stigliani WM, Doelman P, Salomons W, Schulin R, Smidt GRG, Van der Zee SEATM (1991) Chemical time bombs. Environment 33:5–30

Stokes MA, Smiley TL (1968) An introduction to tree-ring dating. University of Chicago Press, Chicago

Stokes S, Walling DE (2003) Radiogenic and isotopic methods for the direct dating of fluvial sediments. In: Kondolf MG, Piegay H (eds) Tools in fluvial geomorphology, John Wiley and Sons, Chichester, pp. 233–267

Stone AT (1987) Microbial metabolites and the reductive dissolution of manganese oxides: Oxalate and pyruvate. Geochimica Coomochimca Acta 51:919–925

Stuiver M, Reimer PJ, Bard E, Beck JW, Burr GS, Hughen KA, Kromer B, McCormac G, vander Plicht J, Spurk M (1998) INTCAL98 Radiocarbon age calibration, 24,000-0 cal BP. Radiocarbon 40:1041–1083

Su SH, Pearlman LC, Rothrock JA, Iannuzzi TJ, Finley BL (2002) Potential long-term ecological impacts caused by disturbance of contaminated sediments: a case study. Environmental Management 29:234–249

Sullivan AN, Drever JI, McKnight DM (1998) Diel variation in element concentrations, Peru Creek, Summit County, Colorado. Journal of Geochemical Exploration 64:141–145

Stumm W (1992) Chemistry of the water-sediment interface. Wiley-Interscience, New York

Taylor MP (1996) The variability of heavy metals in floodplain sediments: a case study from mid Wales. Catena 28:71–87

Taylor MP, Kesterton RGH (2002) Heavy metal contamination of an arid river environment: Gruben River, Namibia. Geomorphology, 42:311–327

Taylor MP, Lewin J (1996) river behaviour and Holocene alleviation: the river Severn at Welshpool, mid-Wales, U.K. Earth Surface Processes and Landforms 21:77–91

Taylor G, Woodyer KD (1978) Bank deposition in suspended-load streams. In: Miall AD (ed) Fluvial sedimentology. Canadian Society of Petroleum Geologists Memoir 5:257–275

Taylor SR (1964) The abundance of chemical elements in continental crust – a new table. Geochimica Cosmochimca Acta 28:1273–1285

Tessier A, Campbell GC, Bison M (1979) Sequential extraction procedure for the speciation of particulate trace metals. Analytical Chemistry 51:844–851

Tessier A, Campbell PGC, Bisson M (1982) Particulate trace metal speciation in stream sediments and relationships with grain size: implications for geochemical exploration. Journal of Geochemical Exploration 16:77–104

Tessier A, Campbell GC (1988) Partitioning of trace metals in sediments. In: Kramer JR, Allen HE (eds) Metal speciation: theory, analysis, and application. Lewis Publishers, Chelsa, Michigan, pp 183–199

Tessier A, Turner DR (1995) Metal speciation and bioavailability in aquatic systems. John Wiley and Sons, Chichester

Thompson A (1986) Secondary flows and the pool-riffle unit: a case study of the processes of meander development. Earth Surface Processes and Landforms 11:631–641

Thompson DM (2002) Long-term effect of instream habitat-improvement structures on channel morphology along the Blackledge and Salmon Rivers, Connecticut, USA. Environmental Management 29:250–265

Thompson DM, Wohl EE, Jarrett RD (1996) A revised velocity-reversal and sediment-sorting model for a high-gradient, pool-riffle stream. Physical Geography 17:142–156

Thompson DM, Nelson JM, Wohl EE (1998) Interactions between pool geometry and hydraulics. Water Resources Research 12:3673–3681

Thorn C, Welford M (1994) The equilibrium concept in geomorphology. Annals of the Association of American Geographers 84:666–696

Thorne CR (1982) Processes and mechanisms of river bank erosion. In: Hey R, Bathurst J, Thorne C (eds) Gravel-bed rivers. John Wiley and Sons, Chichester, pp 1–37

Thorne CR, Tovey NK (1981) Stability of composite river banks. Earth Surface Processes and Landforms 6:469–484

Thorne CR, Hey RD, Newson MD (1997) Applied fluvial geomorphology for river engineering and management. John Wiley and Sons, New York

Thorne CR, Lewin J (1979) Bank processes, bed material movement, and planform development in a meandering river. In: Rhodes DD, Williams GP (eds) Adjustments of the fluvial system. Kendal Hunt, Dubuque, pp 117–137

Thorne CR, Russell APG, Alam MK (1993) Planform pattern and channel evolution of the Brahmaputra River, Bangladesh. In: Best JL, Bristow CS (eds) Braided rivers. Special Publication of the Geological Society of London 75:257–276

Trimble SW (1976) A sediment budget for Coon Creek basin in the Driftless Area, Wisconsin, 1853–1977. American Journal of Science 283:454–474

Turekian KK (1971) Elements, geochemical distribution of. Encyclopedia of Science and Technology, vol 4, McGraw-Hill, New York, pp 627–630

Turekian KK, Wedepohl KH (1961) Distribution of the elements in some major units of the earth's crust. Bulletin of the Geological Society of America 72:175–192

Turner DR (1995) Problems in trace metal speciation modeling. In: Tessier A, Turner DR (eds) Metal speciation and bioavailability in aquatic systems. John Wiley and Sons, Chichester, pp 149–203

Turney CSM, Lowe JJ (2001) Tephrochronology. In: Last WM, Smol JP (eds) Tracking environmental change using lake sediments, vol 1. Basin analysis, coring, and chronological techniques. Kluwer Academic Publishers, Dordrecht, pp 451–471

Twidale CR (1964) Erosion of an alluvial bank at Birdwood, South Australia. Zeitschrift fur Geomorphologie 8:189–211

UKEA (2003) Facts and figures, river quality: an overview. United Kingdom Environmental Agency, http://www. environment—agency.gov.uk/

Ullrich CR, Hagerty DJ, Holmberg RW (1986) Surficial failures of alluvial stream banks. Canadian Geotechnical Journal 23:304–316

UNPD (2001) World population prospects: the 2000 revision. United Nations Department of Economic and Social Affairs, Population Division, United Nations Publications, New York

USACE (1987) Confining disposal of dredged material. US Army Corps of Engineers, Engineering Manual 1110-2-5027

USEPA (1990) Engineering bulletin: Soil washing treatment. EPA/540/2-90/017

USEPA (1988) Guidance for conducting remedial investigations and feasibility Studies under CERCLA. EPA/540/G-89/004

USEPA (1991a) The superfund innovative technology program, technological profiles. EPA540/5-91-008

USEPA (1991b) Engineering Bulletin: In situ soil flushing. EPA/540/2-91/021

USEPA (1992) EPA handbook: vitrification technologies for treatment of hazardous and radioactive waste. EPA/625/R-92/002

USEPA (1993) Selecting remediation techniques for contaminated sediment. EPA-823-B93-001

USEPA (1994) Feasibility study analysis for CERCLA municipal landfill sites. EPA540/R-94/081

USEPA (1995) In situ remediation technology. EPA542-K-94-007

USEPA (1996a) Calculation and evaluation of sediment effect concentrations for the amphipod *Hyalella azteca* and the midge *Chironomus riparius*. EPA 905-R96-008, Great Lakes National Program Office, Region V, Chicago

USEPA (1997) The incidence and severity of sediment contamination in surface waters of the United States, vol 1, The national sediment quality survey. EPA-823-R-97-006

USEPA (1998) EPA's Contaminated sediment management strategy. EPA-823-R98-001

USEPA (2000) National water quality inventory: 2000 report. EPA-841-R-02-001

USEPA (2002) Guidance for comparing background and chemical concentrations in soil for CERCLA sites. EPA 540-R-01-03

USEPA (2005) Contaminated sediment remediation guidance for hazardous waste sites. EPA-540-R-05-012

USGS (1999) The quality of our Nation's waters: nutrients and pesticides. US Geological Survey Circular 1225

USGS (2001) The national water-quality assessment program: informing water-resources management decisions (written by Hamilton PA, Miller, TL), *http://water.usgs.gov/nawqu/docs/xrel*

Vangronsveld J, Cunningham SC (1998) Introduction to the concepts. In: Vangronsveld J, Cunningham SD (eds) Metal-contaminated soils: in situ inactivation and phytorestoration. Springer-Verlag and RG Landes Company, Berlin, pp 1–15

Várallyay G, Salomons W, Czíkós I (1993) Long-term environmental risk for soils, groundwaters and sediments in the Danube catchment area: the Danube chemical time bombs project. Land Degradation and Rehabilitation 4:421–452

Vehik R (1982) The archaeology of the Bug Hill site (34PU116), Pushmataha County, Oklahoma. Archeological Research and Management Center, University of Oklahoma, Norman, OK, Research Series 77

Viganò L et al (2003) Quality assessment of bed sediments of the Po River (Italy). Water Research 37:501–518

Villarroel LF, Miller JR, Lechler PJ, Germanoski D (2006) Lead, zinc, and antimony contamination of the Rio Chilco-Rio Tupiza drainage system, southern Bolivia. Environmental Geology 50:1270–1287

Vink R, Peters S (2003) Modelling point and diffuse heavy metal emissions and loads in the Elbe basin. Hydrological Processes 17:1307–1328

Waal LC, Large ARG, Wade PW (1998) Rehabilitation of rivers: principles and implementation. John Wiley and Sons, Chichester

Wade PW, Large ARG, De Waal LC (1998) Rehabilitation of degraded river habitat: an introduction. In: Waal LC, Large ARG, Wade PW (eds) Rehabilitation of rivers: principles and implementation. John Wiley and Sons, Chichester

Wallach R, Jury WA, Spencer WF (1988) Transfer of chemicals from soil solution to surface runoff: a diffusion-based model. Soil Science Society of America Journal 52:612–618

Wallach R, Grigorin G, Rivlin J (2001) A comprehensive mathematical model for transport of soil-dissolved chemical by overland flow. Journal of Hydrology 247:85–99

Wallach R, Shabtai R (1992) Modeling surface runoff contamination by soil-applied chemical under transient infiltration. Journal of Hydrology 132:263–281

Wallbrink PJ, Murray AS (1996) Determining soil loss using the inventory ratio of excess lead-210 to cesium-137. Soil Science Society of America Journal 60:1201–1208

Wallbrink PJ, Olley JM, Murray AS (1994) Measuring soil movement using ^{137}Cs: implications of reference site variability. IAHS Publication 224:95–105

Walling DE, He Q (1998) The spatial variability of overbank sedimentation on river floodplains. Geomorphology 24:209–223

Walling DE, He Q, Nicholas AP (1996) Floodplains as suspended sediment sinks. In: Anderson MG, Walling DE, Bates PD (eds) Floodplain processes. John Wiley and Sons, Chichester, pp 399–440

Walling DE, Owen PN, Leeks GJL (1998) The role of channel and floodplain storage in the suspended sediment budget of the River Ouse, Yorkshire, UK. Geomorphology 22:225–242

Walling DE, Owens PN, Leeks GJL (1999) Fingerprinting suspended sediment sources in the catchment of the River Ouse, Yorkshire, UK. Hydrological Processes 13:955–975

Walling DE, Owens PN, Carter J, Leeks GJL, Lewis S, Meharg AA, Wright J (2003) Storage of sediment-associated nutrients and contaminants in river channel and floodplain systems. Applied Geochemistry 18:195–220

Walling DE, Quine TA (1990) Calibration of caesium-137 measurements to provide quantitative erosion rate data. Land Degradation and Rehabilitation 2:161–175

Walling DE, Webb BW (1981) The reliability of suspend sediment load data. Erosion and Sediment Transport Measurement Proceedings of the Florence Symposium, IAHS Publication 133:177–194

Walling DE, Webb BW (1986) Solutes in river systems in solute presses. In: Trudgill ST (ed) John Wiley and Sons, New York, pp 251–327

Walling DE, Webb BW (1987) Suspended load in gravel-bed rivers. In: Throne CR, Bathurst JC, Hey RD (eds) Sediment transport in gravel-bed rivers. Wiley, Chichester, pp. 691–723

Walling DE, Webb BW (1992) Water quality I: physical characteristics. In: Calow P, Petts GE (eds) The river's handbook: hydrological and ecological principles. Blackwell, Oxford, pp 48–72

Walling DE, Woodward JC (1992) Use of radiometric fingerprints to derive information on suspended sediment sources. In: Bogen J, Walling DE, Day T (eds) Erosion and sediment transport monitoring programmes in river basins IAHS Publication No 210. IAHS Press, Wallingford, pp 153–164

Walling DE, Woodward JC (1995) Tracing sources of suspended sediment in river basins: a case study of the river Culm, Devon, UK. Marine and Freshwater Research 46:327–336

Ward JV, Stanford JA (1995) Ecological connectivity in aluvial river ecosystems and its disruption by flow regulation. Regulated Rivers: Research and Management 11:105–119

Watterson J, Theobald P (1979) Geochemical exploration, 1978. Proceedings of the 7th International Geochemical Exploration Symposium, Association of Exploration Geochemists, Toronto, Canada,

Wartmough SA, Hughes RJ, Hutchinson TC (1999) ^{206}Pb/^{207}Pb ratios in tree rings as monitors of environmental change. Environmental Science and Technology 33:670–673

Weiss D, Shotyk W, Appleby PG, Kramers JD, Cheburkin AK (1999) Atmospheric lead deposition since the industrial revolution recorded by five Swiss peat profiles: enrichment factors, fluxes, isotopic composition, and sources. Environmental Science and Technology 33:1340–1352

White WR, Paris E, Bettes R (1981) Tables for the design of stable alluvial channels. Report IT208, Hydraulics Research Station, Wallingford

Wieder RK (1994) Diel changes in iron (III)/iron(II) in effluent from constructed acid mine drainage treatment wetlands. Journal of Environmental Quality 23:730–738

Whitehead PG, Neal C, Neale R (1986) Modelling the effects of hydrological changes on stream water acidity. Journal of Hydrology 84:353–364

Wilcock PR, Iverson RM (2003) Prediction in geomorphology. In: Wilcock PR, Iverson RM (eds) Prediction in geomorphology. American Geophysical Union, Geophysical Monograph 135, pp 3–11

Wildavsky A (1995) But is it true? A citizen's guide to environmental health and safety issues. Harvard University Press, Cambridge, MA

Williams GP (1986) River meanders and channel size. Journal of Hydrology 88:147–164

Williams GP (1989) Sediment concentration versus water discharge during single hydrologic events in rivers. Journal of Hydrology 111:89–106

Windom HL, Schropp SJ, Calder FD, Ryan JD, Smith RG Jr, Burney LC, Lewis FG, Rawlinson CH (1989) Natural trace metal concentrations in estuarine and coastal marine sediments of the southeastern US. Environmental Science and Technology 23:314–320

Wischmeier WH, Smith DD (1965) Predicting rainfall erosion losses from cropland east of the Rocky Mountains. In: Agricultural Handbook 282, US Government Printing Office Washington DC

Wissmar RC, Bisson PA (2003) Strategies for restoring river ecosystems: sources of variability and uncertainty in natural and managed systems. American Fisheries Society, Bethesda

Wolf PR (1974) Elements of photogrammetry (with air photo interpretation and remote sensing). McGraw-Hill, New York

Wolfenden PJ, Lewin J (1977) Distribution of metal pollutants in floodplain sediments. Catena 4:309–317

Wolman MG (1955) The natural channel of Brandywine Creek, Pennsylvania. US Geological Survey Professional Paper 271

Wolman MG (1959) Factors influencing erosion of a cohesive river bank. American Journal of Science 257:204–216

Wolman MG, Leopold LB (1957) River flood plains: some observations on their formation. US Geological Survey Professional Paper 282-C

Wolman MG, Miller JP (1960) Magnitude and frequency of forces in geomorphic process. Journal of Geology 68:54–74

Young RA, Onstad CA, Bosch DD, Anderson WP (1989) AGNPS user's guide. USDA Agricultural Research Service, Morris, Minnesota

Younger PL, Coulton RH, Froggatt EC (2005) The contribution of science to risk-based decision-making: lessons from the development of full-scale treatment measures for acidic mine waters at Wheal Jane, UK. Science of the Total Environment 338:137–154

Yu L, Oldfield F (1989) A multivariate mixing model for identifying sediment source from magnetic measurements. Quaternary Research 32:168–181

Yu L, Oldfield F (1993) Quantitative sediment source ascription using magnetic measurements in a reservoir-catchment system near Nijar, S.E. Spain. Earth Surface Processes and Landforms 18 441–454

Zarba CS (1992) Equilibrium partitioning approach. In: Sediment classification methods compendium. EPA 823-R-92-006

Zarull MA, Hartig JH, Maynard L (1999) Ecological benefits of contaminated sediment remediation in the Great Lakes Basin. Sediment Priority Action Committee, Great Lakes Water Quality Board, http://www.ijc.org/php/publications/html/ecolsed/sedremgl.html

Zhang XB, Higgitt DL, Walling DE (1990) A preliminary assessment of the potential for using caesium-137 to estimate rates of soil erosion in the Loess Plateau of China. Hydrological Science Journal 35:243–251

GLOSSARY

Absorption: The incorporation of ions into the crystal structure of a mineral or other substrate.

Accuracy: Degree to which an observation or measurement corresponds to the true value.

Action Levels[1]: (1) The existence of a contaminant concentration in the environment high enough to warrant action or trigger a response.

Acute Exposure[1]: A single exposure to a toxic substance which results in severe biological harm or death. Acute exposures are usually characterized as lasting no longer than a day.

Acute Toxicity[1]: The ability of a substance to cause poisonous effects resulting in severe biological harm or death soon after a single exposure or dose. Can also refer to any severe poisonous effect resulting from a single short-term exposure to a toxic substance.

Adsorption: A process referring to the removal of dissolved species from solution and their accumulation on a particle surface without the formation of a distinct, three-dimensional molecular structure.

Aerobic[1]: Processes or life forms that require, or are not destroyed by, the presence of oxygen.

Aggradation: Long-term accumulation of sediment on the channel bed, forming a new base level.

Alluvial Channel: A channel whose bed and banks are composed of unconsolidated materials.

Alluvial Sediment: Sediment composed of unconsolidated material deposited by a river or running water.

Alluvium: Unconsolidated sediment deposited by a river or running water.

Alpha Particle[1]: A positively charged particle composed of two neutrons and two protons released by some atom undergoing radioactive decay. The particle is identical to the nucleus of a helium atom.

Anaerobic[1]: A process or life form that occurs in, or is not destroyed by, the absence of oxygen.

Aquifer[1]: An underground geological formation, or group of formations, containing usable amounts of groundwater that can supply wells and springs.

Aromatic: Substance characterized by the presence of at least one benzene ring.

Assessment: The means for determining the nature, extent and levels of existing contamination at a site, and the actual or potential risk(s) that the contaminant poses to human or ecosystem health.

Attenuation[1]: The process by which a compound is reduced in concentration over time, through adsorption, degradation, dilution, and/or transformation.

Bacteria[1]: (Singular: bacterium) Microscopic living organisms.

Base Level: Theoretical level to which a river can erode. The ultimate base level is sea level.

Bedload: Material that is transported close to the channel bottom by rolling, sliding, or bouncing (saltating).

Bed Material Load: Particles found in appreciable quantities on the stream bed, but which may be moved as bedload or in suspension.

Benthic Organism (Benthos)[1]: A form of aquatic plant or animal life that is found on or near the bottom of a stream, lake or ocean.

Beta Particle[1]: An elementary particle emitted by radioactive decay that may cause skin burns. It is halted by a thin sheet of paper.

Biodegradable[1]: The ability to break down or decompose rapidly under natural conditions and processes.

Biofilm: A dense, gelatinous coating on sediment produced by a wide range of microorganisms; often highly reactive and composed of hydrated exopolysaccharide and polypeptide polymers.

Biomagnification: A process whereby a substance increases in concentration up the food chain.

Biota[1]: All living material in a given area.

Biotic Community[1]: A naturally occurring assemblage of plants and animals that live in the same environment and are mutually sustaining and interdependent.

Cap[1]: A layer of clay, or other highly impermeable material installed over the top of a landfill or contaminated area to prevent entry of rainwater and minimize production of leachate.

Capacity: Maximum amount of sediment that a river is capable of transporting.

Carcinogen[1]: Any substance that can cause or contribute to the production of cancer.

Cation Exchange Capacity: The propensity for a given substance to exchange cations with those in a chemically defined solution.

Channelization[1]: Straightening and deepening of a stream so water will move faster; typically used as a flood-reduction or marsh-drainage measure that can interfere with waste assimilation capacity and disturb fish and wildlife habitats.

Chronic Toxicity[1]: The capacity of a substance to cause long-term poisonous human health effects.

Cleanup[1]: Actions taken to deal with a release or threat of release of a hazardous substance that could affect humans and/or the environment. The term "cleanup" is sometimes used interchangeably with the terms remedial action, removal action, response action, or corrective action.

Coliform[1]: Microorganisms found in the intestinal tract of humans and animals. Their presence in water indicates fecal pollution and potentially dangerous bacterial contamination by disease-causing microorganisms.

Common Ion Effect: The lowering of the solubility of a mineral as a result of the occurrence of one of its reaction products from another source.

Competence: The size of the largest particle that a river can carrier under a given set of hydraulic conditions.

Composite Sampling: (1) The collection of the water and sediment continuously over a relatively long period of time, (2) The collection of multiple, instantaneous samples which are combined prior to analysis, (3) Collection and mixing of multiple samples from geographically different areas prior to analysis. The intent is to reduce the number of samples that must be analyzed, while obtaining a more accurate, average concentration of the constituent for that particular increment of time.

Contaminant[1]: Any physical, chemical, biological, or radiological substance or matter that has an adverse affect on air, water, or soil.

Curie[1]: A quantitative measure of radioactivity equal to 3.7×10^{10} disintegrations per second.

Degradation: Long-term erosion of channel bed, thereby lowering base level.

Detachment: Surface process that increases the susceptibility of individual particles of a geologic substrate at the ground surface to movement by the breaking of chemical and physical bonds.

Diagenesis[2]: Physical or chemical process affecting a sedimentary unit or rock after it has been deposited, excluding those processes occurring at sufficiently high temperatures to be considered metamorphic.

Diffuse Flow: Movement of water through a network of interconnected pores in the sediment.

Discharge: The volume of water that passes a given cross section in the channel (or subsurface) during a specified time interval. In rivers, it is usually expressed in units of cubic feet per second or cubic meters per second, and can be expressed mathematically as the product of width, depth, and velocity.

Disposal[1]: Final placement or destruction of toxic, radioactive, or other wastes; disposal may be accomplished through use of approved secure landfills, surface impoundments, land farming, deep well injection, ocean dumping, or incineration.

Direct Runoff: The quantity of water that flows into a stream channel in direct response to a precipitation event.

Dolomite[2]: A mineral ($CaMg(CO_3)_2$), or rock composed primarily of that mineral.

Drainage Basin. The discernible, finite land area which is drained by a trunk channel and all its interconnected network of tributaries and streams. Represents a fundamental unit of the landscape through which solutes, sediments, and contaminated particles are collected, transported, and stored. Used interchangeable with watershed.

Dredging[1]: Removal of sediment from the bottom of a water body.

Ecological Impact[1]: The effect that a man-made or natural activity has on living organisms and their non-living (abiotic) environment.

Ecosystem[1]: The interacting system of a biological community and its non-living environmental surroundings.

Effluent[1]: Wastewater – treated or untreated – that flows out of a treatment plant, sewer, or industrial outfall. Generally refers to wastes discharged into surface waters.

Entrainment: All of the processes involved in initiating motion of a particle from a state of rest.

Estuary[1]: Regions of interaction between rivers and nearshore ocean waters, where tidal action and river flow create a mixing of fresh and salt water. These areas may include bays, mouths of rivers, salt marshes, and lagoons.

Facies: The characteristics of a sedimentary package or rock unit that usually reflect the location or conditions of its origin.

Feasibility Study[1]: (1) Analysis of the practicability of a proposal; e.g., a description and analysis of the potential cleanup alternatives for a site or alternatives for a site on the National Priorities List. The feasibility study usually recommends selection of a cost-effective alternative. It usually starts as soon as the remedial investigation is underway; together, they are commonly referred to as the "RI/FS". The term can apply to a variety of proposed corrective or regulatory actions, (2) In research, a small-scale investigation of a problem to ascertain whether or not a proposed research approach is likely to provide useful data.

Fecal Coliform Bacteria[1]: Bacteria found in the intestinal tracts of mammals. Their presence in water or sludge is an indicator of pollution and possible contamination by pathogens.

Felsic Rock[2]: A light-colored rock usually consisting of feldspars and or quartz.

Fluvial: Pertaining to or produced by a river.

Food Chain[1]: A sequence of organisms, each of which uses the next, lower member of the sequence as a food source.

Fresh Water[1]: Water that generally contains less than 1,000 milligrams-per-liter of dissolved solids.

Fungi[1]: (Singular, Fungus) A group organisms that lack chlorophyll (i.e., are not photosynthetic) and which are usually non-mobile, filamentous, and multicellular. Some grow in the ground, others attach themselves to decaying trees and other plants, getting their nutrition from decomposing organic matter. Some cause disease, others stabilize sewage and break down solid wastes in composting. Includes molds, mildews, yeasts, mushrooms, and puff-balls.

Fungicide: Pesticides which are used to control, prevent, or destroy fungi.

Gamma Radiation[1]: Gamma rays are true rays of energy in contrast to alpha and beta radiation. The properties are similar to x-rays and other electromagnetic waves. They are the most penetrating waves of radiant nuclear energy but can be blocked by dense materials such as lead.

Groundwater: Loosely defined as all subsurface water, as distinct from surface water.

Habitat[1]: The place where a population (e.g., human, animal, plant, microorganism) lives and its surroundings, both living and non-living.

Half-life: Time required for a radioactive material to loose half of its radioactivity.

Hazardous Substance[1]: Any material that poses a threat to human health and/or the environment. Typical hazardous substances are toxic, corrosive, ignitable, explosive, or chemically reactive.

Hazardous Waste[1]: By-products of society that can pose a substantial or potential hazard to human health or the environment when improperly managed. Waste possesses at least one of four characteristics (ignitability, corrosivity, reactivity, or toxicity).

Heavy Mineral: A mineral generally possessing a specific gravity of greater than 2.85. Usually found as a minor constituent (<1%) of common rocks. Includes magnetite, ilmenite, zircon, rutile, kyanite, garnet, sphene, apatite, and biotite, among others.

Holocene: An epoch in the Quaternary period extending from approximately ten thousand years ago to the present.

Hydrograph: A graph depicting changes in stage or discharge of water through time.

Humic Substances: Nonliving, partially decomposed and visually unidentifiable materials that are characterized by extreme heterogeneity in chemical composition and structure.

Hydraulic conductivity: A measure of the rate at which water flows through a unit cross section under unit hydraulic gradient.

Hydrogeology[1]: The geology of groundwater, with particular emphasis on the chemistry and movement of water.

Hydrology[1]: The science dealing with the properties, distribution, and circulation of water.

Hydrophilic: Having an affinity for water.

Hydrophobic: Having an attraction or affinity for particulate matter. Generally refers to contaminants that are associated with particles, and for which the dissolved concentrations are extremely low.

Incision: A process in which a channel lowers it bed elevation through downward erosion. Used interchangeably with entrenchment.

Influent Flow: Movement of water into the ground from a surface water body.

Inorganic Chemicals[1]: Chemical substances of mineral origin, not of basic carbon structure.

Infiltration: Portion of precipitation that is absorbed into the underlying materials.

Infiltration capacity: The rate at which infiltration occurs, usually measured in mm per hour.

Interflow: See throughflow.

Interception: Process by which precipitation is caught by vegetation and is returned to the atmosphere before reaching the ground.

Ion: An electrically charged atom or group of atoms.

Isotope[1]: A variation of an element that has the same atomic number but a different weight because of a different number of neutrons.

Isotopic ratio: Ratio of abundance of any two isotopes.

Knickpoint: Point of abrupt change in the longitudinal profile (slope) of a stream. Slope is often nearly vertical.

Lacustrine: Pertaining to, produced by or formed within a lake.

Landfills[1]: (1) Sanitary landfills are land disposal sites for non-hazardous solid wastes at which the waste is spread in layers, compacted to the smallest practical volume, and cover material applied at the end of each operating day, (2) Secure chemical landfills are disposal sites for hazardous waste. They are selected and designed to minimize the chance of release of hazardous substances into the environment.

Landform: A physical, recognizable feature of the Earth's surface possessing a characteristic shape and which is produced by natural processes.

Leachate: A liquid resulting from the percolation of water or other fluid through wastes, contaminated materials, or other substance.

Lithology: Mineral composition of rock or sedimentary unit.

Load: The mass of material that passes a given point in the channel. The timeframe is not explicitly stated, but must be obtained from the context in which it is used. Load is given in units of mass (tons, kg, etc.).

Mafic Rock[2]: A usually dark-colored rock having a high proportion of ferromagnesian minerals.

Mass Movement: Movement of Earth material downslope under the influence of gravity.

Media[1]: Specific environments – air, water, soil – which are the subject of regulatory concern and activities.

Microbes[1]: Microscopic organisms such as algae, animals, viruses, bacteria, fungus, and protozoa, some of which cause diseases (see microorganism).

Mitigation[1]: Measures taken to reduce adverse impacts on the environment.

Model: A working representation of a phenomenon or process that cannot be precisely observed or described.

Monitoring[1]: Periodic or continuous surveillance or testing to determine the level of compliance with statutory requirements and/or pollutant levels in various media or in humans, animals, and other living things.

National Priorities List (NPL)[1]: USEPA's list of the most serious uncontrolled or abandoned hazardous waste sites identified for possible long-term remedial action under Superfund. A site must be on the NPL to receive money from the Trust Fund for remedial action. The list is based primarily on the score a site receives from the Hazard Ranking System.

Non-ionic: Materials possessing no charge on its surface-active ion(s).

Non-point source: The delivery of pollutants to a river from a diffuse area, such as an agricultural field or an urban center.

Oxidation: (1) A chemical process in which electrons are removed from an atom, (2) The addition of oxygen to break down organic waste or chemicals such as cyanides, phenols, and organic sulfur compounds by bacterial and chemical means.

Pathogenic[1]: Capable of causing disease.

Permeability: Capacity of a rock or sedimentary unit to transmit a fluid. Typically used to indicate the ease with which the fluid can move through a material under conditions of unequal pressure.

Phase: Physical or chemical composition of the material with which the constituents are associated.

Pleistocene[2]: A time period roughly corresponding to the ice ages (10,000–2 million years before present). Represents the time period before the recent (Holocene).

Point Source: Discharge of contaminants from a specific location, such as the end of a pipe or canal.

Precision: Describes the ability to reproduce the results of a measurement; if there is little difference between replicate measurements, then the method used in making the measurement is called precise.

Process: Action through which one feature or item is produced from something else.

Quaternary: Period of geologic time extending from approximately two to three million years ago and extending to the present.

Radiation[1]: Any form of energy propagated as rays, waves, or streams of energetic particles. The term is frequently used in relation to the emission of rays from the nucleus of an atom.

Radioactive Substances[1]: Substances that emit radiation.

Radionuclide[1]: Radioactive element characterized according to its atomic mass and atomic number which can be man-made or naturally occurring. Radioisotopes can have a long life as soil or water pollutants, and are believed to have potentially mutagenic effects on the human body.

Reagent: A substance, chemical, or solution involved in a chemical reaction. Often used to detect, measure or otherwise examine other substances, chemicals or solutions.

Recharge[1]: The process by which water is added to a zone of saturation, usually by percolation from the soil surface, e.g., the recharge of an aquifer.

Record of Decision (ROD)[1]: A public document that explains which cleanup alternative(s) will be used at National Priorities List sites where, under CERCLA, Trust Funds pay for the cleanup.

Reduction: A chemical reaction in which an atom gains one or more electrons.

Refractory: A material characterized by a high melting point.

Remedial Action[1]: The actual construction or implementation of site cleanup that follows remedial design.

Remedial Design[1]: A phase of remedial action that follows the remedial investigation/feasibility study and includes development of engineering drawings and specifications for a site cleanup.

Remedial Investigation[1]: An in-depth study designed to gather the data necessary to determine the nature and extent of contamination at a site; investigation establishes criteria for cleaning up the site and identifies preliminary alternatives for remedial actions. Also supports technical and cost analyses of the alternatives. In the case of the U.S. Superfund Program, the remedial investigation is usually done with the feasibility study. Together they are usually referred to as the "RI/FS".

Remediation: The application of methods that prevent, minimize, or mitigate the damage to human health or the environment by a contaminant; it can involve any method that removes, contains, destroys, or reduces the exposure of pollutants to biota.

Residual[1]: Amount of a pollutant remaining in the environment after a natural or technological process has taken place.

Restoration: The return of a riverine ecosystem to a more natural working order that is sustainable over the long-term, while creating a river that is more productive, aesthetically appealing, and valuable.

Return flow: Type of overland flow in which water that has infiltrated the sediment, flows downslope, and subsequently returns to the surface before entering a stream channel.

Riparian Habitat[1]: Areas adjacent to rivers and streams that have a high density, diversity, and productivity of plant and animal species relative to nearby uplands.

Risk Assessment[1]: The qualitative and quantitative evaluation performed in an effort to define the risk posed to human health and/or the environment by the presence or potential presence of specific pollutants.

River Stage: Height of the water surface in a stream above some arbitrary datum.

Saturation Overland Flow: Movement of water over the ground surface where the water table intersects the ground surface, or when water accumulates above a relatively impermeable horizon. It consists of two components: return flow and direct precipitation on saturated areas.

Saturated Zone[1]: A subsurface area in which all pores and cracks are filled with water under pressure equal to or greater than that of the atmosphere.

Sorption: The accumulation of constituents dissolved within river or pore waters on particle surfaces; occurs through a combination of mechanisms, including absorption, adsorption, and precipitation.

Subsurface Stormflow: The combined effects of enhanced groundwater flow and throughflow to a river channel.

Superfund[1]: A program in the U.S. operated under the legislative authority of CERCLA and SARA that funds and carries out the USEPA solid waste emergency and long-term removal remedial activities. These activities include establishing the National Priorities List, investigating sites for inclusion on the list, determining their priority level on the list, and conducting and/or supervising the ultimately determined cleanup and other remedial actions.

Surface Water: All water naturally open to the atmosphere (rivers, lakes, reservoirs, streams, impoundments, seas, estuaries, etc.).

Surfactant: A soluble compound which reduces surface tension in liquids, and/or reduces interfacial tension between liquids or a liquid and a solid.

Suspended Load: The downstream movement of fine-grained particles transported within the water column. Particles are held in suspension by short, but intense, upward deviations in the flow as a result of turbulent eddies.

System: Interconnected group of features or forces.

Tectonic: Pertaining to the forces that deform rocks and sediment, and result in structural features such as faults and joints. Often associated with mountain building processes.

Throughflow: Movement of water above the water table through locally saturated materials during and after a precipitation event.

Toxic Pollutants[1]: Materials contaminating the environment that cause death, disease, birth defects in organisms that ingest or absorb them.

Unsaturated Zone[1]: The area above the water table where the soil pores are not fully saturated, although some water may be present.

Wash Load: Particles so small that they are absent from the channel bed sediment. The implication is that after these particles have been eroded, they are transported through the river system without ever coming to rest within the channel.

Watershed[1]: The land area that drains into a stream.

Volatile: A substance that evaporates readily.

Water Table: Surface along which the groundwater pressure is equal to atmospheric pressure, and which separates the zone of saturation from the zone of aeration.

Zeolite[2]: A group of hydrated aluminosilicate minerals which are chemically similar to feldspars plus water, plus/minus silica. Some zeolites have a high cation-exchange capacity. Some also are used in industry as molecular sieves.

---

[1] Terms from the Glossary of Environmental Terms and Acronym List, U.S. Environmental Protection Agency, August 1988

[2] Terms after from Drever (1988)

CREDITS

CHAPTER 1

Table 1.1: Copyright 2000, from Environmental Chemistry, 7th edn by Manahan SE. Reproduced by permission of Routledge/Taylor and Francis Group, LLC; **Table 1.2:** From USEPA, National Sediment Quality Inventory, 1997; **Figure 1.1**: From Domenico PA, Schwartz FW, Physical and Chemical Hydrology, 2nd edn, Table 11.5, p 246, 1998, reproduced by permission of John Wiley and Sons, Ltd; **Figure 1.3:** From USEPA, National Water Quality Inventory: 2000 Report, 2000, Figs. 2–3, p 12; **Figure 1.4:** From USEPA, National Water Quality Inventory: 2000 Report, 2000, Fig. 2.4, p 13; **Figure 1.5:** USEPA, National Sediment Quality Inventory, 1997, Fig. 4.1, pp 3–4; **Figure 1.6:** From USEPA, National Water Quality Inventory: 2000 Report, 2000, Fig. 2.5, p 14; **Figure 1.7:** Data from Gibbs RJ, Geological Society of America Bulletin, vol 88, Table 1.3, p 836, 1977, reproduced by permission of the Geological Society of America; **Figure 1.9:** From USEPA Remedial Investigation and Feasibility Studies under CERCLA, 1988.

CHAPTER 2

Table 2.1: Copyright 1991, from Primer on Sediment Trace Element Chemistry, 2nd edn by Horowitz AJ. Reproduced by permission of Routledge/Taylor & Francis Group, LLC; **Table 2.2:** Copyright 1991, from Primer on Sediment Trace Element Chemistry, 2nd edn by Horowitz AJ. Reproduced by permission of Routledge/Taylor & Francis Group, LLC; **Figure 2.1:** Copyright 1991 from Primer on Sediment Trace Element Chemistry, 2nd edn by Horowitz AJ. Reproduced by permission of Routledge/Taylor & Francis Group, LLC; **Figure 2.2:** From Förstner U, Witmann GTW, Metal Pollution in the Aquatic Environment; Figs. 39a, b, p 127, 1979, reproduced by permission of Springer Verlag; **Figure 2.3:** From Förstner U, Witmann GTW, Metal Pollution in the Aquatic Environment; Fig. 36, p 122, 1979, copyright by Springer Verlag; **Figure 2.4:** de Groot A, Zshuppe K, Salomons W, Hydrobiologia, vol 92, Fig. 1, p 692, 1982, reproduced by permission of Springer; **Figure 2.5:** From Bourg ACM, Loch JPG, Biogeodynamics of Pollutants in Soil and Sediments: Risk Assessment of Delayed and Non-linear Responses, Figs. 4.4a, b, p 92, 1995, reproduced by permission of Springer; **Figure 2.6:** Copyright 1997, adapted from Deutsch WJ,

Groundwater Geochemistry: Fundamentals and Applications to Contamination, Fig. 3.6, p 58, reproduced by permission of Routledge/Taylor & Francis Group, LLC; **Figure 2.7:** From Langmuir D, Aqueous Environmental Geochemistry, Fig. 10.4, p 349, 1997, reproduced with permission of Pearson Education, Inc; **Figure 2.8:** From Krauskopf KB, Bird DK, Introduction to Geochemistry, 3rd edn, Fig. 6.4, p 141, 1995, reproduced by permission of McGraw-Hill, Inc; **Figure 2.9:** Copyright 1997, adapted from Deutsch WJ, Groundwater Geochemistry: Fundamentals and Applications to Contamination, Fig. 3.8, p 59, reproduced by permission of Routledge/Taylor & Francis Group, LLC; **Figure 2.10:** Copyright 1997, from Deutsch WJ Groundwater Geochemistry: Fundamentals and Applications to Contamination, Fig. 3.5, p 54, reproduced by permission of Routledge/Taylor & Francis Group, LLC; **Figure 2.11:** From Grim RE, Clay Mineralogy, 3rd edn, Figs. 4.1, 4.2, p 52, 1968, reproduced by permission of McGraw-Hill; **Figure 2.12:** Modified from Buckman HO, Brady NC, The Nature and Properties of Soils, 6th edn, 1960, copyright by Prentice/Pearson, Inc; **Figure 2.13:** Copyright 1997, from Deutsch WJ Groundwater Geochemistry: Fundamentals and Applications to Contamination, Fig. 3.2, p 49, reproduced by permission of Routledge/Taylor & Francis Group, LLC; **Figure 2.14:** Copyright 1988, from Gunn AM, Winnard DA, Hunt DTE, Metal Speciation: Theory, Analysis, and Application, Fig. 1, p 262, reproduced by permission of Routledge/Taylor & Francis Group, LLC; **Figure 2.15:** From Bourg ACM, Loch JPG, Biogeodynamics of Pollutants in Soil and Sediments: Risk Assessment of Delayed and Non-Linear Responses, Fig. 4.3, p 91, 1995, Reproduced by permission of Springer.

CHAPTER 3

Figure 3.2: From Ritter DF, Kochel RC, Miller JR, Process Geomorphology, 4th edn, Fig. 5.4, p 139, 2002, reproduced by permission of Waveland Press, Inc; **Figure 3.3:** From Ritter DF, Kochel RC, Miller JR, Process Geomorphology, 4th edn, Fig. 5.3, p 138, 2002, reproduced by permission of Waveland Press, Inc; **Figure 3.4:** From Ritter DF, Kochel RC, Miller JR, Process Geomorphology, 4th edn, Fig. 5.5, p 140, 2002, reproduced by permission of Waveland Press, Inc; **Figure 3.5:** From Dunne T, Leopold LB, Water in Environmental Planning, 1978, reproduced by permission of by Freeman and Company/Worth Publishers; **Figure 3.7a:** From Ritter DF, Kochel RC, Miller JR, Process Geomorphology, 4th edn, Fig. 5.32a, p 165, 2002, reproduced by permission of Waveland Press, Inc; **Figure 3.8:** From Ritter DF, Kochel RC, Miller JR, Process Geomorphology, 4th edn, Fig. 5.33, p 166, 2002, reproduced by permission of Waveland Press, Inc; **Figure 3.9:** From Ritter DF, Kochel RC, Miller JR, Process Geomorphology, 4th edn, Fig. 5.10, p 144, 2002, reproduced by permission of Waveland Press, Inc; **Figure 3.10:** From Burt TP, Pinay G, Progress in Physical Geography, vol 29, Fig. 8, p 308, 2005, reproduced by permission of Sage Publications, Ltd;

Figure 3.11: Modified from Kimball BA, Runkel RL, Walton-Day K, Bencala KE, Applied Geochemistry, vol 17, Table 1, p 1187, 2002, copyright by Elsevier; **Figure 3.12:** From Kimball BA, Runkel RL, Walton-Day K, Bencala KE, Applied Geochemistry, vol 17, Fig. 6a, p 1199, 2002, reproduced by permission of by Elsevier; **Figure 3.14:** Modified from McKergow LA, Prosser IP, Hughes AO, Brodie J, Marine Pollution Bulletin, vol 51, Fig. 1, p 1199, 2005, copyright by Elsevier; **Figure 3.15:** From Vink R, Peters S, Hydrological Processes, vol 17, Fig. 3, p 1313, 2003, reproduced by permission of by John Wiley and Sons, Inc; **Figure 3.16:** From Ritchie JC, McHenry JR, Journal of Environmental Quality, vol 19, Fig. 2, p 217, 1990, reproduced by permission of the American Society of Agronomy; **Figure 3.17:** From Wallbrink PJ, Murray AS, Journal of the Soil Science Society of America, vol 60, Fig. 3, p 1204, 1996, reproduced by permission of the Soil Science Society of America.

CHAPTER 4

Table 4.2: Modified from Knighton D, Fluvial Forms and Processes: A New Perspective, Table 5.2, p 166, 1998, copyright by Hodder Arnold; **Figure 4.1:** From Nagano T, Yanase N, Tsuduki K, Nagao S, Environment International, vol 28, Fig. 6, p 654, 2003, reproduced by permission of Elsevier; **Figure 4.2a:** Modified from Williams GP, Journal of Hydrology, vol 111; Figs. 1–6, 1989, copyright reproduced by permission of Elsevier; **Figure 4.2b:** From Bhangu I, Whitfield PH, Water Resources, vol 9, Fig. 8, p. 2192, 1997, reproduced by permission of Springer; **Figure 4.3:** From Evans C, Davies TD, Water Resources Research, vol 34, Fig. 2, p 131 and Fig. 4, p 132, 1998, reproduced by American Geophysical Union; **Figure 4.4:** From Shiller AM, Geochemica et Cosmochimica Acta, vol 61, Fig. 1, p 4322, 1997, reproduced by permission of Elsevier; **Figure 4.5:** Adapted from Brick CM, Moore JN, Environmental Science and Technology, vol 30, Figs. 2–3, p 1955, 1996, reproduced by permission of the American Chemical Society; **Figure 4.6:** From Sullivan AB, Drever JI, McKnight DM, Journal of Geochemical Exploration, vol 64, Fig. 6, p 144, 1998, reproduced by permission of Elsevier; **Figure 4.7:** From Brick CM, Moore JN, Environmental Science and Technology, vol 30, Fig. 8 p 1958, 1996, reproduced by permission of American Chemical Society; **Figure 4.8:** Modified from Knighton D, Fluvial Forms and Processes: A New Perspective, Fig. 4.8c, p 123, 1998, copyright by Hodder Arnold; **Figure 4.9:** Adapted from Salomons W, Eysink Q, Special Publication of the International Association of Sedimentology, vol 5, 1981, reproduced by permission of Elsevier; **Figure 4.10:** Modified from Gibbs RJ, Geological Society of America Bulletin, vol 88, Fig. 14, p 840, 1977, copyright by Geological Society of America; **Figure 4.11:** From Richards RP (2001) Report to US Environmental Protection Agency, Grant X998397-01-0, Region VIII; **Figure 4.12:** From Richards RP (2001) Report to US Environmental Protection Agency, Grant X998397-01-0, Region VIII, Fig. 2a, p 8; **Figure 4.13:** From Knighton D, Fluvial Forms and Processes: A New Perspective, Fig. 5.6a, p 163, 1998, reproduced by permission of Hodder Arnold.

CHAPTER 5

Table 5.1: From Knighton D, Fluvial Forms and Processes: A New Perspective, Table 4.2, p 101, 1998, reproduced by permission of Hodder Arnold; **Table 5.2:** From Nicholas, AP, Ashworth, PJ, Kirkby MJ, Macklin MG, Murray T, Progress in Physical Geography, vol 19, Table 5.2, p 502, 1995, reproduced by permission of Sage Publications, Ltd; **Table 5.3:** Modified from Mineral Resources Forum, on-line Table, 2002; **Table 5.5:** After Slingerland R, Smith ND, Annual Review of Earth and Planet Science, vol 14, Table 5.1, p 115, 1986, copyright by Annual Reviews; **Table 5.6:** Copyright 1991, modified from Primer on Sediment Trace Element Chemistry, 2nd edn by Horowitz AJ, reproduced by permission of Routledge/Taylor & Francis Group, LLC; **Figure 5.2:** From Hjulström F AAPG© 1939 reprinted by of the AAPG whose permission is required for further use; **Figure 5.3:** From Ritter DF, Kochel RC, Miller JR, Process Geomorphology, 4th edn, Fig. 6.9, p 198, 2002, reproduced by permission of Waveland Press, Inc; **Figure 5.4:** From Hassan MA, Church M, Dynamics of Gravel-bed Rivers, Fig. 8.4, p 169, 1992, reproduced by permission of John Wiley and Sons, Ltd; **Figure 5.5:** From Villarroel LF, Miller JR, Lechler PJ, Germanoski D, Environmental Geology, 2006, reproduced by permission of Springer; **Figure 5.6:** Villarroel LF, Miller JR, Lechler PJ, Germanoski D, Environmental Geology, 2006, reproduced by permission of Springer; **Figure 5.7:** From Leopold LB, Wolman MG, Miller JP, Fluvial Processes in Geomorphology, Fig. 7.13, p 228, 1964, reproduced by permission of Dover Press; **Figure 5.8:** From Miller JR, Lechler PJ, Desilets M, Environmental Geology, vol 33, Figs. 9–10, pp 259–260, 1998, reproduced by permission of Springer; **Figure 5.9:** From Owens PN, Walling DE, Carton J, Meharg AA, Wright J, Leeks GJL, The Science of the Total Environment, vol 266, Fig. 3, p 182, 2001, reproduced by permission of Elsevier; **Figure 5.10:** Extracted from Lewin J, Macklin MG, International Geomorphology, 1987, Copyright by John Wiley and Sons; **Figure 5.12:** From Graf WL, Annals of the Association of American Geographers, vol 82, Fig. 9, p 337, 1990, reproduced by permission of Blackwell Publishing; **Figure 5.13:** From Gallart F, Benito G, Martin-Vide JP, Benito A, Prio JM, Regues D, The Science of the Total Environment, vol 242, Fig. 6, p 20, 1999, reproduced by permission of Elsevier; **Figure 5.14:** From Huggett JR, Fundamental of Geomorphology, Fig. 7.7, p 185, 2003, reproduced by permission of Routledge Publishing Inc; **Figure 5.16:** From Markham AJ, Thorne CR, Dynamics of Gravel-bed Rivers, Fig. 22.2, p 436, 1992, reproduced by permission of John Wiley and Sons, Ltd; **Figure 5.17:** From Thompson A, Earth Surface Processes and Landforms, vol 11, Fig. 4, p 636, 1986, reproduced by permission of John Wiley and Sons, Ltd; **Figure 5.18:** From Andrews ED, 1979, US Geological Survey Professional Paper 1117; **Figure 5.19:** From Reading HG Sedimentary Environments and Facies, Fig. 3.26, p 34, 1978, reproduced by permission of Blackwell Publishing; **Figure 5.21:** From Knighton D, Fluvial Form and Processes: A New Perspective, Fig. 5.23, p 233, 1998, reproduced by permission of Hodder Arnold; **Figure 5.22:** Modified from Guilbert JM, Park CF, Jr, The Geology of Ore Deposits. Fig. 16.1, p 746 and Fig. 16.4, p 749, 1986, copyright by W.H. Freeman and Company;

Figure 5.23: Modified from Bateman AM, Economic Mineral Deposits, 2nd edn, Figs. 5.7–5.8, p 232, 1950, reproduced by permission of the Estate of Alan M. Bateman; **Figure 5.24:** From Graf WL, Clark SL, Kammerer MT, Lehman T, Randall K, Tempe R, Schroeder A, Catena, vol 18, Fig. 2, p 572 and Fig. 6, p 578, 1991, reproduced by permission of Elsevier; **Figure 5.26:** Modified from Viganò L, and 14 others, Water Research, 37, Fig. 3, p 507, 2003, copyright by permission of Elsevier.

CHAPTER 6

Table 6.2: From Rang MC, Schouten CJ, Historical change of large alluvial rivers: Western Europe, Table 13, p 141, 1989, reproduced by permission of John Wiley and Sons, Ltd; **Table 6.3:** From Taylor MP, Catena, vol 28, 1996, reproduced by permission of Elsevier; **Table 6.5:** From Rang MC, Schouten CJ, Historical change of large alluvial rivers: Western Europe, Table 8, p 136, 1989, reproduced by permission of John Wiley and Sons, Ltd; **Figure 6.1:** Modified from Nanson GC, Croke J, Geomorphology, vol 4, Figs. 1-ii, iii, p 471; Figs. 2-i, ii, p 474; Figs. 3-i, ii, p 478, 1992, copyright by John Wiley and Sons, Ltd; **Figure 6.4:** From Everitt BL, Use of cottonwood in an investigation of recent history of a flood plain. American Journal of Science, vol 266, 1968, reprinted by permission of the American Journal of Science; **Figure 6.5:** From Lewin J, Macklin MG, Johstone E, Quaternary Science Reviews, vol 24, Fig. 6, p 1880, 2005, reproduced by permission of Elsevier; **Figure 6.7:** From Brown AG, Alluvial Geoarchaeology: Floodplain Archaeology and Environmental Change, Fig. 1.2, p 22, 1997, reprinted with the permission of Cambridge University Press; **Figure 6.8:** From Brown AG, Alluvial Geoarchaeology: Floodplain Archaeology and Environmental Change, Fig. 1.1, p 18, 1997, reprinted with the permission of Cambridge University Press; **Figure 6.10:** From Smith ND, Cross TA, Duficy JP, Clough SR, Sedimentology, vol 36, Fig. 14, p 14, 1989, reproduced by permission of Blackwell Publishers; **Figure 6.11:** Modified from Brown AG, Alluvial Geoarchaeology: Floodplain Archaeology and Environmental Change, Fig. 1.5, p 29, 1997, reprinted with the permission of Cambridge University Press; **Figure 6.13:** From Leece SA, Geomorphology, vol 18, Fig. 5, p 269 and Fig. 6, p 269, 1997, reproduced by permission of Elsevier; **Figure 6.14:** Miller JR, Lechler PJ, Desilets M, Environmental Geology, vol 33, Fig. 4, p 253, 1998, reproduced by permission of Springer; **Figure 6.18:** From Walling DE, He Q, Geomorphology, vol 24, Fig. 3, 1998, reproduced by permission of Elsevier; **Figure 6.19:** From Shiono K, Knight DW, Journal of Fluid Mechanics, vol 22, Fig. 1, p 618, 1991, reprinted with the permission of Cambridge University Press; **Figure 6.20:** From Ritter DF, Kochel RC, Miller JR, Process Geomorphology, 4th edn, Fig. 7.5, p 237, 2002, reproduced by permission of Waveland Press, Inc; **Figure 6.22:** From Walling DE, He Q, Nicholas AP, Floodplain Processes, Fig. 12.9b, p 418, 1996, reproduced by permission of John Wiley and Sons, Ltd; **Figure 6.23:** From Walling DE, He Q, Nicholas AP, Floodplain Processes, Fig. 12.15b, p 426, 1996, reproduced by permission of John Wiley and Sons, Ltd; **Figure 6.24:** Modified from Macklin MG, Floodplain Processes, Fig. 13.8, p 452, 1996, copyright by John Wiley and Sons, Ltd; **Figure 6.25:** Adapted from Bradley SB, Cox JJ, The Science of the Total Environment vols 97/98, 1990, copyright by Springer; **Figure 6.26:** Modified from UKEA Facts and Figures, River Quality: An Overview, United Kingdom Environmental Agency, 2003, on-line plot; **Figure 6.27:** Modified from Harrison J, Heijnis H, Caprarelli G, Environmental Geology, vol 43, Figs. 2a and 3, p 683, 2003, copyright by Springer; **Figure 6.28:** From Collins AL, Walling DE, Journal of Hydrology, vol 261, Fig. 1, p 219, 2002, reproduced by permission of Elseveir; **Figure 6.29b, c:** From Collins AL, Walling DE, Leeks GJL, Geomorphology, vol 19, Fig. 4, p 161 and Fig. 6, p 163, 1997, reproduced by permission of Elsevier; **Figure 6.31:** From Villarroel LF, Miller JR, Lechler PJ, Environmental Geology, 2006, reproduced by permission of Springer; **Figure 6.33:** From Ritter DF, Kochel RC, Miller JR, Process Geomorphology, 4th edn, Fig. 6.11, p 200, 2002, reproduced by permission of Waveland Press, Inc.

CHAPTER 7

Table 7.2: From Schumm SA, Lichty RW, America Journal of Science, vol 263, Table 1, p 112, 1965, reproduced by permission of the American Journal of Science; **Table 7.4:** Modified from Gilvear D, Bryant R, Tools in Fluvial Geomorphology, Table 6.1 (part A), p 136, 2003, copyright by John Wiley and Sons; **Figure 7.1:** From Langbein WB, Annual Runoff in the United States. US Geological Circular vol 52, 1949; **Figure 7.2:** From Langbein WB, Schumm SA Transactions of the American Geophysical Union, vol 39, p 1077, 1958, reproduced by permission of the American Geophysical Union; **Figure 7.3:** From Ritter DF, Kochel RC, Miller JR, Process Geomorphology, 4th edn, Fig. 1.7, p 22, 2002, reproduced by permission of Waveland Press, Inc; **Figure 7.4:** From Ritter DF, Kochel RC, Miller JR, Process Geomorphology, 4th edn, Fig. 1.3, p 6, 2002, reproduced by permission of Waveland Press, Inc; **Figure 7.5:** From Patton PC, Schumm SA, Geology, vol 3, 1975, reproduced by permission of the Geological Society of America; **Figure 7.6:**. From Knighton A, Fluvial Forms and Processes: A New Perspective, Fig. 6.5, p 284, 1998, reproduced by permission of Hodder Arnold; **Figure 7.7:** From Knighton A, Fluvial Forms and Processes: A New Perspective, Fig. 5.3, p 158, 1998, reproduced by permission of Hodder Arnold; **Figure 7.10:** From James LA, Annals of the Association of American Geographers, vol 79, Fig. 14, p 588, 1989, reproduced by permission of Elsevier; **Figure 7.11:** From Schumm SA, Khan HR, Geological Society of America Bulletin, vol 83, 1972, reproduced by permission of the Geological Society of America; **Figure 7.12:** From Schumm SA, The Fluvial System, Fig. 4.10, p 73, 1977, used with permission of S. Schumm; **Figure 7.14:** Modified from Brown AG, Alluvial Geoarchaeology: Floodplain Archaeology and Environmental change, Fig. 1.8, p 35, 1997, reprinted with the permission of Cambridge University Press; **Figure 7.15:** Modified from Ritter DF, Kochel RC, Miller JR, Process Geomorphology, 4th edn, Fig. 7.11, p 243, 2002, copyright by Waveland Press, Inc; **Figure 7.17:** From Ritter DF,

Kochel RC, Miller JR, Process Geomorphology, 4th edn, Fig. 7.12, p 244, 2002, reproduced by permission of Waveland Press, Inc; **Figure 7.18:** Modified from Brewer PA, Taylor MP, Catena, vol 30, Fig. 2, p 234, 1997, copyright by Elsevier; **Figure 7.19:** From Brewer PA, Taylor MP, Catena, vol 30, Fig. 9, p 246, 1997, reproduced by permission of Elsevier; **Figure 7.20b:** From Miller JR, Rowland J, Lechler PJ, Desilets M, Water, Air and Soil Pollution, vol 86, Fig. 3, p 377, 1996, reproduced by permission of Springer.

CHAPTER 8

Table 8.4: Modified from MacDonald DD, Ingersoll CG, Berger TA, Archives of Environmental Contamination and Toxicology, vol 39, pp 20–31, 2000, copyright by Springer Publishers; **Table 8.5:** Modified from MacDonald DD, Ingersoll CG, Berger TA, Archives of Environmental Contamination and Toxicology, vol 39, pp 20–31, 2000, copyright by Springer; **Figure 8.2:** From Breckenridge RP, Crockett AB, Environmental Monitoring and Assessment, vol 51, Fig. 1, p 625, 1998, reproduced by permission of Springer; **Figure 8.3:** Modified heavily from Reimann C, Filzmoser P, Garrett RG, Science of the Total Environment, vol 346, Fig. 7, p 11, 2005, copyright by Elsevier; **Figure 8.5:** From Loring DH, Rantala RTT, Earth-Science Reviews, vol 32, Fig. 5, p 279, 1992, reproduced by permission of Elsevier.

CHAPTER 9

Table 9.2: Modified from USEPA, Contaminated Sediment Remediation Guidance for Hazardous Waste Sites. EPA-540-R-05-012, 2005; **Figure 9.1:** From USEPA, Contaminated Sediment Remediation Guidance for Hazardous Waste Sites. EPA-540-R-05-012, 2005, Fig. 6.1, p 6.1; **Figure 9.2:** From Cleland J, Results of Contaminated Sediment Cleanups Relevant to the Hudson River, Exhibit 13, p 30, 2000, used with permission of Scenic Hudson; **Figure 9.3:** From Steuer JJ, USGS Water-Resources Investigations. Report 00-4245, 2000, Fig. 1, p 1; **Figure 9.4:** From Romagnoli R, Doody JP, VanDewalker HM, Hill SA, Management of Contaminated Sediments, Fig. 1, p 183, 2002, reproduced by permission of Battelle Press; **Figure 9.5:** Modified from Foth and Van Dyke, 2000, Summary Report: Fox River Deposit N, used with permission of Foth and Van Dyke; **Figure 9.6:** From USEPA, Engineering Bulletin, 1990, EPA/540/2-90/017; **Figure 9.7:** From Lehman D, Soil Science Society of America Proceedings, Fig. 3, p 169, 1963, reproduced by permission of the Soil Science Society of America; **Figure 9.8:** From Krishman R, Parker HW, Tock RW, Journal of Hazardous Materials, vol 48, Fig. 1, p 112, 1996, reproduced by permission of Elseveir; **Figure 9.9:** After Palmer MA, Bernhardt ES, Allan JD, Lake PS, Alexander G, Brooks S, Carr J, Clayton S, Dahm CN, Follstad Shan J, Galat DL, Loss SG, Kondolf GM, Lave R, Meyer JL, O'Donnell TK, Pagano L, Sudduth E, Journal of Applied Ecology, vol 42, Fig. 1, pp 209, 2005, reproduced by permission of Blackwell Publishers; **Figure 9.10:**

From Harmon W, Jennings GD, Patterson JM, Clinton DR, Slate LO, Jessup AG, Everhart JR, Smith RE, Wildland Hydrology, Fig. 2, p 405, 1999, used with permission of American Water Resources Association; **Figure 9.11:** From Shields FD Jr, Copeland RR, Klingeman PC, Doyle MW, Simon A, Journal of Hydraulic Engineering, vol 129, Fig. 1, pp 576, 2003, reproduced by permission of the American Society of Chemical Engineering; **Figure 9.15:** From Bucher B, Wolff CG, Cawlfield L, 2000, Channel remediation and restoration design for Silver Bow Creek, Butte, Montana, used with permission of authors.

CHAPTER 10

Figure 10.1: From USEPA, Engineering Bulletin, 1991, EPA/540/2-91/021; **Figure 10.2:** From Iskandar IK, Adriano DC, Remediation of Soils Contaminated with Metals, Science Reviews, Fig. 4, p 17, 1997, reproduced by permission of Northwood; **Figure 10.3:** Adapted from USEPA, The Superfund Innovative Technology Program, 1991, EPA540/5-91-008; **Figure 10.4:** From Dushenkov S, Kapulnik Y, Phytoremediation of Toxic Metals: Using Plants to Clean Up the Environment, Fig. 7.1, p 91, 2000, reproduced by permission of John Wiley & Sons; **Figure 10.6:** From Houthoofd JM, McCready JH, Roulier MH, Remedial Action, Treatment, Disposal of Hazardous Waste: Proceedings of the 7th Annual RREL Hazardous Waste Research Symposium, 1991; **Figure 10.7:** From Berti WR, Cunningham SD, Phytoremediation of Toxic Metals: Using Plants to Clean Up the Environment, Fig. 6.1, p 74, 2000, reproduced by permission of John Wiley & Sons; **Figure 10.8:** From Brown, Sediment Management Work Group Technical Report, Fig. C-1, p 9, 1999, reproduced by permission of the Sediment Management Working Group.

APPENDIX A

Figure A.1: From Stokes S, Walling DE, In Tools in Fluvial Geomorphology, Fig. 9.10, p 247, 2005, reproduced by permission of John Wiley and Sons, Chichester; **Figure A.2:** From Appleby PG, In Tracking Environmental Change using Lake Sediments, vol 1, Basin Analysis, Coring, and Chronological Techniques, Fig. 8 (a only), p 181, 2001, reproduced by permission of Springer.

INDEX

Printed in the United States
By Bookmasters